YOUR ACCESS TO SUCCESS

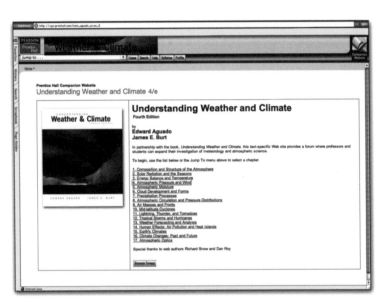

Students with new copies of Aguado/Burt, *Understanding Weather & Climate, Fourth Edition,* have full access to the book's Companion Website with GradeTracker— a 24/7 study tool with Tutorials, quizzes, and other features designed to help you make the most of your limited study time.

Just follow the easy website registration steps listed below...

Registration Instructions for the *Aguado/Burt, Understanding Weather & Climate, Fourth Edition* Companion Website with GradeTracker

1. Go to *www.prenhall.com/aguado*.
2. Click on the cover for *Aguado/Burt, Understanding Weather & Climate, Fourth Edition*.
3. Click on the link to the Companion Website with GradeTracker.
4. Click "Register".
5. Using a coin, scratch off the metallic coating below to reveal your Access Code.
6. Complete the online registration form, choosing your own personal Login Name and Password.
7. Enter your pre-assigned Access Code exactly as it appears below.
8. Complete the online registration form by entering your School Location information.
9. After your personal Login Name and Password are confirmed by e-mail, repeat steps 1-3 above, and enter your new Login Name and Password to log in.

Minimum system requirements

PC Operating Systems:
Windows 2000/XP
Pentium II 233 MHz processor. 64 MB RAM In addition to the minimum memory required by your OS.
Internet Explorer™ 5.5 or 6, Netscape™ 7, Firefox 1.0.

Macintosh Operating Systems:
Macintosh Power PC with OS X (10.2 and 10.3)
In addition to the RAM required by your OS, this application requires 64 MB RAM, with 40MB Free RAM, with Virtual Memory enabled.
Netscape™ 7, Safari 1.3, Firefox 1.0.

Macromedia Shockwave™
8.50 release 326 plugin
Macromedia Flash Player 6.0.79 & 7.0
Acrobat Reader 6.0.1
4x CD-ROM drive
800 x 600 pixel screen resolution

Your Access Code is:

If there is no metallic coating covering the access code above, the code may no longer be valid and you will need to purchase online access using a major credit card to use the website. To do so, go to *www.prenhall.com/aguado*, click the cover for *Aguado/Burt, Understanding Weather & Climate,Fourth Edition*, and follow the instructions to purchase access to the Companion Website with GradeTracker.

Important: Please read the Subscription and End-User License Agreement located on the "Log In" screen before using the *Aguado/Burt, Understanding Weather & Climate, Fourth Edition* Companion Website with GradeTracker. By using the website, you indicate that you have read, understood, and accepted the terms of the agreement.

Technical Support Call 1-800-677-6337. Phone support is available Monday–Friday, 8am to 8pm and Sunday 5pm to 12am, Eastern time. Visit our support site at *http://247.pearsoned.com*. E-mail support is available 24/7.

Understanding Weather and Climate

Fourth Edition

Edward Aguado

San Diego State University

James E. Burt

University of Wisconsin–Madison

PEARSON

Prentice Hall

Upper Saddle River, New Jersey 07458

Library of Congress Cataloging-in-Publication Data

Aguado, Edward.
  Understanding weather and climate/Edward Aguado, James E. Burt.—4th ed.
    p. cm.
  ISBN 0-13-149696-4
  1. Atmospheric physics. 2. Weather. 3. Climatology. I. Burt, James E. II. Title.
QC861.2.A27 2007
551.5—dc22

                        2006000631

Editor: Jeff Howard
Development Editor: Rebecca Strehlow
Production Editor: Tim Flem/PublishWare
Media Editor: Chris Rapp
Associate Editor: Amanda Griffith
Editor in Chief, Science: Daniel Kaveney
Editor in Chief of Development: Carol Trueheart
Executive Managing Editor: Kathleen Schiaparelli
Art Director: John Christiana
Director of Marketing, Science: Patrick Lynch
Senior Managing Editor, Art Production and Management: Patricia Burns
Manager, Production Technologies: Matthew Haas
Managing Editor, Art Management: Abigail Bass
Art Production Editor: Jess Einsig
Illustrations: Precision Graphics
Cartography: MapQuest.com
Creative Director: Juan R. Lopez
Director of Creative Services: Paul Belfanti
Page Composition: PublishWare
Manager of Formatting: Allyson Graesser
Manufacturing Manager: Alexis Heydt-Long
Manufacturing Buyer: Alan Fischer
Managing Editor, Science Media: Nicole M. Jackson
Media Production Editor: Dana Dunn, William Wells
Assistant Managing Editor, Science Supplements: Karen Bosch
Editorial Assistants: Margaret Ziegler; Nancy Bauer
Marketing Assistant: Renee Hanenberg
Cover Designer: John Christiana
Interior Designer: Studio Indigo
Director, Image Resource Center: Melinda Reo
Manager, Rights and Permissions: Zina Arabia
Interior Image Specialist: Beth Boyd-Brenzel
Cover Specialist: Karen Sanatar
Image Permission Coordinator: Debbie Hewitson
Photo Researcher: Jerry Marshall
Cover Photograph: Jim Reed/© Bettmann/CORBIS All Rights Reserved

To Lauren, William,
and Babsie June
—EA
To my parents, Martha F. Burt
and Robert L. Burt
—JEB

© 2007, 2004, 2001, 1999 by Pearson Education, Inc.
Pearson Prentice Hall
Pearson Education, Inc.
Upper Saddle River, New Jersey 07548

Printed in the United States of America

10 9 8 7 6 5 4 3 2

ISBN 0-13-149696-4

Pearson Education Ltd., *London*
Pearson Education Australia Pty., Limited, *Sydney*
Pearson Education Singapore, Pte. Ltd
Pearson Education North Asia Ltd., *Hong Kong*
Pearson Education Canada, Ltd., *Toronto*
Pearson Educación de Mexico, S.A. de C.V.
Pearson Education—Japan, *Tokyo*
Pearson Education Malaysia, Pte. Ltd.

Brief Contents

Contents

Part Two   Water in the Atmosphere   120

5   Atmospheric Moisture   122

Part Four   Disturbances   276

10 Midlatitude Cyclones   278

11 Lightning, Thunder, and Tornadoes   306

12 Tropical Storms and Hurricanes   352

Part Five   Human Activities   386

13 Weather Forecasting and Analysis   388

14 Human Effects: Air Pollution and Heat Islands   428

Part Six   Current, Past, and Future Climates   450

15 Earth's Climates   452

Part Seven   Special Topics and Appendices   510

17 Atmospheric Optics   512

CD-ROM Contents

Tutorials

RADIATION
Nature of Radiation
   Radiation Waves
   Electromagnetic Radiation
   Blackbody Radiation
Quiz
Solar vs. Terrestrial Radiation
Quiz

EARTH-SUN GEOMETRY
Simplifying Assumptions
   Parallel Rays
   Point Source
Quiz
Tilt and Solar Declination
   Solstices and Equinoxes
   Solar Declination
Quiz
Solar Position
   Zenith Angle
Beam Spreading and Depletion
Period of Daylight
Quiz

PRESSURE GRADIENTS
Surface Pressure Maps and Gradients
Pressure Aloft
   Pressure Gradients Aloft
   Isobaric Surfaces and Pressure Gradients
Quiz
Surface Maps and Upper-level Charts
Quiz

CORIOLIS
Fictitious Forces and Moving Frames of
Reference
Rotation Around the Vertical
   Earth Rotation
   Polar vs. Equatorial Rotation
   Local Rotation Rates
Rotation and Apparent Deflection
Quiz
Windspeed and Apparent Deflection
Quiz
Summary: Properties of the Coriolis Force
Grand Quiz

FORCES AND WINDS
Newton's Law and Motion
   Newton's Law
   Accelerations and Decelerations

Equation of Motion
Quiz
Frictionless Flow
   Geostrophic Flow
   Stability of Geostrophic Flow
   Gradient Flow
Quiz
Friction
Combination of Forces
   Interaction of Friction and PGF
   Vertical Changes: 3-D Views
   Cyclonic and Anticyclonic Patterns
Quiz

STABILITY
Buoyancy and Lapse Rates
   Dry Adiabatic Lapse Rate
   Saturated Adiabatic Lapse Rate
   Adiabatic Lapse Rates Compared
Quiz
Environmental Lapse Rates and Stability
   Absolutely Unstable Air
   Absolutely Stable Air
   Conditionally Unstable Air
   Level of Free Convection
Quiz

PRECIPITATION
Why Condensation Alone Does Not Produce
Precipitation
   Condensation Nuclei
   Distance Between Water Droplets
   Droplet Density
   Droplet Size
Quiz
Warm Cloud Precipitation
   When Droplets Collide
   After Droplets Collide
Quiz
Cold Cloud Precipitation
   Saturation Vapor Pressure
   Vapor Deposition
   Riming and Aggregation
Quiz

UPPER LEVEL WINDS
Pressure and Height
Quiz
Ridges and Troughs
   Definition
   Development

   3-D Views
Quiz
Westerly Winds and Waves
   Stationary Waves
   Westerly Waves
   Easterly Waves
   Wave Patterns
Quiz

EL NINO-SOUTHERN OSCILLATION
Air-sea Interactions
Quiz
Composite ENSO Conditions
   El Nino SST
   La Nina SST
   El Nino Teleconnections
   La Nina Teleconnections
   U.S. Regional Responses
Individual Events
   Historical El Niño and
     La Niña Animations
   Individual Event Anomalies
   Risk of Extreme Events
Quiz

MIDLATITUDE CYCLONES
Introduction
Cyclone Basics
   Low-level Pressure and Wind
   Fronts
Quiz
Cyclone Formation and Movement
   Divergence and Flow Aloft
   Movement and Evolution
Quiz
3-D Structure
   Dry Belt
   Warm Belt
   Cold Belt
Quiz

DOPPLER RADAR
Radar Pulses
Doppler Shift
Radar Rotation
Radar Sweeps
Examples
Quiz
Lessons from a Live Meteorologist

Weather in Motion

Weather Images

Media Library

Preface

Meteorology is perhaps the most dynamic of all the earth sciences. In no other sphere do events routinely unfold so quickly, with so great a potential impact on humans. Some of the most striking atmospheric disturbances (such as tornadoes) can take place over time scales on the order of minutes—but nevertheless have permanent consequences. Wind speeds of several hundred kilometers per hour accompany the most violent storms, and large-scale extreme events with attendant widespread destruction are common. Furthermore, even the most mundane of atmospheric phenomena influence our lives on a daily basis (for instance, the beauty of blue skies or red sunsets, rain, the daily cycle of temperature).

Atmospheric processes, despite their immediacy on a personal level and their importance in human affairs on a larger level, are not readily understood by most people. This is probably not surprising, given that the atmosphere consists primarily of invisible gases, along with suspended, frequently microscopic particles, water droplets, and ice crystals. In this book, our overriding goal is to bridge the gap between abstract explanatory processes and the expression of those processes in everyday events. We have written the book so that students with little or no science background will be able to build a nonmathematical understanding of the atmosphere.

That said, we do not propose to abandon the foundations of physical science. We know from our own teaching experience that physical laws and principles can be mastered by students of widely varying backgrounds. In addition, we believe one of meteorology's great advantages is that reasoning from fundamental principles explains so much of the field. Compared to some other disciplines, this is one in which there is an enormous payoff for mastering a relatively small number of basic ideas.

Finally, our experience is that students are always excited to learn the "why" of things, and to do so gives real meaning to "what" and "where." For us, therefore, the idea of forsaking explanation in favor of a purely descriptive approach has no appeal whatsoever. Rather, we propose merely to replace mathematical proof (corroboration by formal argument) with qualitative reasoning and appeal to everyday occurrences. As the title implies, the goal remains understanding atmospheric behavior.

Understanding Weather and Climate is a college-level text intended for both science majors and non-majors taking their first course in atmospheric science. We have attempted to write a text that is informative, timely, engaging to students and easily used by professors.

Distinguishing Features

Scientific Literacy and Currency. We have emphasized scientific literacy throughout the book. This emphasis gives students an opportunity to build a deeper understanding about the building blocks of atmospheric science and serves as tacit instruction regarding the workings of all the sciences. For instance, in Chapter 2 we cover the molecular changes that occur when radiation is absorbed or emitted, items that are often considered a "given" in introductory texts. In Chapter 3 these basic ideas are used to help build student understanding of why individual gases radiate and absorb particular wavelengths of radiation and illustrate how processes operating at a subatomic level can manifest themselves at global scales.

An emphasis on scientific literacy can be effectively implemented only if it is accompanied by careful attention to currency. We believe that two kinds of currency are required in a text: an integration of current *events* as they relate to the topic at hand, and an integration of current *scientific thinking*. For instance, the reader will find discussion of both recent hurricane activity and the most recent theories regarding the mechanisms that generate severe storms. Scientific literacy also calls for attention to language—after all, precision of language is an important distinguishing characteristic of science, one that sets it apart from other intellectual activities. With that in mind, we have tried to avoid some common statements of dubious accuracy, such as "warm air is able to hold more water vapor than cold air."

Media.   A fundamental feature of this book is the integration of the classic textbook model with the emerging areas of instructional technology. These nontraditional resources are delivered through the CD provided with the book and via the Internet. The software on the accompanying CD consists of several components. Perhaps most fundamental to our approach, the CD features 13 computer tutorials covering basic principles of atmospheric science. The software modules have undergone considerable testing and have been used successfully by thousands of students. They rely heavily on three-dimensional diagrams and animations to present material not easily visualized using conventional media. In choosing topics for the modules, we have emphasized material that is both difficult to master and has the potential to benefit from computer technology. We made no attempt to cover every chapter in the modules.

The software modules follow a tutorial style, with explanations and new vocabulary introduced incrementally, building on what was presented earlier in the modules and what was presented in the text. They have been reworked to substantially increase opportunities for student interaction with the material and they have also been redesigned to employ a more modern, visually attractive look and feel. The tutorials are best used in conjunction with the assigned readings. Students and professors will notice that the book and the tutorials are linked. First, the tutorials are described in Media Enrichment sections found at the end of every chapter. In addition, CD icons and screenshots in the book margins indicate that the topic under discussion is covered in a tutorial as well. We advise that you first view a tutorial in its entirety. If additional review is needed, you can easily move within a tutorial to the section under discussion. The tutorials are also linked back to the text. Icons are used to locate places where the book provides more detailed or background information about the topic at hand.

In addition to the tutorials, the CD for the fourth edition of the book contains other useful resources, including:

- *Weather in Motion* movies, depicting events and phenomena discussed in the text. Examples include a satellite movie showing clouds and temperature across the globe, three-dimensional simulations of thunderstorm development, and animations depicting variations in Earth's orbit. Like the tutorials, each movie is described in the Media Enrichment section at the end of every chapter.

- *Weather Images*, providing additional illustration of weather phenomena. These include photographs, satellite images, and computer diagrams complementing the text. Each is described in a Media Enrichment section.

- *Additional Media* resources, consisting of additional images, movies, and animations. These are intended for self-guided browsing by students and are therefore not explicitly mentioned in the text. A short description of each is found on the CD.

The CD-ROM for this fourth edition of the text has been completely revamped. Students will find more tutorials (13 instead of 8), a modern look to the content, and much more interactivity than before. The new navigation scheme is more intuitive and provides for easier movement from section to section. In addition, the software has been ported to a Flash™ environment. Flash is now a mature, stable technology and has become the most widely used platform for delivery of animated elements. Flash does not depend on other software, and no CD installation is required. Most importantly, the move to Flash allowed us to decouple animations from tutorials, so that instructors can play the animations in class by themselves, without any distracting textual material.

The Internet site **http://www.prenhall.com/aguado** includes review exercises, quantitative exercises, and other materials that allow users to query the Internet for timely atmospheric data and the ability to file one's exercise electronically. The site also offers instructors access to Syllabus Manager, Prentice Hall's online syllabus creation and management tool. The GradeTracker functionality allows students to answer study questions and track their progress in the gradebook (which can be imported by an instructor any time during the semester).

Instructor Flexibility.   During the writing process, we have enjoyed interacting with many of our colleagues who teach courses in weather and climate on a regular basis. It was especially interesting to see how little consensus exists regarding topic order (truth be told, the authors of this book don't agree on the optimal sequence). With this in mind, we tried to minimize the degree to which individual chapters depend on material presented earlier. Thus, instructors who prefer a chapter order different than the one we ultimately chose will not be disadvantaged. In this fourth edition we are using a novel approach to the sequencing of chapters on atmospheric moisture and pressure. The introductory chapter on pressure now resides ahead of the chapters on moisture. This facilitates student learning with regard to vapor pressure, saturation, and the importance of surface convergence into low pressure regions as a mechanism for cloud formation. Chapters on the distribution of pressure and wind patterns follow the discussion of moisture, as they have in previous editions.

Emphasis on Forecasting.   In addition to a comprehensive chapter on the topic, this text contains numerous examples of how physical principles are employed in weather forecasting. We have included several discussions of the use of thermodynamic diagrams in weather forecasting and analysis. These charts are extremely valuable but not immediately comprehensible to most students. To alleviate this problem, we introduce thermodynamic diagrams in a sequential fashion. That is, their use for plotting vertical temperature profiles is presented in the chapter on temperature. We expand upon this in the chapter on atmospheric moisture to show how various measures of humidity can also be determined with the aid of the charts. Thus, instructors can teach their students how to use these diagrams without inundating them with excessive detail all at once.

Readability.   In contrast to the more formal scientific style used in many science textbooks, we have chosen to adopt more casual prose. Our goal is to present the material in language that is clear, readable, and friendly to the student reader. We employ frequent headings and subheadings to help students follow discussions and identify the most important ideas in each chapter. As a rule, we keep technical language to a minimum.

Focus on Learning.   The chapters offer a number of study aids:

- **NEW!** *Did You Know.* This feature highlights interesting meteorological facts in every chapter.
- *Key Terms.* Key terms in each chapter are printed in boldface when first introduced. Most are also listed at the end of each chapter, along with the page number on which each first appears. All key terms are defined in the glossary at the end of the book and are available on the CD as well.
- *Focus on the Environment Boxes.* These boxes highlight environmental issues as they relate to the study of the atmosphere.
- *Physical Principles Boxes.* More mathematical in nature than the rest of the text, these boxes accommodate students who have a more quantitative interest in the topic. An understanding of the material in these boxes is not essential to an understanding of the material presented in the body of the text.
- *Special Interest Boxes.* These boxes highlight interesting topics related to the discussion at hand.
- *Forecasting Boxes.* These describe how the principles discussed in the chapter can be used in forecasting and often include simple "rules of thumb" that help students make their own forecasts.

- *Chapter Summary.* Each chapter concludes with a chapter summary that highlights the main points in the chapter.
- *Review Questions.* At the end of each chapter, you will find a list of questions about the subject of that chapter. These review questions test reading comprehension and can be answered from information presented in the chapter.
- *Quantitative Problems.* The *Understanding Weather and Climate* Web site features quantitative exercises to accompany each chapter. If you choose to work these problems, the computer will grade them and provide you with immediate feedback.
- *Critical Thinking Questions.* Each chapter concludes with questions that require students to use the thought process and material presented in the chapter to work out answers relevant to real-world questions.
- *Problems and Exercises.* These questions encourage students to work out solutions to numerical questions to gain a better understanding of chapter material.
- *Useful Web Sites.* These describe relevant and interesting Web sites with valuable information or current data at the end of each chapter.
- *Software Tutorial Icons.* Throughout the book, the software icons (see the list above) will appear where the topic under discussion is reviewed in the tutorial modules that accompany the book. We suggest that you use the tutorial if you are having trouble understanding this topic using the text alone.

Supplements

The authors and publisher have been pleased to work with a number of talented people to produce an excellent supplements package for the text. This package includes the traditional materials that students and professors have come to expect from authors and publishers, as well as some new kinds of supplements that involve electronic media.

FOR THE STUDENT

- *Integrated CD Resources.* The CD includes software tutorials that are explicitly linked to the book. They contain interactive exercises, animations, three-dimensional diagrams, and review quizzes. As mentioned, the tutorials cover the most difficult material presented. In addition, the CD contains *Weather in Motion* animations, *Weather Images*, and *Additional Media*, as described previously.
- *Internet Support.* The *Understanding Weather and Climate* Web site gives students the opportunity to further explore the book's topics using the Internet. The site contains numerous review exercises (from which students get immediate feedback), exercises to expand one's understanding of atmospheric science, and resources for further exploration. Please visit the site at **http://www.prenhall.com/ aguado**
- *Rand McNally Atlas of World Geography* (0-13-959339-X). This atlas includes 126 pages of up-to-date regional maps and 20 pages of illustrated world information tables. It is available FREE when packaged with *Understanding Weather and Climate*. Please contact your local Prentice Hall representative for details.

FOR THE PROFESSOR

- *Transparencies* (0-13-154783-6). More than 200 full-color acetates of illustrations from the text are available to qualified adopters.
- *Instructor Resource Center on CD* (0-13-154784-4). This resource provides high-quality electronic versions of photos and illustrations from the book, as well as customizable PowerPoint lecture presentations, images and animations extracted from the student tutorials, and the instructor's manual and test item file presented in Microsoft Word. Images are high-resolution, low-compression, 16-bit jpeg files. To

further guarantee classroom projection quality, all images are manually adjusted for color, brightness, and contrast. For easy reference and identification all images are organized by chapter. This powerful presentation tool is available at no cost to qualified adopters of the text.

- *Test Item File.* An extensive array of test questions accompanies the book. These questions are available in hard copy (0-13-154786-0) and also on disks formatted for Windows or Macintosh (0-13-154785-2).

Acknowledgments

The authors would like to extend our deepest appreciation to our editor and friend, Dan Kaveney, for his astute management of the entire project and for the careful thought he gave to the numerous issues that arose along the way. His support has been unfailing and his judgments superb, and he demonstrated preternatural resilience in the face of the authors' inclination to extend every deadline to the last possible minute. Our media editor Chris Rapp deserves special thanks for guiding what became a near-total revamping of the CD and Web-based digital resources. We benefited immensely from his exceptional managerial abilities, technical savvy, and intuitive understanding of instructional design. Numerous other people at Prentice Hall were particularly supportive, including Carol Trueheart, Editor-in-Chief of Development; Margaret Ziegler, Editorial Assistant for Geography; and Amanda Griffith, Associate Editor for Geography.

We greatly appreciate the assistance of our development editor, Rebecca Strehlow, who provided excellent feedback on revised and new material for this edition. Jerry Marshall very ably researched the photos for the text. Bobbie Dempsey, our copy editor, did a meticulous job of improving the final drafts of each chapter. Tim Flem of PublishWare did a terrific job of tracking details and putting the project into its final form. He was a delight to work with on a day-to-day basis throughout the production stages of the book.

We also benefited greatly from the advice of many professional educators and meteorologists. Mark Moede of the San Diego office of the National Weather Service provided valuable information on the day-to-day activities of weather forecasters.

We thank researchers and staff at government agencies and institutions throughout the world for creating many of the images and movies on the accompanying CD and for their willingness to make them freely available to use in projects like this. The following deserve special mention in this regard: Space Science and Engineering Center, University of Wisconsin–Madison; Dr. Richard Orville, Texas A&M University; Dr. Gary Huffines, University of Northern Colorado; Diana DeRubertis, University of California, Berkeley; and Paul Morin, University of Minnesota. We also thank CADRE design for their creative work in proposing novel solutions for our sometimes ill-formed teaching objectives, and for their skill in addressing a number of difficult technical issues related to the CD.

We must also offer special thanks to the many colleagues who spent valuable time and energy preparing in-depth reviews of our early efforts, many of whom have continued in this role through multiple revisions. In that regard we are particularly grateful to Robert Rohli of Louisiana State University, who read the entire manuscript with exceptional care and made many excellent suggestions. Additionally, we thank the following reviewers:

Rafique Ahmed, *University of Wisconsin–LaCrosse*
Greg Bierly, *Indiana State University*
Mark Binkley, *Mississippi State University*
Gerald Brothen, *El Camino College*
David P. Brown, *University of Arizona*
Adam W. Burnett, *Colgate University*
Gregory Carbone, *University of South Carolina*

R. E. Carlson, *Iowa State University*
Donna J. Charlevoix, *University of Illinois, Urbana–Champaign*
Christopher R. Church, *Miami University of Ohio*
John H. E. Clark, *Penn State University*
Andrew Comrie, *University of Arizona*
Eugene Cordero, *San Jose State University*
Mario Daoust, *Southwest Missouri State University*
Arthur (Tim) Doggett, *Texas Tech University*
Dennis M. Driscoll, *Texas A & M University*
Christopher M. Godfrey, *University of Oklahoma*
Rex J. Hess, *University of Utah*
Jay S. Hobgood, *Ohio State University*
Edward J. Hopkins, *University of Wisconsin–Madison*
Scott A. Isard, *University of Illinois*
Eric Johnson, *Illinois State University*
Scott Kirsch, *University of Memphis*
Daniel James Leathers, *University of Delaware*
Gong-Yuh Lin, *California State University–Northridge*
Anthony Lupo, *University of Missouri*
Thomas L. Mote, *University of Georgia*
Gerald R. North, *Texas A & M University*
Jim Norwine, *Texas A & M–Kingsville*
John E. Oliver, *Indiana State University*
Stephen Podewell, *Western Michigan University*
David Privette, *Central Piedmont Community College*
Azizur Rahman, *University of Minnesota, Crookston*
Robert V. Rohli, *Louisiana State University*
Steven A. Rutledge, *Colorado State University*
Arthur N. Samel, *Bowling Green State University*
Hans Peter Schmid, *Indiana University*
Brent Skeeter, *Salisbury State University*
Stephen Stadler, *Oklahoma State University*
S. Elwynn Taylor, *Iowa State University*
Mingfang Ting, *University of Illinois*
Graham Tobin, *University of South Florida*
Paul E. Todunter, *University of North Dakota*
Liem Tran, *Florida Atlantic University*
Donna Tucker, *University of Kansas*
Barry Warmerdam, *Kings River Community College*
Thompson Webb III, *Brown University*
Jack Williams, *University of Wisconsin-Madison*
Thomas B. Williams, *Western Illinois University*
Morton Wurtele, *University of California–Los Angeles*
Douglas Yarger, *Iowa State University*
Charlie Zender, *University of California, Irvine*

Energy and Mass

Summer solstice at Deadhorse in Prudhoe Bay, Alaska. The Sun appears as a series of glowing stars in a time elapsed photo.

```
SEVERE WEATHER STATEMENT

NATIONAL WEATHER SERVICE NORMAN OK

7:31 PM CDT MON MAY 3 1999

LARGE DAMAGING TORNADO MOVING THROUGH OKLAHOMA CITY METRO.
A LARGE TORNADO HAS CAUSED EXTENSIVE DAMAGE IN SOUTHERN
PORTIONS OF THE OKLAHOMA CITY METRO AREA. AT 7:31 PM THE
TORNADO WAS ENTERING SOUTHERN OKLAHOMA COUNTY JUST EAST OF
INTERSTATE 35 AND SOUTH OF CROSS ROADS MALL. PERSONS IN
SOUTHEAST OKLAHOMA CITY AND MIDWEST CITY ARE IN DANGER! IF
YOU LIVE NEAR THESE AREAS TAKE IMMEDIATE TORNADO PRECAU-
TIONS! THE TORNADO WAS MOVING NORTHEAST. THIS IS AN EX-
TREMELY DANGEROUS AND LIFE-THREATENING SITUATION. IF YOU
ARE IN THE PATH OF THIS LARGE AND DESTRUCTIVE TORNADO ...
TAKE COVER IMMEDIATELY.

BULLETIN - EAS ACTIVATION REQUESTED

TORNADO WARNING

NATIONAL WEATHER SERVICE NORMAN OK

10:17 PM CDT MON MAY 3 1999

THE NATIONAL WEATHER SERVICE IN NORMAN HAS ISSUED A

*TORNADO WARNING FOR NOBLE COUNTY IN NORTHEAST OKLAHOMA
UNTIL 10:45 PM CDT.

*AT 10:17 PM CDT A ONE-MILE WIDE TORNADO WAS REPORTED NEAR
MULHALL, MOVING NORTHEAST AT 30 MPH.

THIS TORNADO WILL CROSS INTERSTATE 35 NEAR OR SOUTH OF
PERRY. RESIDENTS IN NOBLE COUNTY NEED TO TAKE COVER NOW.
PERRY IS IN THE PATH OF THIS STORM.

THIS IS AN EXTREMELY DANGEROUS AND POTENTIALLY LIFE-THREAT-
ENING SITUATION. IF YOU ARE IN THE PATH OF THIS LARGE AND
DESTRUCTIVE TORNADO, TAKE COVER IMMEDIATELY.

TORNADOES ARE ESPECIALLY DANGEROUS AT NIGHT BECAUSE THEY
ARE HARD TO SEE. TAKE COVER NOW. IF A BASEMENT IS NOT
AVAILABLE, MOVE TO AN INTERIOR ROOM OR HALLWAY ON THE LOW-
EST FLOOR. LEAVE MOBILE HOMES AND VEHICLES FOR REINFORCED
SHELTER. STAY AWAY FROM WINDOWS.
```

During the afternoon and evening of May 3, 1999, dozens of emergency advisories like these two were issued to residents of the south central United States as one of the worst tornado outbreaks in decades punished the area. Worst hit was central Oklahoma, where at least 57 tornadoes killed 44 people and injured 748 others. In Kansas another 5 people were killed and 11,000 homes or businesses were damaged or destroyed. Over the next few days, the same storm that spawned the tornadoes slowly worked its way to neighboring states, killing 5 more people in Texas and Tennessee.

◀ Looking north at Atigun Pass, Alaska, on July 4, 1977, just before midnight. Because Atigun Pass is located north of the Arctic Circle, the midnight sun can be seen much of the summer.

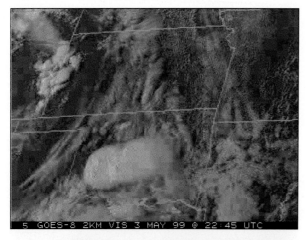

WEATHER IN MOTION
Satellite Movies of the May 3–4, 1999, Tornado Cluster

The storm system contained 11 major storm elements, called *supercells*, that followed a southwest-to-northeast path. Each supercell produced at least 1 tornado, with the most destructive supercell yielding 22 in a 3-hour period while covering a distance of 61 kilometers (km), or 38 miles (mi). The largest of its tornadoes was categorized as an F5, the most destructive possible, measuring as much as 1 km (0.6 mi) wide and staying on the ground for about 4 hours.

As terrible as the outbreak was, things would have been very much worse had it not been for the extensive forecasting, tracking, and warning capabilities of the National Weather Service. One study based on long-term observations suggests that as many as 700 people might have perished had it not been for the recent modernization of the Weather Service. People such as Air Force Captain John Millhouse would readily agree with that notion. Having heard one of the tornado warnings, he and his family moved to the basement of their home. When they emerged unscathed, only two walls of the house were left standing; the rest was totally destroyed.

Not all weather phenomena are as dramatic as tornadoes, but it is indisputable that weather exerts a tremendous impact on our day-to-day lives: We revel in the beauty of a sunrise, marvel at the power of hurricanes, rely on the rain to nourish our gardens, and complain about the heat and cold. And weather can throw a wrench into our travel plans—in the year 2000, the whims of the atmosphere caused more than 300,000 U.S. aviation delays! Table 1–1 summarizes recent estimates of annual financial and human costs of various types of extreme weather events within the United States for the 1990s. Though hurricanes and tornadoes can exact a terrible human cost, extreme heat and cold are the biggest killers in the United States, with floods and hurricanes producing the greatest financial costs. Table 1–1 actually understates the toll exerted by weather. Though severe weather kills about 900 people annually, that figure is compounded by an estimated 6000 weather-related traffic fatalities that accrue mostly from nonsevere events.

Despite its immediacy, most of us know relatively little about how and why the atmosphere behaves as it does. In the pages that follow, we hope to provide an account of both the how and the why, in ways that will lead you to understand the underlying physical processes. This chapter introduces the most basic elements of meteorology, laying the foundation for much of the rest of the book.

TABLE 1–1	**Averages of Annual Fatalities and Financial Costs of Weather Events in the United States During the 1990s**	
EVENT	**ANNUAL MEAN NUMBER OF FATALITIES**	**ANNUAL MEAN LOSS (ADJUSTED TO 1999 DOLLARS)**
Floods	98	$5,300,000,000
Hurricanes	21	$5,400,000,000
Winter Storms	57	$329,000,000
Tornadoes	56	$777,000,000
Extreme Heat	282	$85,000,000
Extreme Cold	292	$368,000,000
Lightning	69	$38,000,000
Hail	1	$938,000,000
Total	876	$13,000,000,000

Source: Data from Pielke and Carbone, *Bulletin of American Meteorological Society*, March 2002.

The Atmosphere, Weather, and Climate

The **atmosphere** is a mixture of gas molecules, microscopically small suspended particles of solid and liquid, and falling precipitation. **Meteorology** is the study of the atmosphere and the processes (such as cloud formation, lightning, and wind movement) that cause what we refer to as the "weather." **Weather** is distinct from **climate** in that the former deals with short-term phenomena and the latter with long-term patterns. Take a look outside your window and what you will see is the current weather. The current temperature, humidity, wind conditions, amount and type of cloud cover, and presence or absence of precipitation—these are all elements of weather.

    Climatology concerns itself with the same elements of the atmosphere that meteorology does, but on a different time scale. Understanding the average values of the atmospheric characteristics across Earth's surface is a major component of climatology, but it is not the only aspect of the discipline. Climatologists also want to know the variability of the weather elements. So, for example, it is useful to know that Boulder, Colorado, may have an average April temperature of 7 °C (45 °F), but this figure becomes more meaningful when one understands just how far the temperature might depart from the value on any given day. Frequencies of occurrence of weather events—such as hail or lightning—are aspects of climates. Climatology also incorporates the values of weather extremes, such as the lowest and highest recorded temperatures, maximum one-day or one-hour rainfall amounts, and the largest snow accumulation. The distinction between weather and climate might well be summarized by the old saying, "Climate is what you expect but weather is what you get."

Thickness of the Atmosphere

Every child has wondered, "How high is the sky?" There is no definitive answer to that question, however, because Earth's atmosphere becomes thinner at higher altitudes. A person in a rising hot-air balloon would be surrounded by an atmosphere that gradually becomes less dense. At some height, the air becomes so thin that the balloonist would pass out from a shortage of oxygen—but there would still be an atmosphere. At an altitude of 16 km, or 10 mi, the density of the air is only about 10 percent of that at sea level, and at 50 km (30 mi) it is only about 1 percent of what it is at sea level. Even at heights of several hundred kilometers above sea level, there is some air and, hence, an atmosphere. However, because no universally accepted definition exists of how much air in a given volume constitutes the presence of an atmosphere (is there an atmosphere if there is only, say, one molecule of air per cubic kilometer?), we have no way to establish its upper boundary.

    Viewed from Earth's surface, the atmosphere appears to be extremely deep. In reality, however, most of the atmosphere is contained within a relatively shallow envelope surrounding the oceans and continents. Let's assume for the sake of discussion that the upper limit of the atmosphere occurs at 100 km (60 mi) above sea level (in fact, 99.99997 percent of the atmosphere is below this height). By comparing this 100-km thickness with the 6500-km (4000-mi) radius of Earth, we see that the depth of the atmosphere is less than 2 percent of Earth's thickness from the center. This is evident in Figure 1–1, an image of Earth and its atmosphere taken from space. The top of the thunderstorm cloud

FIGURE 1–1

Despite its appearance from the surface, the atmosphere is extremely thin relative to the rest of Earth.

The total mass of the atmosphere, 5.14×10^{15} kilograms (kg),[1] is equivalent to 5.65 billion million tons—the amount of water that would fill a lake the size of California to a depth of 13 km (7.7 mi).

---

[1] 5,140,000,000,000,000 kg

probably has an altitude of about 12 km (7.5 mi), but when viewed from space, it appears to hug the ground. Though impressive when we look up at them, those clouds are no thicker than the skin of an apple, comparatively speaking.

Given the shallowness of the atmosphere, its motion over large areas must be primarily horizontal. Indeed, with some very notable exceptions, horizontal wind speeds are typically a thousand times greater than vertical wind speeds. However, we cannot overlook the wind's vertical motions. As we shall see, even small vertical displacements of air have an impact on the state of the atmosphere. Paradoxically, the least impressive motions—vertical—despite being hardest to detect and forecast, turn out to determine much of atmospheric behavior.

Composition of the Atmosphere

The atmosphere is composed of a mixture of invisible gases and a large number of suspended microscopic solid particles and water droplets. Molecules of the gases can be exchanged between the atmosphere and Earth's surface by physical processes, such as volcanic eruptions, or by biological processes, such as plant and animal respiration. Molecules can also be produced and destroyed by purely internal processes, such as chemical reactions between the gases.

Consider a gas that is constantly being cycled between the atmosphere and Earth's surface. If we think of the atmosphere as a reservoir for this gas, the gas concentration in the reservoir will remain constant so long as the input rate (the rate at which the gas moves from ground to atmosphere) is equal to the output rate (the rate at which the gas moves from atmosphere to ground). Under such conditions, we say that the concentration of the gas exists in a *steady state*.

Although the atmospheric concentration of a gas remains constant under steady-state conditions, individual molecules stay in the atmosphere for only a finite period of time before they are removed by whatever output processes are active. The average length of time that individual molecules of a given substance remain in the atmosphere is called the *residence time*. The residence time is found by dividing the mass of the substance in the atmosphere (in kilograms) by the rate at which the substance enters and exits the atmosphere (in kilograms per year). Thus, everything else being equal, gases that are rapidly exchanged between Earth's surface and the atmosphere have brief residence times, as do gases that have relatively low atmospheric concentrations.

Figure 1–2 illustrates the concept of a steady state and residence time. Parts (a) and (b) show the same mass constituting the "atmospheric reservoir," with the length of the input and output arrows indicating the rates at which gases are put into and removed from the reservoir. In (a), the input and output rates are equal, indicating a steady state. Both rates are small, however, meaning that any particular molecule of the gas has a long residence time. In (b), there is again a steady state, but the greater rate of input and output relative to reservoir size leads to a shorter residence time.

Atmospheric gases are often categorized as being permanent or variable, depending on whether or not their concentration is stable. **Permanent gases** are those that form a constant proportion of the atmospheric mass, whereas **variable gases** are those whose distribution in the atmosphere varies in both time and space.

Permanent gases account for the greater part of the atmospheric mass—99.999 percent—and occur in nearly constant proportion throughout the atmosphere's lowest 80 km (50 mi). Because of its chemical homogeneity, this region within 80 km of Earth's surface is called the **homosphere**. For most purposes, we consider the homosphere virtually the entire atmosphere.

Above the homosphere is the **heterosphere**, where lighter gases (such as hydrogen and helium) become increasingly dominant with increasing altitude. Because its composition varies with altitude, the heterosphere contains no truly permanent gases. Given the small mass at these altitudes, we won't pursue these and other details of the heterosphere but instead will deal almost exclusively with the homosphere.

FIGURE 1–2

The atmosphere can be thought of as a gas reservoir experiencing constant input and output by surface exchange and/or internal processes. If the inputs and outputs occur at the same rate, there is no net change in the content of the gas. In (a) the arrows depict a slow rate of exchange of a hypothetical gas. As a result, any molecule of that gas can be expected to remain for a long time before being cycled out of the atmosphere. In (b) the reservoir size is the same, but the exchange rate is much more rapid, so the gas has a shorter residence time.

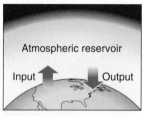

(a)

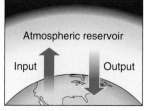

(b)

TABLE 1–2	Permanent Gases of the Atmosphere		
CONSTITUENT	FORMULA	PERCENT BY VOLUME	MOLECULAR WEIGHT
Nitrogen	N_2	78.08	28.01
Oxygen	O_2	20.95	32.00
Argon	Ar	0.93	39.95
Neon	Ne	0.002	20.18
Helium	He	0.0005	4.00
Krypton	Kr	0.0001	83.8
Xenon	Xe	0.00009	131.3
Hydrogen	H_2	0.00005	2.02

THE PERMANENT GASES

The homosphere is composed mostly of nitrogen and oxygen, with small amounts of the inert gases argon and neon and even smaller amounts of several other gases (Table 1–2). Atmospheric **nitrogen** occurs primarily as paired nitrogen atoms bonded together to form single molecules denoted N_2. Nitrogen gas has a molecular weight of 28.02, meaning that on average the mass of one N_2 molecule is slightly in excess of that of a combined total of 28 protons and neutrons. Almost all N atoms contain 7 protons and 7 neutrons, but some are heavier, having 8 neutrons. As a result, the average molecular weight of N_2 is slightly greater than 28.

Nitrogen is a stable gas that accounts for 78 percent of the volume of all the permanent gases, or 75.5 percent of their mass (the difference arises because the gases have different molecular weights but occupy the same volume under similar conditions). The processes that add and remove nitrogen from the atmosphere occur very slowly, so it has a very long residence time—42 million years.

Despite the fact that it makes up more than 75 percent of the atmosphere, N_2 has relatively little effect on most meteorological and climatological processes. (However, as we shall discuss later, nitrogen-bearing gases, such as nitrous oxide, are quite important to Earth's climate.)

The second most dominant gas, **oxygen** (O_2), constitutes 21 percent of the volume of the atmosphere and 23 percent of its mass. Oxygen is crucial to the existence of virtually all forms of life. Like nitrogen, the oxygen molecules of the atmosphere consist mostly of paired atoms, called *diatomic oxygen*. Their residence time is about 5000 years. Together, nitrogen and oxygen account for 99 percent of all the permanent gases, with **argon** making up most of the remainder. Removal processes are so slow for argon that its residence time is extremely long.

VARIABLE GASES

The variable gases account for only a small percentage of the total mass of the atmosphere (Table 1–3). Despite their relative scarcity, some of these gases affect the behavior of the atmosphere—and even your own physical comfort.

TABLE 1–3	Variable Gases of the Atmosphere		
CONSTITUENT	FORMULA	PERCENT BY VOLUME	MOLECULAR WEIGHT
Water Vapor	H_2O	0.25	18.01
Carbon Dioxide	CO_2	0.038	44.01
Ozone	O_3	0.01	48.00

WATER VAPOR

The most abundant of the variable gases, **water vapor**, accounts for about one-quarter of 1 percent of the total volume of the atmosphere. Because the source of water vapor in the atmosphere is evaporation from Earth's surface, its concentration normally decreases rapidly with altitude, and most atmospheric water vapor is found in the lowest 5 km (3 mi) of the atmosphere.

Water is constantly being cycled between the planet and the atmosphere in what is called the **hydrologic cycle**, which is described in Chapter 5. Earth is sometimes referred to as the "water planet" because three-quarters of its surface is covered by oceans, ice sheets, lakes, and rivers. In addition, much water exists beneath Earth's surface both in saturated (groundwater) formations and in partially saturated upper soil layers. Water continuously evaporates from both open water and plant leaves into the atmosphere, where it eventually condenses to form liquid droplets and ice crystals. These liquid and solid particles are removed from the atmosphere by precipitation as rain, snow, sleet, or hail. Because of the rapidity of global evaporation, condensation, and precipitation, water vapor has a very short residence time of only 10 days.

We all know how damp and muggy the air feels when the water vapor content is high and how parched our skin can get when the air is dry (not to mention "bad hair days" when the moisture content is extremely high or low). Despite the wide range of physical comfort levels we experience in response to variations in water vapor content, the actual range of water vapor content is really quite limited. Near Earth's surface, the water vapor content ranges from just a fraction of 1 percent of the total atmosphere over deserts and polar regions to about 4 percent in the tropics. This means that at most we would find that 4 out of every 100 air molecules are water vapor. (Outside of the tropics, water vapor content does not usually exceed 2 percent.) At higher altitudes, water vapor is even rarer.

Despite being a relatively small portion of the atmosphere, water vapor is extremely important. Not only is it the source of the moisture needed to form clouds, it is a very effective absorber of energy emitted by Earth's surface. (We describe radiant energy in the next chapter.) Its ability to absorb Earth's thermal energy makes water vapor one of the "greenhouse gases" we discuss in Chapter 3.

Keep in mind that water vapor is not the same as small droplets of liquid water. Water vapor exists as individual gas molecules. Unlike the molecules of liquids and solids, water vapor molecules are not bonded together. In that regard, water vapor is similar to N_2, O_2, and other atmospheric gases. Unlike other gases, however, it readily changes phase into liquid and solid forms both at Earth's surface and in the atmosphere.

Although water vapor is an invisible gas, some satellite systems can detect the amount of water vapor in the air and display its varying content on images (Figure 1–3). These images typically show that water vapor exhibits large changes over very short distances, more so than other variable gases. Displays of water vapor content can be quite valuable to forecasters, who use the images to determine broad-scale wind patterns in the middle and upper atmosphere. The images also help meteorologists identify the boundaries between adjoining bodies of air.

CARBON DIOXIDE

Another variable gas is **carbon dioxide**, CO_2. We do not physically sense variations in the amount of carbon dioxide present in the atmosphere, as we do with water vapor; and yet, as you will see shortly, we can't ignore these variations. Increases in the carbon dioxide content of the atmosphere may have some important climatic consequences that could greatly affect human societies.

Carbon dioxide currently accounts for about 0.038 percent of the atmosphere's volume. When a gas occupies such a small proportion of the atmosphere, we often express its content as parts per million (ppm) rather than percent (parts per hundred). Thus, the current atmospheric concentration of CO_2 is about 380 ppm. It is supplied to the atmosphere by plant and animal respiration, the decay of organic

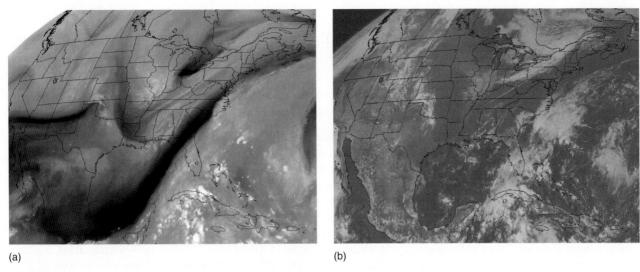

(a)   (b)

FIGURE 1–3

Paired satellite images based on a system that detects the presence of water vapor (a) and one that gives a visible rendition of cloud tops (b). Notice that though the two patterns are somewhat similar, the water vapor image shows a broader distribution of moisture than does the image showing actual clouds.

material, volcanic eruptions, and natural and *anthropogenic* (human-produced) combustion. Carbon dioxide is removed from the atmosphere by **photosynthesis**, the process by which green plants convert light energy to chemical energy. Photosynthesis uses energy from sunlight to convert water absorbed by the roots of plants and carbon dioxide taken in from the air into the chemical compounds known as carbohydrates, which both support plant processes and ultimately nourish the animal kingdom. In this process, oxygen molecules are released into the atmosphere as a byproduct (see *Box 1–1, Focus on the Environment: Photosynthesis, Respiration, and Carbon Dioxide*).

In recent decades, the rate of carbon dioxide input to the atmosphere has exceeded the rate of removal, leading to a global increase in concentration. Figure 1–4 plots the record of carbon dioxide content obtained from the Mauna Loa Observatory. (Taken from a 3400-m, or 11,150-ft, elevation on one of the Hawaiian Islands, Mauna Loa's observations are considered representative for the Northern Hemisphere.) Since the 1950s, the concentration of CO_2 has increased at a

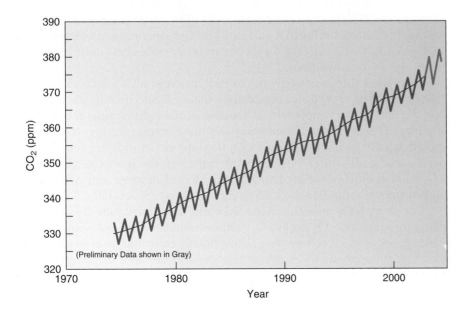

FIGURE 1–4

The carbon dioxide content of the atmosphere has been steadily increasing over the last half-century due to human activities. The data are obtained from the Mauna Loa Observatory, with the zig-zag line showing the seasonal cycle in the growth and decay of plants. The smooth line corresponds to annual average values, illustrating the long-term trend.

Photosynthesis, Respiration, and Carbon Dioxide

Without the process of photosynthesis, Earth would have an entirely different atmosphere—and would probably be without life as we know it. Through photosynthesis, plants utilize light energy from the Sun and make food. Because it requires sunlight, photosynthesis occurs only during the day. It also requires chlorophyll, an organic substance found in green plants and some single-celled organisms. Photosynthesis converts solar energy, water, and carbon dioxide into simple carbohydrates. These can then be converted to complex carbohydrates, starches, and proteins, all of which supply plants with the material for their own growth. Plants in turn provide the basic nutrients for grazing and browsing animals.

In addition to photosynthesis, an exchange of gases takes place through the leaves of plants—**respiration**. Respiration provides the mechanism by which plants obtain the oxygen they need to perform their metabolic processes. (For animals, respiration is synonymous with breathing. Though plants don't have lungs, they take in oxygen through their leaves.) Unlike photosynthesis, respiration occurs during both day and night. A plant must "fix" carbon into its cells in order to grow; to obtain this carbon, the photosynthesis rate in the plant must exceed the respiration rate. Thus, during periods of plant growth more carbon dioxide is being removed from the air than is being added through respiration. After a plant dies, however, it no longer takes in CO_2 from the air. Instead, its stored carbon is oxidized and released back to the atmosphere as CO_2 as the plant decomposes. If no major changes occur in the amount and distribution of vegetation, CO_2 intake (for photosynthesis) balances CO_2 output (from respiration and decay) over the course of a year, and the total store of atmospheric CO_2 is unaffected by plant growth.

There is, however, another factor governing the CO_2 balance. Under some circumstances, dead plant material is quickly buried beneath the surface, does not decompose, and therefore does not release its stored carbon back into the atmosphere. Instead, over millions of years the material is transformed into fossil fuels, such as petroleum or coal, which we extract from the ground and burn for heat and energy, yielding CO_2 as a combustion product. In doing so, we release carbon that otherwise would have remained underground for eons.

Although we speak of "adding" to the atmosphere, it's probably more accurate to think of "moving" carbon back to the atmosphere. Regardless, the result is certainly an increase in atmospheric CO_2, with potentially global consequences.

Recognizing the role of fossil fuel combustion on the carbon balance of the atmosphere, 160 of the world's nations met in 1997 at Kyoto, Japan, to formally agree on a multiyear plan to lower CO_2 emissions. Under this plan, averaged over the 5-year period from 2008 to 2012, industrialized countries would reduce CO_2 emissions by 6 to 8 percent of 1990 values. To be legally binding, the plan would not only have to be signed by the representatives of the participating countries, but also formally ratified by their governments. In March 2002 the European Union formally ratified the treaty, followed by Canada in December of that year. Ratification by Russia in November 2004 satisfied the remaining condition for approval and allowed the Kyoto Agreement to come into force on February 16, 2005. The United States, which has not signed on to this agreement, is the only major industrialized nation not to have done so.

rate of about 1.8 ppm per year. The increase has occurred mainly because of an increase in anthropogenic combustion and, to a lesser extent, deforestation of large tracts of woodland (Figure 1–5). The increase has received considerable scientific and media attention because CO_2 (like water vapor) effectively absorbs radiation emitted by Earth's surface. The CO_2 increases projected to the year 2100 would double existing levels of carbon dioxide and could lead to a warming of the lower atmosphere.

Figure 1–4 shows not only the overall increase in CO_2 levels in the Northern Hemisphere but also shows seasonal oscillations. Specifically, the amount of carbon dioxide in the atmosphere is greatest in the early spring and lowest in late summer. The springtime maximum occurs because during the winter plant growth is slow, so plants take less CO_2 from the air. In addition, leaf litter has been decomposing all winter, which means carbon in the litter has been oxidized to CO_2 and has entered the air. During the period of summer regrowth, carbon is removed from the atmosphere, and so carbon dioxide levels fall.

For every CO_2 molecule removed from the atmosphere by photosynthesis, plants produce one O_2 molecule. Atmospheric O_2 does not show any seasonal variation, however, because the photosynthetic contribution is so small relative to the huge atmospheric O_2 reservoir. Carbon dioxide has a residence time of about 150 years.

FIGURE 1–5
Burning rain forest in Guatemala. Deforestation by humans has contributed to the increase of carbon dioxide in the atmosphere because it reduces the amount of plant material that can carry out photosynthesis.

OZONE

The form of oxygen in which three O atoms are joined to form a single molecule is called **ozone**, O_3. This substance is something of a paradox. The small amount of it that exists in the upper atmosphere is absolutely essential to life on Earth; near Earth's surface, however, it is a major component of air pollution, causing irritation to lungs and eyes and damage to vegetation. Fortunately, ozone occurs only in minute amounts in the lower atmosphere, so that even in highly polluted urban air its concentration may be only about 0.15 ppm (that is, 15 out of every 100 million molecules are ozone). Aloft, at altitudes of 25 km, its concentration might be 50 to 100 times higher (up to about 15 ppm).

Ozone in the part of the upper atmosphere called the *stratosphere* is vital to life on Earth because it absorbs lethal ultraviolet radiation from the Sun. Why is ozone mostly found in the stratosphere rather than at Earth's surface? The answer is complicated, as literally hundreds of chemical reactions govern its abundance. To simplify, it can be said that ozone forms when atomic oxygen (O) collides with molecular oxygen (O_2). Atomic oxygen is produced in the very upper reaches of the atmosphere, but little ozone forms there because of low air density at those altitudes. Nearer Earth's surface (but still high up in the atmosphere), the chances of O atoms colliding with O_2 molecules are greater, so the highest ozone values are found there.

When it absorbs ultraviolet radiation, ozone splits into its constituent parts ($O + O_2$), which can then recombine to form another ozone molecule. Through these reactions, ozone is continually being broken down and re-formed to yield a relatively constant concentration in the ozone layer (see *Box 1–2, Focus on the Environment: Depletion of the Ozone Layer*).

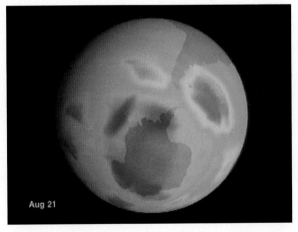

Aug 21

WEATHER IN MOTION
The Ozone Hole

METHANE

Another variable gas is **methane**, CH_4. For hundreds of thousands of years the concentration of methane has varied cyclically, rising and falling on roughly a 23,000-year basis. The pattern was broken about 5000 years ago, when methane began increasing from the expected downward trend. With industrialization the rate of increase accelerated, so that values have more than doubled over the

Depletion of the Ozone Layer

In 1972 Sherwood Rowland and Mario Molina, atmospheric chemists then working at the University of California at Irvine, proposed that certain human-produced chemicals called chlorofluorocarbons (CFCs) could be carried naturally into the stratosphere and damage the ozone layer. CFCs are widely used in refrigeration and air conditioning, in the manufacture of plastic foams, and as solvents in the electronics industry. Although banned in the United States and Canada as propellants in aerosol cans, they are still used in this capacity elsewhere.

Knowing that CFCs do not easily react with other molecules in the lower atmosphere, Rowland and Molina proposed that these molecules could reach the stratosphere intact, where they would break down and release free atoms of chlorine (Cl). Under certain circumstances, chlorine atoms can effectively destroy ozone molecules. In the first step of this process, a chlorine atom reacts with an ozone molecule to produce O_2 and chlorine monoxide (ClO). Next, an oxygen atom (O) reacts with ClO, creating another O_2 molecule while freeing the chlorine atom (Cl). Note that the chlorine atom that first reacted with the ozone molecule is still present and capable of reacting again with another ozone molecule. This fact makes chlorine atoms able to repeatedly break down ozone molecules. In fact, as many as 100,000 ozone molecules can be removed from the atmosphere for every chlorine molecule present. Rowland and Molina's theory is now accepted as fact, and the two scientists were rewarded for their work with a Nobel prize in 1995.

The most severe ozone depletion occurs every October (spring in the Southern Hemisphere) and persists for several months. Why is the ozone hole found over the Antarctic during the spring? The answer is fairly complex, but a variety of factors can be cited here. First, air currents surrounding Antarctica isolate the region from the rest of the hemisphere, so there is less mixing with ozone-rich air from the north. Another factor is the unusual chemistry of clouds found in the Antarctic stratosphere. At the very low temperatures found there, clouds are largely made up of nitric acid and water, rather than ordinary water ice. Processes involving these clouds allow certain chlorine compounds to accumulate. When

the dark Antarctic winter ends, the burst of ultraviolet radiation breaks apart these compounds to create free chlorine atoms (Cl). These Cl free radicals readily destroy ozone, as described previously. A very recent discovery shows that depletion begins at the rim of Antarctica, where sunlight arrives first, and works its way poleward. Thus depletion begins in June at 65° S but not until late August at 75° S. We should emphasize that chlorine is not usually abundant in the Antarctic stratosphere. Rather, what chlorine exists is in a form able to destroy ozone, thanks to the unique conditions that take place in the spring (see Figure 1).

Significant stratospheric ozone depletion has also been detected over much of Europe and North America, including a less pronounced hole over the Arctic. Nobody knows how long the ozone will continue to decrease, or how depleted it may become, but in November 1999 satellite and land-based data revealed abnormally low ozone levels over northwestern Europe. Stratospheric zone values were 30 percent below normal for that

time of year, resulting in levels nearly as low as those normally observed over the Antarctic.

Government and private industry have taken action to help reduce the amount of CFCs entering the atmosphere. In accordance with the Montreal Protocol of 1987 and subsequent conferences, the world's developed countries have ceased production of CFCs, and the developing nations are scheduled to do likewise by 2006. CFCs have lifetimes in the atmosphere of about 100 years, so an immediate reduction in the ozone hole will not occur. But progress has been made; since 1997 there has been a decline in the amount of chlorine in the stratosphere, and the size of the ozone hole appears to have stabilized. A recent study predicts that with full compliance with the Montreal Protocol, actual decreases in the size of the ozone hole might become detectable sometime between 2015 and 2045. Thus, curbing CFC emissions illustrates how solid science, combined with international cooperation, can have a major impact on the protection of the environment.

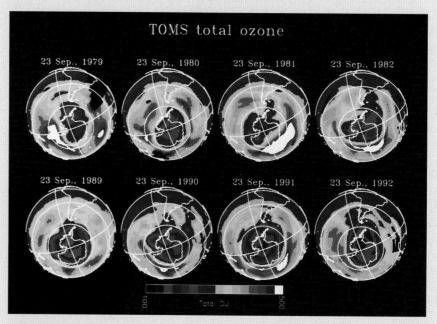

FIGURE 1

A series of satellite images showing a reduction of the stratospheric ozone over Antarctica. The area in red indicates the extent of the "ozone hole."

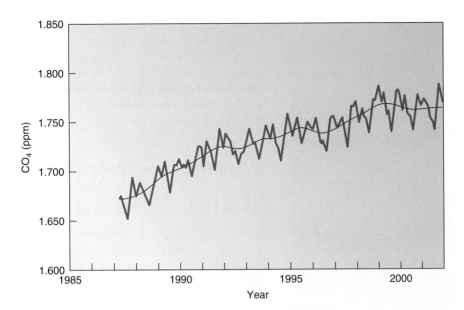

FIGURE 1–6
Methane is a gas that has varied greatly over long time scales. The most recent increase began about 5000 years ago and continued until roughly 1999. Since then its concentration has remained nearly constant.

last 200 years, reaching about 1.7 ppm in 1999 (Figure 1–6). Most of this increase is attributed to rice cultivation, biomass burning, and fossil fuel extraction (coal and petroleum mining). In the half-dozen years since 1999 atmospheric methane values stabilized, apparently because of the decline in oil exploration in Russia since 1990. The residence time for methane is about 10 years, and the current input of the gas into the atmosphere now approximately equals its natural removal.

Despite its low concentration in the atmosphere, methane is an extremely effective absorber of thermal radiation emitted by Earth's surface. Thus, increases in atmospheric methane levels could play a role in the warming of the atmosphere.

AEROSOLS

Small solid particles and liquid droplets in the air (excluding cloud droplets and precipitation) are collectively known as **aerosols**.[1] Although we associate them with polluted urban atmospheres, aerosols are formed by both human and natural processes. Thus, they are ever present, even in areas far from human activity. They normally occur at concentrations of about 10,000 particles per cubic centimeter (cm^3) over land surfaces, which is equivalent to roughly 17,000 particles per cubic inch ($in.^3$).

The smallest of these particles have radii on the order of 0.1 micrometer (μm; one-millionth of a meter) and are believed to form primarily from the chemical conversion of sulfate gases to solids or liquids. Larger particles are introduced into the air directly as wind-generated dust, volcanic ejections, sea spray, and combustion byproducts. Because these particles are extremely small, most fall so slowly that even the smallest of vertical motions keep them suspended in the atmosphere. (The most effective mechanism for removing aerosols is their capture by falling precipitation.) Aerosols typically have life spans of a few days to several weeks.

Aerosols have some noticeable effects on the atmosphere. Urban smog, which includes aerosols, severely reduces visibility, and dust storms can reduce visibility

DID YOU KNOW?

On average, each breath a person takes brings into the lungs about 1000 cm^3 (1 liter, or 64 $in.^3$) of air. Given the average size and concentration of aerosols, each of us draws about 1 trillion aerosols into our lungs several times each minute, or about two tablespoons of solids each day.

---

[1]The term **particulate** is often used interchangeably with *aerosol*. This terminology may be confusing for some people, who would take *particulate* to mean solid particles only. However, at these microscopically small sizes, it is difficult to make a distinction between the liquid and solid states; thus, there is little reason to draw a fine distinction.

FIGURE 1–7
A dust storm in Australia.

to near zero when large volumes of soil are dislodged from the surface (Figure 1–7). Aerosols also play a major role in the formation of cloud droplets because virtually all cloud droplets that form in nature do so on suspended aerosols called **condensation nuclei**. Condensation nuclei are discussed in more detail in Chapter 5.

Vertical Structure of the Atmosphere

As we have seen, the atmosphere has no distinct upper boundary; the air simply becomes less and less dense with increasing altitude. We also know that its composition remains nearly constant up to a height of about 80 km (50 mi)—that is, within the homosphere. Yet despite this gradual change in density and nearly constant chemical composition with height, meteorologists still find it convenient to divide the atmosphere vertically into several distinct layers. Some layers are distinguished by electrical characteristics, some by chemical composition (this is the homosphere/heterosphere distinction discussed earlier), and some by temperature characteristics. Together with the change in density with height, this layering of the atmosphere gives it its **structure**. In this section, we look first at the changing density of the atmosphere, then at its temperature changes, and finally at the electrical properties of the layer called the *ionosphere*.

DENSITY

The **density** of any substance is the amount of mass of the substance (expressed in kilograms)[2] contained in a unit of volume (1 cubic meter, m^3). In gases, individual molecules have no attachment to one another and move about randomly. One characteristic of gases is that no definite limit exists to the amount of mass that can fill a given volume—molecules can always be added to or removed from the volume, resulting in a density change. Alternatively, the volume containing a fixed mass of gas can decrease (as happens in a bicycle pump as the piston moves inward and compresses the air), resulting in a density increase as the constant mass is squeezed into a smaller and smaller volume.

Like any other assemblage of gases, therefore, the atmosphere is compressible. When we feel the weight of something, we are being subjected to the downward gravitational force exerted by the overlying mass. At lower altitudes, there is more overlying atmospheric mass than at higher altitudes. Because air is compressible and subjected to greater compression at lower elevations, the density of the air at lower levels is greater than that aloft. Figure 1–8 illustrates this principle. Despite its simplicity, this concept is a key to understanding concepts presented later.

Air density decreases gradually with increasing altitude. At sea level, the air density is normally about 1.2 kg/m^3. By comparison, at Denver, Colorado, the "Mile High City," the air density is only about 85 percent of that. As a result,

FIGURE 1–8
Because of compression, the atmosphere is more dense near its base and progressively "thins out" with altitude.

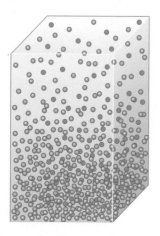

[2]Although a kilogram corresponds to about 2.2 pounds in the English system of measurement used in the United States, there is a difference between the two. A kilogram is a unit of mass. A pound, however, is a unit of weight equal to mass times the acceleration of gravity. On Earth 1 kilogram of a substance has a weight of 2.2 pounds. If the same kilogram goes to the Moon, its mass will not change but its weight will be only one-sixth of what it was on Earth. This is true because the acceleration of gravity on the Moon is one-sixth as strong as the acceleration of gravity on Earth.

The Three Temperature Scales

FAHRENHEIT AND CELSIUS

At one time, the temperature scale used all over the world was the Fahrenheit scale. Invented in the early 1700s by Gabriel Fahrenheit, it assigns values of 32° and 212° to the freezing and boiling points of water.* There are thus 180 Fahrenheit degrees between freezing and boiling. Although the Fahrenheit scale has been replaced in Canada and nearly every other country around the world, in the United States it is still the scale used by the general public.

The other familiar scale for measuring temperature is the Celsius scale, named for Anders Celsius, who formulated it in 1742. The Celsius scale assigns values of 0° and 100° to the freezing and boiling points of water, so there are only 100 Celsius degrees between the two points. This means that a Celsius degree is larger than a Fahrenheit degree. Thus, for example, a 2 °C change is larger than a 2 °F change. However, this is *not* to say a temperature expressed in °C is always higher than the same temperature expressed in °F; it can be higher or lower. To convert from Celsius to Fahrenheit, we use the following formula:

$$°F = 9/5 \ °C + 32$$

To convert from Fahrenheit to Celsius, we use

$$°C = 5/9(°F - 32)$$

You can use these formulas to verify that $-40 \ °F = -40 \ °C$.

KELVIN

Though useful in everyday applications, both the Fahrenheit and Celsius scales have a significant shortcoming—they allow for negative values. This

*To be precise, these values apply to pure water at sea level. In addition, water doesn't freeze spontaneously at 32 °F, so "melting point" is a better term than "freezing point."

is not the case for other units of measurement. For example, buildings don't have negative heights and weights, cars don't travel at negative speeds, and children don't have negative ages. But negative temperature values give the impression that substances can have negative heat contents—a situation that is physically impossible. To overcome this problem, scientists use a different scale for the measurement of temperature, called the **Kelvin scale**. In this system, the temperature 0 K is the lowest possible temperature that can exist in the universe. (Notice that we omit the degree notation with this scale and just refer to the number of kelvins,† K). A temperature of 0 K implies no heat, and it is therefore impossible for subzero temperatures to exist with this scale.

†The Kelvin scale is abbreviated with a capital K; the unit of measurement is spelled with a lowercase k.

The Kelvin scale is really a modified form of the Celsius scale insofar as the increments of the two are equal. Thus, if the temperature increases 1 degree Celsius, it also increases 1 kelvin. The only difference between the two is the starting point; 0 K corresponds to −273.16 °C. Therefore, conversion from Celsius to Kelvin is simply

$$K = °C + 273.16$$

To convert from Kelvin to Celsius, we use

$$°C = K - 273.16$$

Figure 1 shows the Kelvin, Celsius, and Fahrenheit scales.

It is vitally important to use Kelvin units in the radiation formulas presented in this chapter. The use of Fahrenheit or Celsius temperatures would result in erroneous or, at worst, nonsensical results.

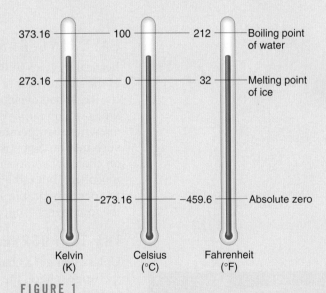

FIGURE 1
The Kelvin, Celsius, and Fahrenheit scales.

punted footballs and batted baseballs experience a corresponding decrease in air resistance and travel farther than they would at sea level.

One way to think about density is in terms of the average distance a molecule travels before colliding with another, called the **mean free path**. Near Earth's surface, the mean free path is a mere 0.0001 millimeter (mm). In contrast, at 150 km (93 mi) above sea level, a molecule travels about 10 m (3.3 ft) before colliding with another; at 250 km (155 mi), a molecule is likely to travel a full kilometer (0.62 mi) before meeting another.

FIGURE 1–9
The temperature profile of the atmosphere results in four
layers based on thermal characteristics.

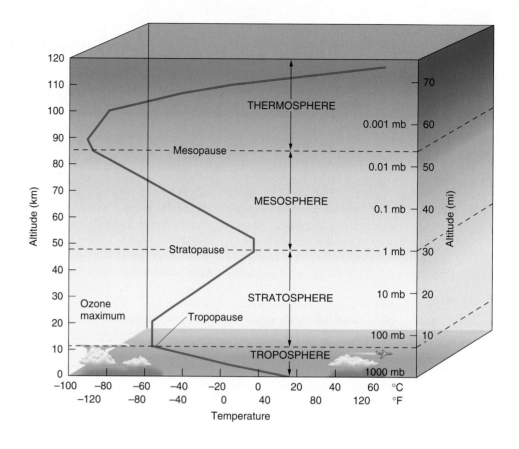

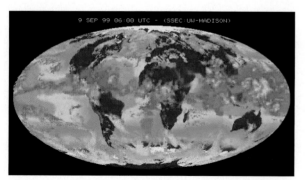

WEATHER IN MOTION
A Global Montage of Clouds

LAYERING BASED ON TEMPERATURE PROFILES

Vertical motions in the atmosphere are usually slow and limited in range because of the shallowness of the atmosphere. Nonetheless, such motions greatly influence the likelihood of cloud development, precipitation, and thunderstorm activity. Air whose temperature decreases rapidly with height is easily moved vertically, while air whose temperature either decreases slowly or increases with height resists such motion. Scientists therefore divide the atmosphere into four layers based not on chemical composition (which is relatively constant throughout most of the atmosphere) but rather on how mean temperature varies with altitude. The average temperature profile shown in Figure 1–9, called the **standard atmosphere**, shows the four layers: troposphere, stratosphere, mesosphere, and thermosphere.

THE TROPOSPHERE

Unless you get a chance to fly in a military jet, you will spend your entire life in the **troposphere**, the lowest of the four temperature layers. The name is derived from the Greek word *tropos* ("turn") and implies an "overturning" of air resulting from the vertical mixing and turbulence characteristic of this layer. The troposphere is where the vast majority of weather events occur and is marked by a general pattern in which temperature decreases with height. Despite being the shallowest of the atmosphere's four layers, the troposphere contains 80 percent of its mass. This is possible, of course, because air is compressible.

The depth of the troposphere varies considerably, ranging from 8 to 16 km (3.6 to 10 mi), with a mean of about 11 km (6.6 mi). The altitude at which the troposphere ends depends largely on its average temperature, being highest where the air is warm and lowest in cold regions. The troposphere is therefore thicker over the tropics than over the polar regions and thicker during the summer than during the winter.

Temperatures vary greatly from bottom to top in the troposphere. The average global temperature is about 15 °C (59 °F) near the ground but only about –57 °C

(–71 °F) at the top of the troposphere, an average decrease of about 6.5 °C/km (3.6 °F/1000 ft). Thus, you might feel perfectly comfortable inside an airplane at an altitude of 10 km (33,000 ft), but the temperature outside would likely be in the neighborhood of –50 °C (–58 °F). At the top of the troposphere, a transition zone called the **tropopause** marks the level at which temperature ceases to decrease with height.

An apparent paradox is associated with the trend toward decreasing temperature with height in the troposphere. We know that Earth is heated almost entirely by the Sun, which might make you think temperature should *increase* with height during the day (because as we move upward from the surface, we get closer to the Sun). The explanation for this paradox is that the atmosphere is relatively transparent to most types of radiant energy emitted by the Sun. In other words, a large portion of the sunlight passing through the atmosphere is not absorbed and therefore does not contribute greatly to its warming. As we shall see in Chapter 3, the most important direct source of energy for the atmosphere is not downward-moving solar radiation but rather energy emanating upward from Earth's surface.

Despite the strong tendency for temperature to decrease with altitude in the troposphere, it is not uncommon for the reverse situation to occur. Such situations, where temperature increases with height, are known as **inversions**. An inversion is significant because it inhibits upward motion and thereby allows high concentrations of pollutants to be confined to the lowest parts of the atmosphere.

THE STRATOSPHERE

Above the tropopause is the **stratosphere**, a name derived from the Latin word for *layer*. Except for the penetration of some strong thunderstorms into the lower stratosphere, little weather occurs in this region (Figure 1–10). In the lowest part of the stratosphere, the temperature remains relatively constant at about –57 °C (–71 °F) up to a height of about 20 km (12 mi). From there to the top of the stratosphere (called the **stratopause**), about 50 km (30 mi) above sea level, the temperature increases with altitude until it reaches a mean value of −2 °C (28 °F).

In the upper stratosphere, heating is almost exclusively the result of ultraviolet radiation being absorbed by ozone. Therefore, as solar energy penetrates downward through the stratosphere, there is less and less ultraviolet radiation available and a resultant decrease in temperature. In the part of the stratosphere where temperature does not vary with height, heating is the result of both absorption of solar ultraviolet radiation and absorption of thermal radiation from below. Thus, as we move up or down in this region, the reduction in solar heat is offset by the increase in heat given off by Earth. The net result is the straight vertical line of no temperature change shown in Figure 1–9.

The stratosphere contains about 19.9 percent of the total mass of the atmosphere. Thus, the troposphere and stratosphere together account for 99.9 percent of the total mass of the atmosphere. The fact that gases are so compressible allows the major portion of the mass of the atmosphere to be contained in the two lowest layers. Note that despite being more than triple the depth of the troposphere, the stratosphere contains only one-quarter the mass: troposphere = 80 percent of total atmospheric mass; stratosphere = 20 percent.

Within the stratosphere is the **ozone layer**, a zone of increased ozone concentration at altitudes between 20 and 30 km (12 and 18 mi). Despite its name, the ozone layer is not composed primarily of ozone. In fact, at 25 km (15 mi) above sea level, where the percentage of ozone in this region is greatest, its concentration

FIGURE 1–10
Most clouds exist in the troposphere, but the tops of strong thunderstorm clouds can extend kilometers into the stratosphere when violent updrafts occur. The flattened area at the top of this cumulonimbus cloud is in the stratosphere.

might be only about 10 ppm. Despite its scarcity, ozone is an extremely important constituent of the stratosphere. It is largely responsible for absorbing the solar energy that warms the stratosphere, and it also protects life on Earth from the lethal effects of ultraviolet radiation.

Being well removed from Earth's surface, where evaporation supplies water vapor to the atmosphere, the stratosphere has a very low moisture content. Moreover, the temperature characteristics of the stratosphere inhibit vertical motions that favor the formation of clouds (to be discussed in Chapter 6). These conditions inhibit precipitation, and consequently particulates from volcanic eruptions can remain in the stratosphere for many months. Furthermore, the strong winds of the stratosphere can cause its aerosol content to be distributed across the globe, creating a veil of material that can affect the penetration of sunlight to the surface. For a couple of years after the 1991 eruption of Mount Pinatubo in the Philippines, for example, the Northern Hemisphere experienced redder-than-normal sunrises and sunsets as a result of aerosols in the stratosphere.

THE MESOSPHERE AND THERMOSPHERE

Of the 0.1 percent of the atmosphere not contained in the troposphere and stratosphere, 99.9 percent exists in the **mesosphere**, which extends to a height of about 80 km (50 mi) above sea level. As in the troposphere, temperature in the mesosphere decreases with altitude. Scientists believe that the absorption of solar radiation near the base of the mesosphere provides most of the heat for the layer, which is dispersed upward by vertical air motions.

Above the mesosphere is the **thermosphere**, where temperature increases with altitude to values in excess of 1500 °C. These high temperatures can be misleading, however, if we overlook the distinction between high temperature and high heat content. The temperature of the air is an expression of its kinetic energy, which is related to the speed at which its molecules move. The amount of heat contained in the air reflects not only its temperature but also its mass and *specific heat* (the amount of energy needed to change its temperature by a certain amount). Because there are so few gas molecules in this layer, the air cannot have a high heat content no matter what its temperature is. In fact, the atmosphere is so sparse in the upper reaches of the thermosphere that a gas molecule will normally move as much as several kilometers before colliding with another. Thus, an ordinary thermometer in this part of the atmosphere would have little contact with the surrounding air. Under these circumstances, the concept of temperature loses meaning and cannot be associated with everyday terms such as *hot* and *cold*. (See *Box 1–3, Physical Principles: The Three Temperature Scales*, for important information on temperature scales.)

A LAYER BASED ON ELECTRICAL PROPERTIES: THE IONOSPHERE

The four layers of the atmosphere described previously are delineated on the basis of temperature profiles. An additional layer, called the **ionosphere**, can be defined based on its electrical properties. This layer, which extends from the upper mesosphere into the thermosphere, contains large numbers of electrically charged particles called *ions*. **Ions** are formed when electrically neutral atoms or molecules lose one or more electrons and become positively charged ions or gain one or more electrons and become negatively charged ions. In the ionosphere, atoms and molecules lose electrons as they are bombarded by solar energy, thus creating positively charged ions and free electrons.

The ionosphere is important, among other reasons, for reflecting AM radio waves back toward Earth, thereby increasing the distance at which broadcasts can be received. As shown in Figure 1–11, the ionosphere is divided into several sublayers, including the D-, E-, and F-layers. The lowest of the three, the D-layer, exists only during the daylight hours and absorbs AM radio waves. When

DID YOU KNOW?

Although Venus is approximately the same size as Earth, its atmosphere is vastly different. The surface density of Venus's atmosphere is some 50 times greater than Earth's and it is 96.5% carbon dioxide. Our planetary neighbor has a vastly different temperature as well—averaging about 475 °C, or 890 °F!

FIGURE 1–11

The ionosphere influences the distance to which AM radio waves can be transmitted. During the day (a), the D-layer absorbs the radio waves. The D- and E-layers break up at night (b), allowing the waves to reach the F-layer. The F-layer reflects the radio waves back toward Earth's surface. Repeated reflections between the E-layer and Earth's surface enable the waves to be picked up by receivers at greater distances at night.

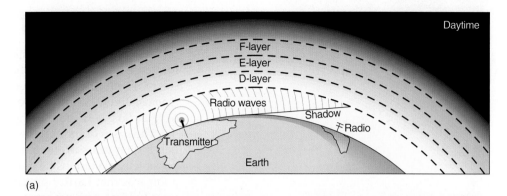
(a)

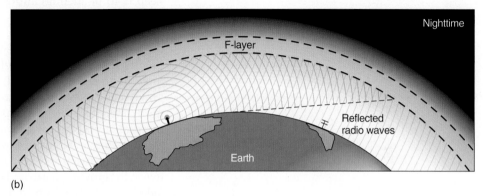
(b)

night comes, the D-layer begins to disappear and the E-layer weakens as their free electrons recombine with positively charged ions. The radio waves are then able to reach the F-layer, which reflects radio waves rather than absorbing them and thus redirects the transmission back toward the surface. Earth's surface also reflects these waves, so they rebound back toward the ionosphere. The repeated reflection of the radio waves between the F-layer and the surface allows them to overcome the effect of Earth's curvature and greatly extends the distance at which they can be "heard." Perhaps you have listened to an AM radio at night and happened to pick up a distant radio station, only to be surprised by its disappearance the following day. This results from the breakup and re-formation of the D- and E-layers.

Because of these changes in the range of transmission, radio stations reduce the power of their transmitters at night so that stations on the same frequency do not interfere with one another. Moreover, the Federal Communications Commission (FCC) has designated certain radio stations as "clear channels." Such stations are given unique frequencies (each is the only one in the United States allowed at a given point on the radio dial) and allowed to broadcast from very powerful transmitters. The purpose of such stations is to guarantee that all parts of the country, no matter how far from a city, have access to at least one radio station. Clear channel stations are indeed heard over great distances, and a truck driver going through eastern Colorado may have no difficulty picking up stations from Chicago, Illinois, or Ft. Worth, Texas.

The ionosphere is also responsible for the **aurora borealis**, or northern lights, and **aurora australis**, the southern lights (Figure 1–12). In the ionosphere, subatomic particles from the Sun are captured by Earth's magnetic field (the same field that makes compass needles point to the north). The captured energy causes the gases of the atmosphere to become *excited* (meaning the electrons of the atoms jump to greater orbital distances from their nuclei). When electrons fall back to lower orbits, radiation is emitted. Thus, the aurora does not reflect radiant energy as do clouds but rather emits light that is much like a neon lamp.

FIGURE 1–12

An aurora borealis. Subatomic particles from the Sun are captured by Earth's magnetic field, causing an agitation of molecules and the emission of light with different colors.

Evolution of the Atmosphere

It is generally believed that Earth was formed 4.6 billion years ago. If an atmosphere formed with Earth, it must have consisted of the gases most abundant in the early solar system—including large amounts of hydrogen and helium, the two lightest elements. Today's atmosphere is much different, composed mostly of nitrogen and oxygen. So where did the original gases go and how were they replaced?

The gases of the earliest atmosphere were simply lost to space during the first half-billion years of the planet's existence. We are not sure exactly how, but two possible explanations have been offered. For many years, the conventional viewpoint was that Earth's gravitational field would not have been strong enough to keep hold of such an atmosphere, and the light elements therefore escaped into space. To understand why, however, we need a bit of background. Gases surrounding a planet would easily disperse into outer space (much the same way air disperses from a bicycle tire when the valve is opened) were it not for the fact that gravity exerts an opposing effect and prevents their escape. However, if molecules move with sufficient speed, known as their **escape velocity**, they can overcome gravity and leave the atmosphere. Gases having lower molecular weights are more likely to achieve escape velocity; thus, the hydrogen and helium of the primordial atmosphere were most readily lost.

Recently, an alternate explanation has gained acceptance. According to this view, gases drawn to Earth during its formation would have been removed by collisions between the growing Earth and other large bodies. Some of these other objects, so-called failed planets, would have been as large as Mars. Tremendous energy would have been released by collisions with failed planets, ejecting large amounts of planetary material along with any atmosphere that might be present. (In fact, the Moon is believed to have formed by condensation of matter dislodged by just such a collision.) If this hypothesis is correct, removal by slow leakage would have been insignificant.

Over time a new, secondary atmosphere formed, made up of gases released from Earth's interior by volcanic eruptions—a process called **outgassing**. The gases spewed out during volcanic events are predominantly water vapor and carbon dioxide, with lesser amounts of sulfur dioxide, nitrogen, and other gases. Volcanic outgassing was possibly augmented, maybe even dwarfed, by material brought to Earth in small comets on the order of 15 m (45 ft) in diameter. Recent satellite observations suggest that there is a steady rain of these water-bearing "cosmic snowballs," with from 5 to 30 striking Earth at any one time. If they've been falling at this rate since the formation of Earth, they might well be the source of most of the world's water vapor. However it arose, the secondary atmosphere has been greatly transformed, because water vapor and carbon dioxide now constitute only a small portion of the atmosphere.

The removal of the water vapor is easily explained: Upon cooling in the prehistoric atmosphere, it readily condensed and precipitated to the surface to form the oceans. Precipitation also contributed to the removal of carbon dioxide from the atmosphere. As raindrops fell through the CO_2-rich atmosphere, some CO_2 was dissolved in every drop. (Soft drinks provide an everyday example of how CO_2 is dissolved in water.) As they fell, the CO_2-enriched drops transferred carbon from the atmosphere to the oceans, where it combined with material eroding from the continents and was buried in seafloor sediments.

The transformation to an atmosphere high in oxygen depended on the advent of primitive, anaerobic bacteria (those that survive in the absence of oxygen) about 3.5 billion years ago. These primitive life-forms were the first in a long line of organisms that removed carbon dioxide from the air and replaced it with oxygen (these gases are exchanged freely between the oceans and the atmosphere). Ultimately, plant and later animal material sank to the ocean floor (as it continues to do today), where the organic carbon was locked away in sediments. As a matter of fact, the vast majority of carbon released by volcanoes exists in neither the atmosphere nor the ocean; it is held in carbonate rock formations.

All of these processes gradually led to an increase in atmospheric oxygen at the expense of carbon dioxide. But one other transformation had to take place to create an atmosphere that could support life at the surface. Recall that without an ozone layer in the upper atmosphere, sunlight reaching the surface would contain lethal levels of ultraviolet radiation. Thus, a protective ozone layer had to develop before life could exist outside the oceans. Fortunately, this occurs naturally when the ultraviolet radiation breaks down diatomic oxygen molecules into individual oxygen atoms. The oxygen atoms could then recombine with O_2 in the upper atmosphere to form the ozone layer. Once this happened, the amount of ultraviolet radiation reaching the surface was reduced sufficiently to allow plants to occur on land. This increase in plant cover in turn accelerated the rate at which photosynthesis replaced atmospheric carbon dioxide with oxygen.

The last thing we need to account for is the high concentration of nitrogen in the atmosphere. Although it constitutes a small portion of the material released by outgassing, nitrogen is removed from the atmosphere very slowly. As a result, its concentration has gradually increased to the point that it is now the main constituent of the atmosphere.

Some Weather Basics

All of us are familiar with daily weather forecasts, from which we routinely receive information on the present and predicted state of the atmosphere. In recent years, the amount of weather information available to the public has exploded, most notably via the Internet. Detailed maps and weather reports that were once available only to professional meteorologists can now be accessed by anybody with a computer and a modem. Such access makes learning the fundamentals of weather and climate far more enjoyable, because we can now look at map, satellite, radar, and other resources and see how the principles of meteorology play out on a daily basis. We now present an overview of the fundamentals of weather, along with an introduction to weather maps.

ATMOSPHERIC PRESSURE AND WIND

Atmospheric **pressure** is one of the most fundamental weather characteristics, but at the same time it is one you don't directly feel. Unlike temperature and humidity, it is impossible to sense whether the air pressure at the surface is high or low. Yet atmospheric pressure affects all other aspects of weather. Two generalizations about surface pressure are particularly important. First, air tends to blow away from regions of high pressure toward areas of lower pressure. In other words, it is the horizontal variation in air pressure that generates **winds**. If the pressure were uniform from place to place, the air would be continually calm. The second generalization is that air tends to rise in areas of low surface pressure and sink in zones of high pressure. This is important because rising motions favor the formation of clouds, while sinking motions promote clear skies.

Atmospheric pressure is routinely plotted on maps with lines called **isobars**. Each isobar connects points having equal air pressure, with the pressure being expressed in units of **millibars** (abbreviated as mb) in the United States and **kilopascals** (kPa) in Canada. Figure 1–13 illustrates how isobars depict the distribution of pressure. Notice that an isobar labeled 1024 encircles most of Idaho. Any point on that line has a pressure that is equal to 1024 mb (which corresponds to 102.4 kPa). Another isobar, indicating a pressure of 1020 mb, surrounds the 1024 mb isobar. Not only do the two isobars tell us what the pressure is at any point along those lines, but they also allow us to infer what the pressure is anywhere between them. Thus, over nearly all of the states of Washington and Wyoming, which are situated between the two isobars, the air pressure is between 1020 and 1024 mb. The fact that higher pressure exists over Idaho implies

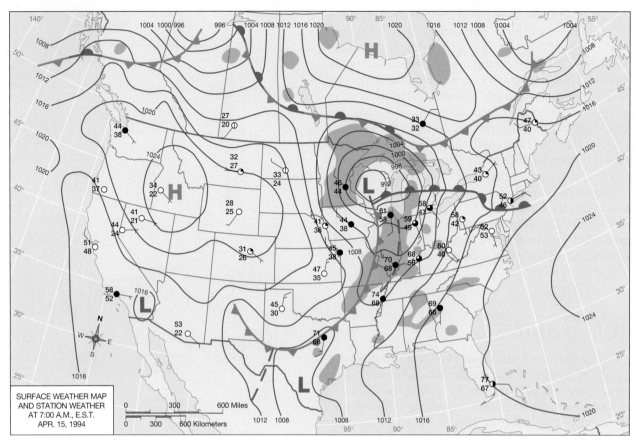

FIGURE 1–13

A typical surface weather map.

that the air flows outward from that region. Just west of the Great Lakes in Figure 1–13 is an area of lower pressure, and we can expect that air is flowing into that region from surrounding areas.

Information regarding wind speed and direction can be obtained on weather maps by looking at the **station models**, which contain symbols and numbers giving detailed weather information for particular locations. Notice the station model for Tucson, Arizona, in the southwestern United States. The station model in Tucson contains an open circle with a line pointing toward the southeast, which indicates the direction that the wind is blowing *from*. At the end of the southeastward-pointing line is a single tick mark. The number and type of tick marks at the end of the line give an approximate wind speed, using the conventions shown in Figure 1–14a. In this example, Tucson has winds coming from the southeast at 14 to 19 km per hour (9–14 mph).

Station models also give cloud cover information. Open circles, such as that shown for Tucson, indicate clear skies, while fully shaded circles indicate overcast conditions. Intermediate amounts of cloud cover are indicated by the patterns shown in Figure 1–14b.

In Figure 1–13, the isobar pattern and station model information verify the generalizations about pressure distributions, wind movement, and cloud cover. The wind barbs show that the air does indeed move outward from the region of high pressure, while air is flowing into the low-pressure area west of the Great Lakes. Furthermore, the station models show that much of the area around the high-pressure system has clear skies, in contrast to the general overcast near the low-pressure system.

TEMPERATURE

Temperature is one of the most obvious weather components. Everyday experience indicates that temperatures vary from place to place systematically. In other words, as you take a 50-mile car trip you are not likely to experience temperatures that vary wildly. Instead, you will probably observe only gradual changes in temperature as you drive along. On the other hand, there are times when substantial temperature differences appear over short distances, or when major shifts in temperature occur over short time periods at a particular location. These major changes in temperature often occur due to the presence of **fronts**, fairly narrow boundary zones separating relatively warm and cold air. As you drive across the frontal zones, you will experience notable temperature shifts. Likewise, if a front passes over the place you are, the temperature will change substantially.

Four types of fronts exist, which are discussed later in this text. For now, let's concern ourselves with two particular types: *cold fronts* and *warm fronts*. Figure 1–13 maps the presence of a cold front (shown as a blue line with triangles) extending southward from the low-pressure region over Wisconsin, and a warm front (red line with semicircles) extending eastward from the same location. To the east of the cold front, temperatures are higher than the area to the west of the front. The exact temperatures are shown in degrees Fahrenheit, on the upper left portion of the station models. In this illustration, Chicago, Illinois, has a temperature of 61 °F, while Des Moines, Iowa, to the west, has a temperature of 44 °F.

Figure 1–13 also shows the distribution of precipitation, indicated by the areas in green. Notice that precipitation is occurring to the northwest of the low-pressure system over Wisconsin and also along the cold front. Precipitation is very common along fronts, so when a front approaches it is not unusual to experience rain or snow in addition to a change in temperature.

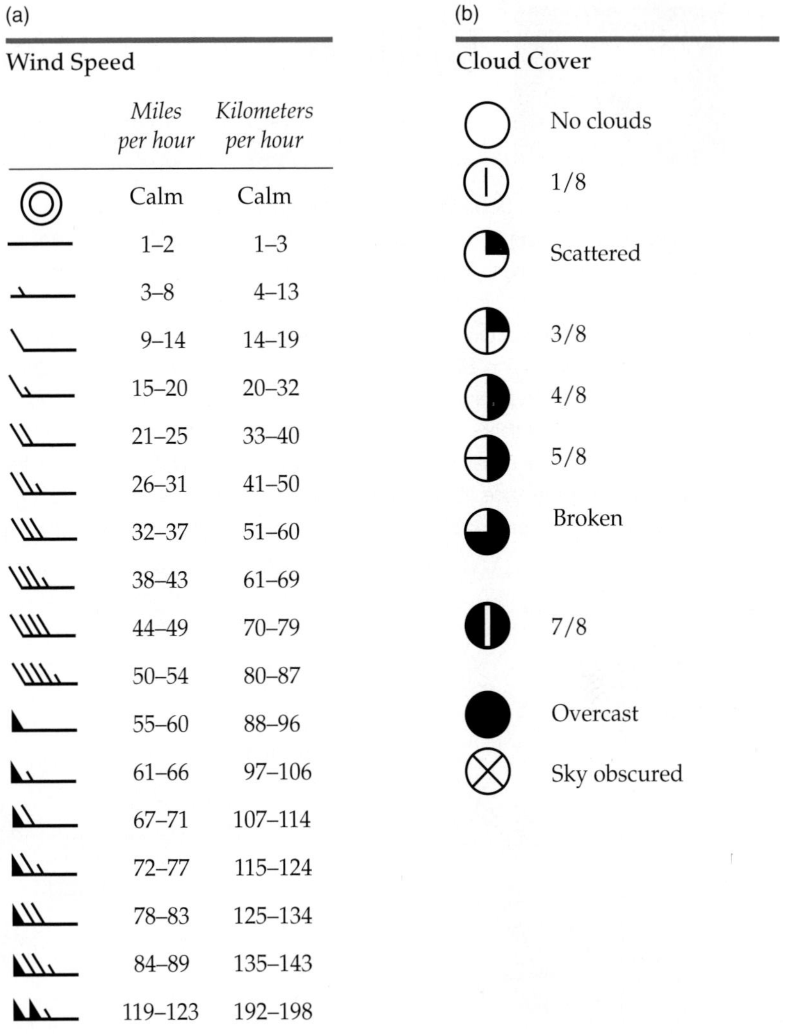

HUMIDITY

You have undoubtedly heard the term **relative humidity**. Relative humidity is just one of several ways of expressing the amount of water vapor in the air. (Remember, water vapor is a gas!) It indicates the amount of water vapor present relative to the maximum possible; thus, it is usually reported as a percentage. Though a commonly used indicator of water vapor content, relative humidity has some serious shortcomings. For this reason, another index called the **dew point temperature** (or simply, the dew point) is often preferred. For now let's just say that the higher the dew point, the greater the amount of water vapor in the air. Dew points above about 15 °C (59 °F) or so indicate humid air, and dew points above about 20 °C (68 °F) are very uncomfortable. Dew points less than about 5 °C (41 °F) are relatively dry. Dew point values are given at the bottom left of the station models, just below the temperature readings. In Figure 1–13, much of the southeastern United States has humid air, while the northeastern United States and eastern Canada are somewhat drier. The driest air in this example occurs over the West.

(a)

Wind Speed

	Miles per hour	Kilometers per hour
◎	Calm	Calm
	1–2	1–3
	3–8	4–13
	9–14	14–19
	15–20	20–32
	21–25	33–40
	26–31	41–50
	32–37	51–60
	38–43	61–69
	44–49	70–79
	50–54	80–87
	55–60	88–96
	61–66	97–106
	67–71	107–114
	72–77	115–124
	78–83	125–134
	84–89	135–143
	119–123	192–198

(b)

Cloud Cover

○	No clouds
◔	1/8
◕	Scattered
◑	3/8
◐	4/8
⊖	5/8
◒	Broken
◧	7/8
●	Overcast
⊗	Sky obscured

FIGURE 1–14
Station model symbols describing (a) wind speed and (b) percentage of cloud cover.

A Brief History of Meteorology

The amount of information available to even casual weather observers is enormous compared to just a couple of decades ago. The Internet now allows anybody with access to a computer and a network connection to obtain better information than a professional weather forecaster had available in the late 1900s. But the information explosion did not happen overnight; for centuries the advances in meteorological knowledge came about in piecemeal increments.

For atmospheric science to become quantitative, basic instrumentation was required. Galileo led the way in 1593 with a prototype version of the basic thermometer, which ultimately led to the development of the Fahrenheit and Celsius scales in 1714 and 1736, respectively. In 1643, Evangelista Torricelli invented the

Billion-Dollar Weather Disasters

Between 1980 and 2004 the United States experienced 62 separate billion-dollar weather-related disasters, amounting to $390 billion in total losses (adjusted to 2002 dollars). Four such events—all hurricanes—occurred in 2004, causing $41 billion in damages and 152 fatalities. Hurricanes Charley, Frances, Ivan, and Jeanne were part of an unusually active Atlantic hurricane season made all the more unusual because all four hurricanes went through the state of Florida. Only once before had four hurricanes gone through a particular

state in a given year, and three of the 2004 storms crossed over the same part of central Florida.

A year later, Floridians were still rebuilding when the 2005 hurricane season got off to a remarkable start: Hurricane Dennis set two significant milestones on July 5. It marked the earliest date ever that a fourth named storm had formed in a given season (hurricane season in the Atlantic doesn't usually become active until August or September), and it was the earliest date that a rare Category 4 Atlantic hurricane had ever oc-

curred. In August 2005, Hurricane Katrina moved through southern Florida as a Category 1 hurricane, depositing rainfall amounts approaching 50 cm (20 in.) in places and bringing its share of wind damage. Within a few days, Katrina intensified to an extremely rare Category 5 hurricane in the Gulf of Mexico, turned northward toward the Gulf Coast, and then ravaged southern Louisiana (including New Orleans), Mississippi, and Alabama. Though reduced to a Category 4 hurricane just prior to landfall, Katrina is one of

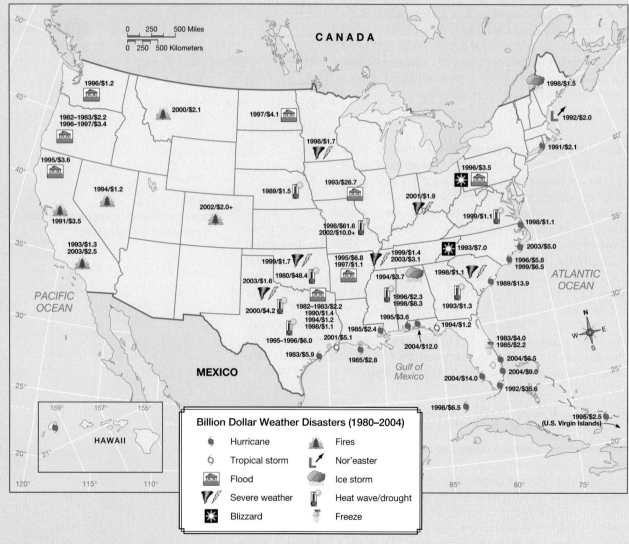

FIGURE 1
Map of billion-dollar disasters.

24

the biggest natural disasters ever to hit the United States. Chapter 12 contains analyses of the major hurricanes of 2004 and 2005.

Hurricanes were not the only major weather problem for North America in recent years. In spring 2003 numerous tornadoes hit the midwestern and southeastern United States, and severe thunderstorms and hail pummeled the southern Great Plains and Texas. These storms brought $5 billion in damages and 50 fatalities.

Several months later, prolonged drought and dry, hot, windy conditions contributed to large fires in southern California that destroyed 3700 homes and killed 22 people.

By far the costliest and deadliest disasters to hit the United States in the last quarter-century were the heat/drought events of 1980 and 1988, which resulted in nearly $100 billion in economic losses and perhaps 20,000 fatalities from heat stress. Another heat wave in 1999

killed an additional 500 people. Clearly, more Americans and Canadians die from severe heat episodes than from hurricanes and tornadoes. This may portend a particularly ominous situation because atmospheric scientists believe that global warming may lead to a more frequent occurrence of severe heat waves. Human activities have almost certainly contributed to global warming, but we are still powerless to mitigate many of its consequences.

FIGURE 2
A casino barge sits among residential homes in Biloxi, Mississippi, after Hurricane Katrina passed through the area.

WEATHER IN MOTION
Tropical Storm Allison

FIGURE 3
Tornado damage at Grandview, Missouri, May 2003. The twister demolished homes on one side of the street while only causing roof damage across the street.

FIGURE 4
Part of the outbreak of fires that ravaged southern California in October 2003.

mercury barometer, the instrument still used as a standard for measuring atmospheric pressure, but it was not until the late 1700s that an instrument became available for measuring water vapor.

Instruments alone can't discern atmospheric patterns or make forecasts—there must also exist a network of observers to operate the instruments and maintain permanent records of weather elements. The first network of this type in North America was authorized in 1847 when the Smithsonian Institution allocated the considerable sum of $1000 to purchase instruments for a network of volunteer weather observers across the United States. These observers submitted their subjective observations and instrument data to the Institution each month, providing useful climatological information. Still, the inability to collect, map, and rapidly disseminate the data made operational forecasting impossible. At about the same time, however, a new invention—the telegraph—enabled the rapid transmission of data from sites across the country to a central collecting station, thus making possible the forecasting of weather by simply noting the location and movement of current weather conditions.

The birth of U.S. weather forecasting occurred in 1870, when President Ulysses S. Grant authorized the establishment of the Army Signal Service. In 1891 the agency was renamed the Weather Bureau and transferred to the Department of Agriculture, where it remained until 1940. The agency was then transferred to the Department of Commerce and in 1970 was renamed the **National Weather Service** as part of the newly established National Oceanic and Atmospheric Administration.

The ever-growing network of surface stations was important to understanding and forecasting weather, but the greatest amount of the atmosphere lies well above the surface. Perhaps the most famous attempt to gain insights into what happens in the atmosphere well above the ground was Ben Franklin's kite experiment in 1752. But the balloon ascents taken by J. L. Gay-Lussac and Jean Biot in 1804 were landmark events in scientific understanding and human courage. The first ride, taken by both scientists, rose to a height of 3,962 m (13,000 ft). A second ascent taken solo by Gay-Lussac went up to 7000 m (23,000 ft). At that height, the average temperature is some 30 degrees below zero Celsius (−22 °F) and the air density is only 40 percent of its value at sea level. Gay-Lussac continued to make temperature and pressure observations until he passed out from oxygen deficiency. Unmanned weather balloons became a valuable source of data in the 1920s, and since the 1940s the tracking of balloon-carried instrument packages called *radiosondes* became a regular source of climatological and forecasting data.

The era following World War II brought with it enormous advances in meteorological knowledge and technology. Weather radars, first developed during the war, have evolved into units that can peer into clouds and observe their internal characteristics and motions. Weather satellites have likewise become essential to meteorologists since the launching of Tiros I (Television and Infrared Observation Satellite) in 1960. And technological advances in computer hardware and software that have revolutionized most other aspects of society have had an incredible impact on atmospheric understanding and forecasting since they were introduced to meteorology in the 1950s.

Summary

Meteorology and climatology study the elements of the atmosphere and the causes of atmospheric behavior. While these two sciences are closely intertwined, the former places a greater emphasis on the short-term events that we think of as weather, while the latter looks at longer-term characteristics. Earth's atmosphere is composed of a mixture of gases and contains an enormous number of suspended solids and liquids called *aerosols*. Three of the gases—nitrogen, oxygen, and argon—occur in nearly constant proportions and constitute the vast majority of the atmospheric mass. Other gases occur in slight amounts and vary considerably in their concentration. These variable gases, especially water vapor and carbon dioxide, can be extremely important to life on Earth.

Four layers of the atmosphere—the troposphere, stratosphere, mesosphere, and thermosphere—can be distinguished on the basis of their temperature profiles, and one—the ionosphere—is designated for its electrical characteristics. Of the first four, the lowest two (the troposphere and stratosphere) contain almost all atmospheric mass. The other two layers, the mesosphere and thermosphere, account for less than 0.1 percent of the atmosphere's mass and are relatively unimportant to most of the processes described in this book. The ionosphere spans the upper mesosphere and thermosphere. It affects the transmission of AM radio waves and provides the locale for the aurora borealis and aurora australis.

The present atmosphere did not arrive with the formation of the planet some 4.6 billion years ago but rather evolved after the primordial atmosphere was lost to space. The process of outgassing released water vapor and carbon dioxide (along with other gases) from Earth's interior. Ultimately, photosynthesis reduced carbon dioxide levels and contributed oxygen to the atmosphere.

This chapter introduced some significant weather elements and demonstrated how they are depicted on surface weather maps, and provided an overview to the development of meteorology as we know it today.

Key Terms

atmosphere page 5
meteorology page 5
weather page 5
climate page 5
climatology page 5
permanent gases page 6
variable gases page 6
homosphere page 6
heterosphere page 6
nitrogen page 7
oxygen page 7
argon page 7
water vapor page 8
hydrologic cycle page 8

carbon dioxide page 8
photosynthesis page 9
respiration page 10
ozone page 11
methane page 11
aerosols page 13
particulate page 13
condensation nuclei page 14
structure page 14
density page 14
Kelvin scale page 15
mean free path page 15

standard atmosphere page 16
troposphere page 16
tropopause page 17
inversion page 17
stratosphere page 17
stratopause page 17
ozone layer page 17
mesosphere page 18
thermosphere page 18
ionosphere page 18
ions page 18
aurora borealis page 19
aurora australis page 19

escape velocity page 20
outgassing page 20
pressure page 21
wind page 21
isobar page 21
millibar page 21
kilopascal page 21
station model page 22
front page 23
relative humidity page 23
dew point temperature page 23
National Weather Service page 26

Review Questions

1. Explain why television newscasts have weather segments but not climate segments.

2. Why is it difficult to define an absolute top of the atmosphere?

3. What are the homosphere and the heterosphere?

4. What is the difference between the permanent and variable gases of the atmosphere? Which gases contribute most to the total mass of the atmosphere?

5. Given that variable gases are so rare, why consider them at all?

6. Why has the concentration of carbon dioxide in the atmosphere been increasing over the last century?

7. What is ozone, and why is it both beneficial and harmful to life on Earth?

8. What are aerosols? Are they formed only by human activities or do they occur naturally?

9. Convert the following Fahrenheit temperatures to Celsius: −22 °F, 50 °F, 113 °F.

10. Convert the following Celsius temperatures to Fahrenheit: −20 °C, 10 °C, 40 °C.

11. How do photosynthesis, respiration, and decay affect the carbon dioxide balance of the atmosphere?

12. In what way does the density of the atmosphere vary with altitude?

13. What are the distinguishing characteristics of the troposphere, stratosphere, mesosphere, and thermosphere?

14. What is the tropopause?

15. In which thermal layer of the atmosphere is the ozone layer found? Why is the term "ozone layer" somewhat misleading?

16. What percentage of the total mass of the atmosphere is contained in the troposphere and the stratosphere?

17. Why does the troposphere contain much more mass than the stratosphere, despite the fact that the troposphere is a thinner layer than the stratosphere?

18. How is the ionosphere distinct from the layers of the atmosphere defined by their temperature profiles?

19. What is outgassing and why was it important?

20. Why were anaerobic bacteria important to the evolution of the atmosphere?

21. Briefly describe the effect that variations in pressure exert on other weather elements.

22. What are isobars?

23. What are station models and what useful information do they depict?

Critical Thinking

1. The Kyoto Treaty that limits carbon dioxide emissions has not been ratified by all countries that originally signed the treaty. What are the pros and cons of doing so?

2. Volcanic eruptions continue to occur and outgas water vapor, carbon dioxide, and other gases. Do you think that this will be a significant factor in increasing the concentration of these gases over the next century? Why or why not?

3. Temperatures usually decrease with height in the troposphere but increase with height in the stratosphere. Why do the two layers have such different profiles?

4. The thermosphere has extremely high temperatures, but a person exposed to the thermosphere would rapidly freeze. Explain the apparent contradiction in terms of what you know about heat and temperature.

Problems and Exercises

1. The National Climatic Data Center (NCDC) has a Web site at **http://lwf.ncdc.noaa.gov/oa/climate/severeweather/extremes. html** that provides a wealth of information on extreme weather and climate events. Examine the site to see how much information is available. Do any of the topics relate to weather events in your hometown?

2. Examine the map of National Weather Service Offices at **http://www.wrh.noaa.gov/wrhq/nwspage.html**. Is there a Weather Service office near you? If so, consider visiting the office and seeing firsthand what goes into producing a forecast.

3. Keep a record of daily weather in your area and download current weather maps from one of the available Web sites. Do you notice any patterns on the maps that tend to be associated with particular weather conditions?

4. Look at today's weather map and observe the contrasting weather conditions across the United States and Canada. Do any areas exhibit significant changes in weather from adjacent regions? How well defined are the boundary zones? Repeat this exercise for several days and see if these transition regions show movement.

Quantitative Problems

Your comprehension of a great many concepts can be enhanced by working out numerical solutions to questions. Go to this book's Web site at **http://www.prenhall.com/aguado/** and click on the Begin button. This will take you to the main page for Chapter 1, which offers a variety of valuable resources. Take a look at the options available. Along with the self-quizzes, links, and other resources on the page, the quantitative exercises should prove particularly valuable.

Useful Web Sites

http://www.weather.com/
Home page for the Weather Channel. Contains current weather information along with reports related to travel, health, severe weather, and many other topics. A good first stop.

http://www.nws.noaa.gov/organization.php
Listing of all local National Weather Service offices, with links connecting the viewer directly to sites. Also lists Uniform Resource Locators (URLs)—Internet addresses—for regional and national support offices such as the National Hurricane Center. An excellent site for local information anywhere in the United States and for advisories on severe weather.

http://weatheroffice.ec.gc.ca/canada_e.html
Home page for the Meteorological Service of Canada. Provides forecasts and links to satellite, radar, and many other useful resources.

http://www.wunderground.com/ and http://www. wunderground. com/global/CN_ST_Index.html

Multipurpose sites that provide detailed information for the United States and Canada, respectively. Offer standard and unusual weather maps and local weather data. Also good sources for weather information all over the world.

http://www.ozonelayer.noaa.gov/

Up-to-date information on the status of the ozone layer.

Media Enrichment

Weather in Motion
Satellite Movies of the May 3–4, 1999, Tornado Cluster

The large weather system that produced this severe weather outbreak was tracked by weather satellites. Two movies on this book's CD-ROM provide different perspectives on this storm. The black-and-white movie is based on visible light, and the other is based on infrared radiation (discussed in Chapter 2). The most severe portions of the storm system show up dramatically in both movies. Note that the movie based on visible energy ends at sunset, while the infrared movie continues past sunset.

Weather in Motion
A Global Montage of Clouds and Sea Surface Temperature

Although this chapter hasn't emphasized atmospheric motion, it's not too soon to stress the ceaseless movement of Earth's fluid envelope. This movie depicts weather for an entire year (1999), obtained by merging information acquired every 6 hours from three satellite systems. It contains a wealth of information and will be used repeatedly in the text. For now, we draw your attention to just a couple of features. First, notice the distinctly different movement of major cloud systems at various latitudes. In particular, you can see tropical systems drifting mostly from east to west, whereas systems at higher latitudes show west-to-east movement. There are also strikingly different shapes to the cloud systems, with tropical versions more or less circular compared to the elongated storm clouds at higher latitudes.

There is a pronounced north–south migration of the band of clouds over the tropics during the year, especially over Africa. During January the clouds are centered south of the equator, but as the movie progresses into July and August, the clouds move into the Northern Hemisphere. The latitudinal movement of the cloud bands is directly related to the seasonal changes caused by Earth–Sun geometry, discussed in Chapter 2. You see this also reflected in sea surface temperature, as the zone of high temperatures (colored red) moves north and south. Finally, pay close attention to the Atlantic coast of North America during the August and September portions of the movie. Some notable hurricanes that hit the eastern United States show up very well. We will discuss hurricanes and other major weather events throughout this book.

Weather in Motion
The Ozone Hole

This movie shows the progressive development of the Antarctic ozone hole during the Southern Hemisphere spring of 1991. The movie starts in late September, when the hole normally begins to develop. By mid-October most of the Antarctic continent is covered by a zone of greatly reduced stratospheric ozone. Notice that the ozone hole is not stationary; instead, its size, shape, and intensity vary throughout the next few months.

Weather in Motion
Tropical Storm Allison

Tropical Storm Allison brought huge amounts of rain and extreme flooding to Texas and Louisiana, as it hovered over the region for several days. The system then moved to the east coast where it brought floods as far north as New England. This movie tracks the storm from June 4–14, 2001.

Weather Image
First Look at the New Millennium

If you were wondering what North America looked like at the very beginning of the new millennium, here is the first weather satellite image of the year 2000. The image was collected at 15 minutes past midnight, Greenwich mean time (also called UTC, for Universal Time Coordinated). We see a band of clouds that has recently passed over the eastern United States, along with generally cloudy conditions over eastern Canada. Another extensive region of cloud cover exists over much of the western United States and Canada.

SOLAR RADIATION AND THE SEASONS

Over the period of January 5–9, 1998, residents of a large part of the eastern United States and Canada encountered a rainfall event that greatly affected their lives. Unlike most rain events, the falling drops froze immediately on hitting the surface to form a continuous layer of smooth, white ice. The ice became so thick that it caused tree branches to break, power lines to snap, and roads to become impassable. The freezing rain was so widespread and the damage so pervasive that millions of residents in New England, New York State, and the Canadian province of Quebec lost electrical power—for more than 2 weeks in some places. In Quebec alone, 3 million people—40 percent of the population—were without power, and damage estimates were in the neighborhood of $1 billion. One of those who endured the outage, Mr. Andre Champagne of Montreal, described it thusly: "It's so dark, it's like I'm on Mars."

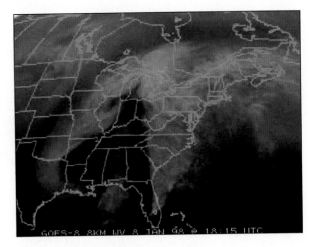

WEATHER IN MOTION
January 1998 Ice Storm

The storm had numerous other consequences—all of them bad. In the southeastern and Mid-Atlantic states, the rain caused extensive flooding, in part because of direct runoff into rivers. The percolation of rainwater into the existing snowpack caused rapid melt, which further exacerbated the flooding problem. Four states called in the National Guard. And there was no shortage of unhappy cattle, as the loss of electricity made it difficult for farmers to milk their cows, causing more than a million dollars in agricultural losses. More than 30 people in the United States and Canada died as a result of the storm.

The cold winter storm of January 1998 offers a vivid contrast to the severe weather outbreak that hit the northern United States and southern Canada on May 30 and 31, when tornadoes, hail, and lightning killed at least 17 people. Such storms, though possible any time during the year, are most prevalent in spring. And then there was the summer heat wave of July 1999, which spread its share of misery from eastern Canada down to the southern plains and southeastern portion of the United States. More than 250 people died across the United States over the 2-week episode.

As these examples show, many sorts of severe weather can create widespread havoc, but each is more likely to occur at some particular time of the year. Even in the absence of severe weather, our lives are profoundly influenced by the seasonal cycle that occurs each year. But as important as the changes of the seasons are to everyday activities, many of us do not really understand what causes them. In many cases we incorrectly think we know what causes them. For example, many people believe that variations in the distance between Earth and the Sun are responsible for changes in the seasons, with summer occurring when Earth and the Sun are closest together. But Earth and the Sun are closest to each other on or about January 4—in the midst of the Northern Hemisphere winter! Likewise, the Sun is farthest from Earth on July 4, a time Northern Hemisphere residents associate not with winter, but with long, hot days, barbecues, and fireworks. Obviously, then, there must be another explanation. In this chapter, we describe exactly how Earth's orbit about the Sun produces our seasons. We begin with some of the basics.

Energy

Energy is traditionally defined as "the ability to do work." This definition isn't entirely accurate and raises it own questions (What's work?), but it's impossible to do better in just a few words. Rather than travel far afield in search of a precise definition, we'll assume everyone has at least a vague idea of energy as an agent capable of setting an object in motion, warming a teapot, or otherwise manifesting itself in everyday events. The standard unit of energy in the International System (SI) used

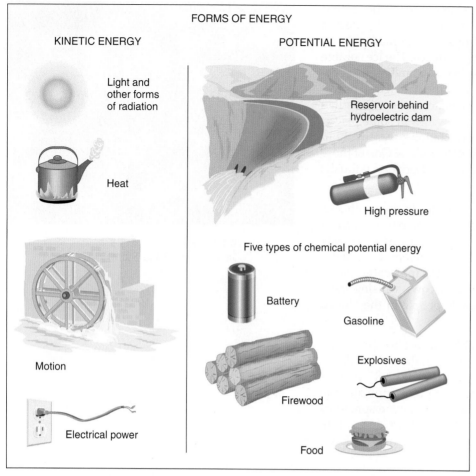

FORMS OF ENERGY

KINETIC ENERGY

Light and other forms of radiation

Heat

Motion

Electrical power

POTENTIAL ENERGY

Reservoir behind hydroelectric dam

High pressure

Five types of chemical potential energy

Battery

Gasoline

Firewood

Explosives

Food

FIGURE 2–1

Energy assumes several different forms, but each of these is a form of either kinetic energy (the energy of motion) or potential energy.

in scientific applications is the **joule** (abbreviated as J). Although students may be more familiar with the calorie as the unit for energy, the joule is preferred in this context (1 joule = 0.239 calories). A related term, **power**, is the rate at which energy is released, transferred, or received. The unit of power is the **watt** (W), which corresponds to 1 joule per second.

Even the simplest activity requires a transfer of energy. In fact, while you read these words, an energy transfer is occurring as the chemical energy from food you have eaten is converted into the kinetic energy (energy of motion) needed to move your eyes across this line of type. But your body, like any other machine, is not perfectly efficient; it loses some thermal energy (heat) as chemical energy is converted into kinetic energy. Thus, your eye muscles give off heat as they contract and relax.

The same concept applies to our atmosphere. About one two-billionth of the energy emitted by the Sun is transferred to Earth as **electromagnetic radiation**, some of which is directly absorbed by the atmosphere and surface. This radiation provides the energy for the movement of the atmosphere, the growth of plants, the evaporation of water, and an infinite variety of other activities.

KINDS OF ENERGY

Energy can occur in a variety of forms. We often speak of radiant, electrical, nuclear, and chemical energy; but, strictly speaking, all forms of energy fall into the general categories of **kinetic energy** and **potential energy**. These are illustrated in Figure 2–1.

Kinetic energy can be viewed as energy in use and is often described as the energy of motion. Motion can occur on a large scale, as in the movement of an object from one place to another. Examples that occur in nature include falling raindrops (Figure 2–2a), water flowing through a river channel, and grains of dust transported by the wind. The motion of kinetic energy can also occur at a microscale, as in the case of molecular vibration or rotation (illustrated for water in Figure 2–2b).

A solid object may seem to be standing still, but its atoms or molecules undergo a certain amount of vibration. Gas and liquid molecules, in contrast, are not fixed in space but move about randomly (Figure 2–3). In solids, liquids, or gases, the rate of vibration or random movement determines the temperature of the object.

If kinetic energy is energy in use, potential energy is energy that hasn't yet been used. Potential energy can assume many forms. For example, a plant's carbohydrates have potential energy that can be consumed by animals (or by the plant itself) and then metabolized to yield the energy needed for all of its biological activity. When our own bodies metabolize food, we are using this potential energy, converting it to kinetic energy, and releasing heat as a byproduct.

Another form of potential energy results from an object's position. Consider, for example, a cloud droplet that occupies some position above Earth's surface. Like all other objects, the droplet is subject to the effect of gravity. As it falls toward Earth's surface, the object's potential energy is converted to kinetic energy. Obviously, the higher the droplet's elevation, the greater the distance it is capable of falling and the greater its potential energy. It is important to recognize that the droplet did not attain its height by magic, because energy was used to elevate its mass in the first place.

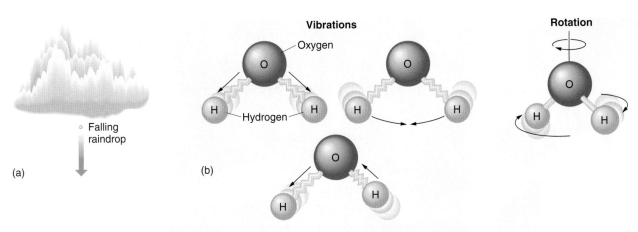

FIGURE 2–2
Kinetic energy can occur as the motion associated with moving objects, such as the falling raindrop in (a), or as molecular vibration or rotation, as depicted for water molecules in (b). The greater the rate of vibration or rotation, the higher the temperature of the substance.

ENERGY TRANSFER MECHANISMS

Energy can be transferred from one place to another by three processes: conduction, convection, and radiation.

CONDUCTION

Conduction is the movement of heat through a substance without the movement of molecules in the direction of heat transfer. A simple example is a metal rod, one end of which is placed over a campfire. The part of the rod above the flame is warmed, and molecules there gain energy. Some of this energy is passed to neighboring molecules, which in turn heat adjacent molecules. (The exact mechanism of molecular "passing" depends on the substance—in metals, it is mainly accomplished by electrons.) This process occurs throughout the length of the rod so that after a few moments the entire piece of metal becomes too hot to handle. The transfer of heat from the warmer to the colder part of the rod is conduction. Note that although heat travels through the rod, the molecules that make up the rod do not move. Conduction is most effective in solid materials, but as we will see in Chapter 3, it also is an important process in a very thin layer of air near Earth's surface.

CONVECTION

The transfer of heat by the mixing of a fluid is called **convection**. Unlike conduction, convection is accomplished by movement of the liquid or gas in which the process occurs. You can observe this process by watching a pot of water boil on a kitchen stove. The water at the bottom of the pot is closest to the source of energy and warms most rapidly. In warming, the water expands ever so slightly, becomes less dense, and rises to the surface. The rising water must, of course, be replaced from above, so water formerly at the surface sinks to the base of the pot. These rising and sinking motions cause a rapid movement not only of mass, but also of the thermal energy within the circulating water.

Convection in the atmosphere is not much different from that within a pot of boiling water. During the daytime, heating of Earth's surface warms a very thin layer of air (on the order of 1 mm thick) in contact with the surface. Above this thin *laminar layer*, air heated from below expands and rises upward because of the inherent **buoyancy** of warm air (the tendency for a light fluid [liquid or gas] to float upward when surrounded by a heavier fluid). Unlike water in a pot, the atmosphere can undergo convection even in the absence of buoyancy through a process called *forced convection*, the vertical mixing that happens as the wind blows. We discuss these processes in more detail in Chapter 3.

FIGURE 2–3
Air molecules move about in random motion.

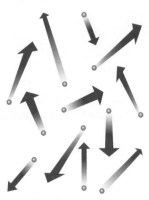

Randomly moving
air molecules

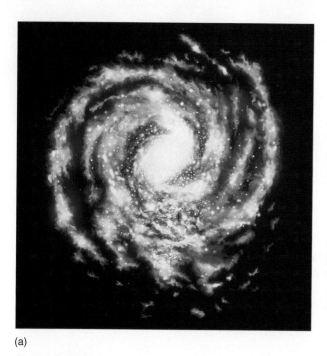

(a)

(b)

FIGURE 2–4

Our solar system is part of the Milky Way galaxy that contains more than 100 billion stars. (a) An artist's rendition. (b) A wide-angle infrared image of the plane and bulge of our Milky Way.

RADIATION

Of the three energy transfer mechanisms, **radiation** is the only one that can be propagated without a transfer medium. In other words, unlike conduction or convection, the transfer of energy by radiation can occur through empty space. Virtually all the energy available on Earth originates from the nearby (in astronomical terms) star we call the Sun, a member of the Milky Way galaxy (Figure 2–4). The atmosphere also has other sources of energy: Minute amounts of radiation are received from the billions of other stars in the universe, and some energy reaches the surface from Earth's interior. However, the contribution of these sources is minuscule compared to the energy from the Sun.

We will now examine the characteristics of radiation and the way Earth's orientation affects the radiation received. The spatial and seasonal variations in the receipt of solar energy are not mere abstractions; they are, in fact, the driving force for virtually all the processes discussed in the rest of this book.

Radiation

Radiation is emitted by all matter. Thus, *everything*—including the stars, Earth, ourselves, and this book—is constantly emitting electromagnetic energy. We are all familiar with electromagnetic energy in many of its forms. We see the environment around us because a type of radiation we call *visible light* impinges on our eyes, which then send signals to our brains to produce visual images. A different type of electromagnetic energy is used when we warm a meal in a microwave oven; the radiation agitates the molecules of the food and thereby increases its temperature. Other types of radiation may be less beneficial or even harmful, such as ultraviolet radiation, which can lead to sunburns, malignancies, or even death. Although different types of radiation have different effects, they are all very similar in that they are transmitted as a sequence of waves.

TUTORIAL
Radiation

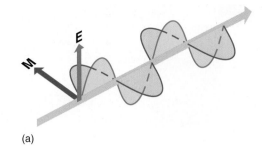

(a)

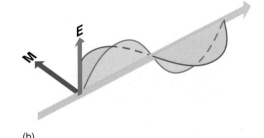

(b)

FIGURE 2–5

Electromagnetic radiation consists of an electric wave (E) and a magnetic wave (M). As radiation travels, the waves migrate in the direction shown by the pink arrow. The waves in (a) and (b) have the same amplitude, so the radiation intensity is the same. However, (a) has a shorter wavelength, so it is qualitatively different from (b). Depending on the exact wavelengths involved, the radiation in (a) might pass through the atmosphere, whereas that in (b) might be absorbed.

Think of a wave created by a rock tossed into a pond. The wave is revealed by an oscillation in the water surface with alternating crests (high points in the ripple) and troughs (low points). When you observe the regular rise and fall of the surface as the wave passes, you know energy is being transferred.

In the case of radiation, the waves are electrical and magnetic oscillations. That is, radiation consists of both an electrical and a magnetic wave. With the proper instruments, we would detect these electrical and magnetic variations—hence the term *electromagnetic radiation*. To put it differently, when an object emits radiation, both an electrical field and a magnetic field radiate outward. At a fixed point in space, the strength of both fields rises and falls rhythmically, thereby forming electric and magnetic waves, each with its own crest-to-trough pattern. The electric and magnetic waves are perpendicular to one another, as shown in Figure 2–5. More importantly, the electric and magnetic components are closely coupled—the two rise and fall in unison.

RADIATION QUANTITY AND QUALITY

To describe electromagnetic radiation completely, we need to provide information about the amount of energy transferred (quantity), and the type, or quality, of the energy. This is similar to describing someone's weight, where we might state quantity in pounds and indicate quality using words such as "mostly flab." In the case of radiation, quantity is associated with the height of the wave, or its *amplitude*. Everything else being equal, the amount of energy carried is directly proportional to wave amplitude.

The quality, or "type," of radiation is related to another property of the wave, the distance between wave crests. Figure 2–5 shows waves of electromagnetic radiation moving in the same direction. All have the same amplitude, but the distance between the individual wave crests is smaller for the waves depicted at the top. The upper waves therefore have a shorter **wavelength**, which is the distance between any two corresponding points along the wave (crest-to-crest, trough-to-trough, etc.). Because of their shorter wavelengths, the waves in Figure 2–5a are qualitatively different, and might produce different effects, from the waves in Figure 2–5b. For example, X-ray radiation has an extremely short wavelength and is able to penetrate soft tissues. On the other hand, ordinary light, having a somewhat longer wavelength, is absorbed by the skin. Compared to everyday objects, the radiation of interest here has very small wavelengths. It is therefore convenient to specify wavelengths using small units called **micrometers** (or **microns**). One micrometer—signified by μm—equals one-millionth of a meter, or one-thousandth of a millimeter (0.00004 in.).

All forms of electromagnetic radiation, regardless of wavelength, travel through space at the speed of light, which is about 300,000 km (186,000 mi) per second. At that speed, it takes 8 minutes for energy from the Sun to reach Earth. The energy received from the other, more distant stars takes even longer to arrive at Earth. For instance, radiation from the next nearest star, Proxima Centauri, must travel through space for 4.3 years before reaching us. Though this may seem like a long time, it is minuscule compared to the *billions* of years needed for light from a distant star to arrive at Earth.

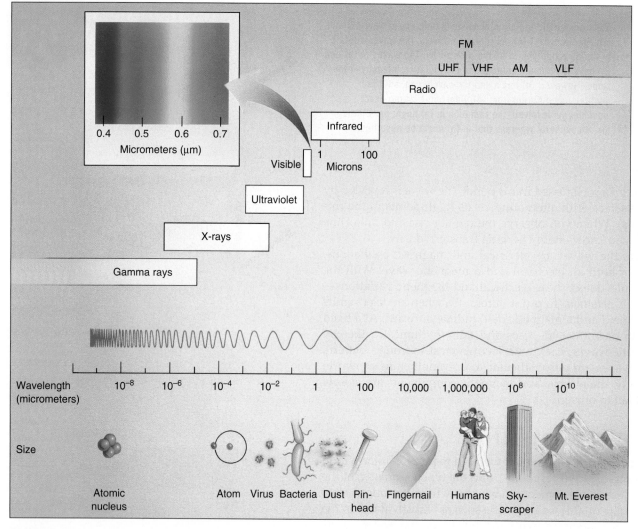

FIGURE 2–6
Electromagnetic energy can be classified according to its wavelength.

TABLE 2–1	Wavelength Categorizations
TYPE OF ENERGY	**WAVELENGTH (MICROMETERS)**
Gamma	<0.0001
X-ray	0.0001 to 0.01
Ultraviolet	0.01 to 0.4
Visible	0.4 to 0.7
Near Infrared (NIR)	0.7 to 4.0
Thermal Infrared	4 to 100
Microwave	100 to 1,000,000 (1 meter)
Radio	>1,000,000 (1 meter)

Electromagnetic energy comes in an infinite number of wavelengths, but we can simplify things by categorizing wavelengths into just a few individual "bands," as indicated in Figure 2–6 and Table 2–1. The band with the shortest wavelengths consists of gamma rays, with a maximum wavelength of 0.0001 μm. Successively longer wavelength bands include X-rays, ultraviolet (UV), visible, near-infrared (NIR), thermal infrared (IR), microwave, and radio waves. Note that there is nothing unique or special about the visible portion of this electromagnetic spectrum other than the fact that our eyes and nervous systems have evolved to be able to sense this type of energy. Except for their wavelengths, visible rays are just like any other form of electromagnetic energy.

INTENSITY AND WAVELENGTHS OF EMITTED RADIATION

All objects radiate energy, not merely at one single wavelength but over a wide range of wavelengths. Figure 2–7a graphs the intensity of radiation emitted at all wavelengths every second by a square meter of the surface of the Sun (in red) and Earth (in blue). We can readily see that a unit of area on the Sun emits much more radiation (about 160,000 times more) than does the same amount of surface area on Earth (notice that the curve showing Earth's emission is actually greatly exaggerated—if it were drawn to true scale, it would be too small to see). The shape of the curve showing the intensity of energy emitted by Earth at different wavelengths

(Figure 2–7b) is similar to that of the Sun, but the total energy released is much less, and the peak of the curve corresponds to a longer wavelength.

Of course, the amount of radiation emitted and its wavelengths are not the result of mere chance; they obey some fundamental physical laws. Strictly speaking, these laws apply only to perfect emitters of radiation, so-called **blackbodies**. Blackbodies are purely hypothetical bodies—they do not exist in nature—that emit the maximum possible radiation at every wavelength. Earth and the Sun are close to blackbodies and therefore nearly follow the laws described shortly. Other materials may or may not approximate blackbodies. In particular, the atmosphere, composed mainly of gases, is especially far from a blackbody, so we will not treat it as one.

STEFAN-BOLTZMANN LAW

The single factor that determines how much energy a blackbody radiates is its temperature. Hotter bodies emit more energy than do cooler ones; thus, not surprisingly, a glowing piece of hot iron radiates more energy than an ice cube. Interestingly, though, the amount of radiation emitted by an object is more than proportional to its temperature. In other words, a doubling of temperature produces *more* than a doubling of the amount of radiation emitted. Specifically, the intensity of energy radiated by a blackbody increases according to the fourth power of its absolute temperature. This relationship, the blackbody version of the **Stefan-Boltzmann law**, is expressed as

$$I = \sigma T^4$$

where I denotes the intensity of radiation in watts per square meter, σ (Greek lowercase sigma) is the Stefan-Boltzmann constant (5.67×10^{-8} watts per square meter per K^4), and T is the temperature of the body in kelvins (see *Box 1–3, Physical Principles: The Three Temperature Scales, page 15*).

Because the intensity of radiation depends on the temperature raised to the fourth power, a doubling of temperature leads to a sixteenfold increase in emission. Solving the Stefan-Boltzmann equation using the mean temperature of Earth's surface (about 288 K, 15 °C, or 59 °F) reveals that a square meter emits about 390 watts of power. In contrast, the surface of the Sun, with its temperature of about 6000 K (5700 °C, or 10,300 °F), emits about 73 million watts per square meter.

Although true blackbodies do not exist in nature, they provide a useful model for understanding the maximum amount of radiation that can be emitted. Most liquids and solids can be treated as **graybodies**, meaning that they emit some percentage of the maximum amount of radiation possible at a given temperature. Whereas some substances (for example, water) are highly efficient at radiating energy, others (for example, aluminum) are less efficient. The percentage of energy radiated by a substance relative to that of a blackbody is referred to as its **emissivity**. Emissivities range from just above zero to just below 100 percent and are denoted by the Greek letter epsilon (ε). By incorporating the emissivity of any body, we derive the complete version of the Stefan-Boltzmann law:

$$I = \varepsilon \sigma T^4$$

Including the emissivity factor means that the electromagnetic energy emitted by any graybody will be some fraction of what would be emitted by a blackbody. Note that even though the graybody form of the Stefan-Boltzmann law shows radiation intensity to be a function of both emissivity and temperature, most natural surfaces have emissivities above 0.9 (that is, above 90 percent of blackbody emission). In most cases, therefore, differences in emission are governed by temperature differences. The atmosphere is an exception to this because emission depends

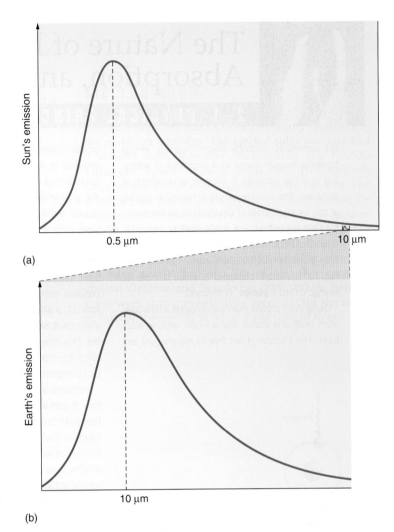

(a)

(b)

FIGURE 2–7
Energy radiated by substances occurs over a wide range of wavelengths. Because of its higher temperature, emission from a unit of area of the Sun (a) is 160,000 times more intense than that of the same area on Earth (b). Solar radiation is also composed of shorter wavelengths than Earth radiation.

DID YOU KNOW?

To be comfortable, humans need to maintain a skin temperature of about 306 K (33 °C or 91 °F). At this temperature an average-sized person with surface area of 1.8 square meters emits 895 watts of power—about the same as that used by nine household lightbulbs. Of course, the wavelengths emitted are in the thermal infrared range, so we don't light up a dark room.

The Solar Constant

We all know that the Sun is extremely hot and we are protected from its great heat by our distance from the solar surface. But the electromagnetic energy moving through space is not depleted as it moves toward Earth. Radiation traveling through space carries the same amount of energy and has the same wavelength as when it left the solar surface. However, at greater distances from the Sun, it is distributed over a greater area, which reduces its intensity.

Consider a sphere completely surrounding the Sun, whose radius is equal to the mean distance between Earth and the Sun, or 1.5×10^{11} m (Figure 2–9). As the distance from the Sun increases, the intensity of the radiation diminishes in proportion to the distance squared. This relationship is known as the **inverse square law**. By dividing total solar emission (3.865×10^{26} W) by the area of our imaginary sphere surrounding the Sun (the area of any sphere is given as $4\pi r^2$), we can determine the amount of solar energy received by a surface perpendicular to the incoming rays at the mean Earth–Sun distance. This incoming radiation is equal to

$$\frac{3.865 \times 10^{26}\ \text{W}}{4\pi(1.5 \times 10^{11}\ \text{m})^2} = 1367\ \text{W/m}^2$$

We refer to the value 1367 W/m^2 as the **solar constant** (although minor variations in solar output and other factors allow for some minor departures from this "constant"). For the sake of comparison, using the same procedure, we can determine that the solar constant for Mars (2.25×10^{11} m from the Sun) is 445 watts per square meter.

WEATHER IN MOTION
A Solar Eclipse

FIGURE 2–9
The intensity of a beam of solar radiation does not weaken as it travels away from the Sun. However, its intensity is reduced when it is distributed over a larger area. Imagine two spheres encompassing the Sun (such as the one with a radius equal to the mean Earth–Sun distance). All the radiation emitted from the Sun would be captured by this surrounding sphere. Now imagine that the surrounding sphere has a radius equal to the mean distance between the Sun and Mars. This sphere is larger than the previous one, so the energy emitted must be distributed over a greater area.

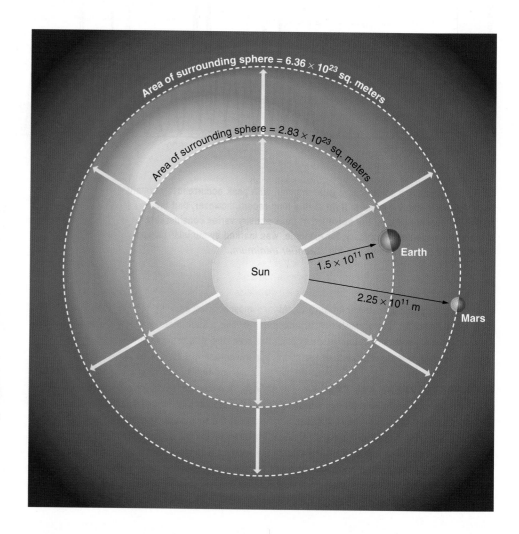

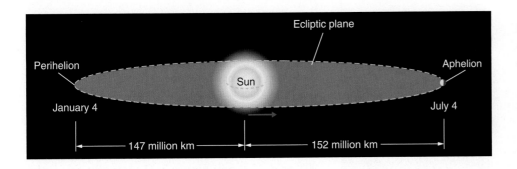

FIGURE 2–10

Earth's orbit around the Sun is not perfectly circular but is an ellipse. Earth is nearest to the Sun (perihelion) on about January 4 and farthest away (aphelion) on July 4.

The Causes of Earth's Seasons

Although the Sun emits a nearly constant amount of radiation, on Earth we experience significant changes in the amount of radiation received during the course of a year. These variations in energy manifest themselves as the seasons. We also know that the low latitudes (for example, the tropics and subtropics) receive more solar radiation per year at the top of the atmosphere than do regions at higher latitudes (for example, the Arctic and Antarctic). In this section, we will discuss how Earth's orbit around the Sun and its orientation with respect to incoming radiation influence the seasonal and latitudinal receipt of incoming solar radiation (often called **insolation**[1]).

EARTH'S REVOLUTION AND ROTATION

As we know, and as Figure 2–10 shows, Earth orbits the Sun once every $365\frac{1}{4}$ days as if it were riding along a flat plane. We refer to this imaginary surface as the **ecliptic plane** and to Earth's annual trip about the plane as its **revolution**.

The orbit is not quite circular but instead sweeps out an elliptical path, so that the distance between Earth and the Sun varies over the course of the year. Earth is nearest the Sun—at a point called **perihelion**—on or about January 4, when Earth–Sun distance is about 147,000,000 km (91,000,000 mi). Earth is farthest from the Sun—at a point called **aphelion**—on or about July 4, when Earth–Sun distance is about 152,000,000 km (94,000,000 mi). Thus, on perihelion Earth is 3 percent closer to the Sun than on aphelion. But because the intensity of incoming radiation varies inversely with the square of Earth–Sun distance (recall the inverse square law), the radiation is almost 7 percent more intense. (As mentioned at the outset of this chapter, however, this variation in intensity is not what causes the change of seasons.)

In addition to its revolution, Earth also undergoes a spinning motion called **rotation**. Rotation occurs every 24 hours (23 hours and 56 minutes, to be exact) around an imaginary line, called Earth's axis, connecting the North and South Poles. The axis is not perpendicular to the plane of the orbit of Earth around the Sun but is tilted 23.5° from it. Moreover, no matter what time of year it is, the axis is always tilted in the same direction and always points to a distant star called **Polaris** (the North Star). The constant direction of the tilt means that for half the year the Northern Hemisphere is oriented somewhat toward the Sun, and for half the year it is directed away from the Sun. The changing orientation of the hemispheres with regard to the Sun is the true cause of the seasons—not the varying distance between Earth and the Sun.

It is easier to visualize how the tilt of the axis influences the seasons if we consider a hypothetical situation in which the axis is tilted not 23.5°, but rather a full

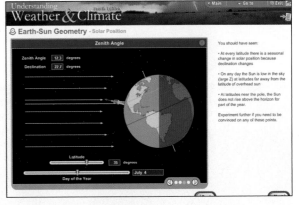

TUTORIAL
Earth-Sun Geometry

---

[1]It is sometimes reported that the term *insolation* is an acronym for incoming solar radiation. Such is not the case; it is from a Latin word and was first used in the 1600s.

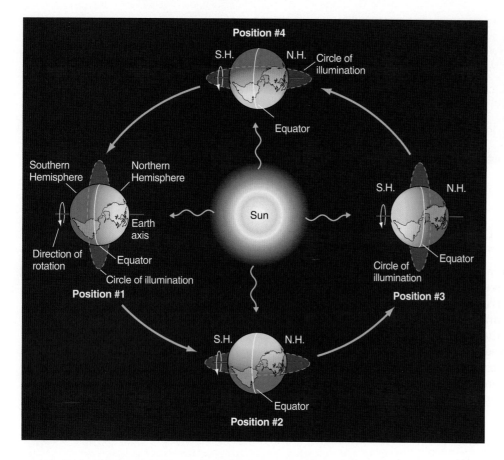

FIGURE 2-11

A hypothetical situation wherein Earth's axis is aligned along the ecliptic plane. In position #1, the Northern Hemisphere receives much energy from the Sun while the Southern Hemisphere is in constant darkness. The situation reverses six months later (position #3). In positions #2 and #4, both hemispheres receive equal amounts of solar energy.

90°, as depicted in Figure 2–11. Actually, this is the case for Uranus, so what we describe is not entirely hypothetical. Examine the situation when Earth is in position #1. The Northern Hemisphere is oriented directly toward the Sun so that it is fully illuminated over the entire 24-hour period of rotation. Meanwhile, the Southern Hemisphere undergoes 24 hours of continual darkness. This situation favors greater warmth in the Northern than in the Southern Hemisphere. Furthermore, a person standing at the North Pole would observe the Sun as being directly overhead during the entire day. Moving away from the North Pole toward the equator, there is a gradual reduction in the angle of the Sun above the horizon, until at the equator the Sun appears to be right on the horizon. South of that line, the Sun is below the horizon and nighttime covers the Southern Hemisphere.

Now refer to position #3, which occurs 6 months after position #1. In this situation, the Southern Hemisphere is in continual sunlight while the Northern Hemisphere is subjected to 24 hours of darkness. Furthermore, someone standing at the South Pole would see the Sun directly overhead, and the apparent position of the Sun would shift toward the horizon for viewers located closer to the equator.

Finally, observe the intermediate positions, #2 and #4. In these two situations, the 90° tilt of the axis is neither toward nor away from the Sun, and the tilt becomes irrelevant to the receipt of insolation. Moreover, in positions #2 and #4 every place on Earth receives 12 hours of daylight and 12 hours of darkness because every latitude is half sunlit and half dark. Finally, note that at noon (when the longitude of any place in question is aligned directly toward the Sun), a person standing at the equator would observe the Sun to be directly overhead. Thus, Earth's revolution causes seasonal changes in the amount of heating of the surface. When either the Northern or Southern Hemisphere is oriented toward the Sun, that hemisphere receives a greater amount of insolation and therefore warms more effectively. Whichever hemisphere is oriented away from the Sun receives less radiation.

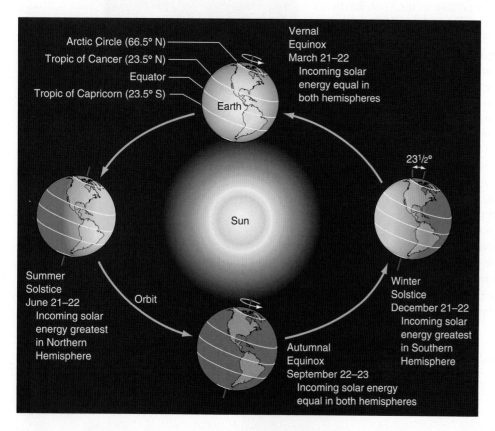

FIGURE 2-12

Earth's revolution around the Sun.

SOLSTICES AND EQUINOXES

Figure 2–12 shows the true seasonal change in orientation of Earth with respect to the Sun, based on the actual 23.5° tilt of the axis. Although the axis is tilted only 23.5°, and not 90°, the principle just described still applies. During 6 months of the year, the Northern Hemisphere receives more sunlight than does the Southern Hemisphere; during the other 6 months, the Southern Hemisphere receives a greater amount of insolation. The four positions shown in the diagram represent 4 days that have particular significance.

In the farthest left position in Figure 2–12, the Northern Hemisphere has its maximum tilt toward the Sun. This occurs on or about June 21, which we refer to as the **June solstice** (this is also called the *summer solstice* according to the corresponding Northern Hemisphere season). Although we designate this as the first day of summer, it actually represents the day on which the Northern Hemisphere has its greatest availability of insolation. Six months later (on or about December 21), the Northern Hemisphere has its minimum availability of solar radiation on the **December solstice** (*winter solstice* in the Northern Hemisphere), which is called the first day of winter in the Northern Hemisphere and the first day of summer in the Southern Hemisphere. Intermediate between the two solstices are the **March equinox** (often called the *vernal* or *spring equinox* for the Northern Hemisphere) on or about March 21, and the **September equinox** (called the *autumnal equinox* in the Northern Hemisphere) on or about September 21. On the equinoxes, every place on Earth has 12 hours of day and night (the word *equinox* refers to "equal night"), and both hemispheres receive equal amounts of energy.

Of course, the transitions between the four positions shown in Figure 2–12 do not occur in sudden leaps; instead, a steady progression occurs from one position to the next. As shown in Figure 2–13, the 23.5° tilt of the Northern Hemisphere toward the Sun on the June solstice causes the *subsolar point* (the point on Earth where the Sun's rays meet the surface at a right angle—and where the Sun appears directly overhead) to be located at 23.5° N. This is the most northward latitude at which the subsolar point is located. The fact that the Sun never appears directly overhead

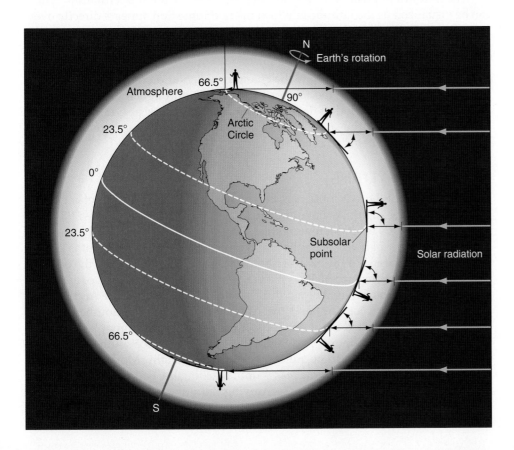

FIGURE 2–13
Because Earth's axis is tilted 23.5°, the subsolar point is at 23.5° N during the summer solstice.

FIGURE 2-14
The solar declination gradually migrates north and south over the course of the year. On the June solstice the subsolar point marks its most northward extent, 23.5° N. Solar declination is 23.5° S on the December solstice.

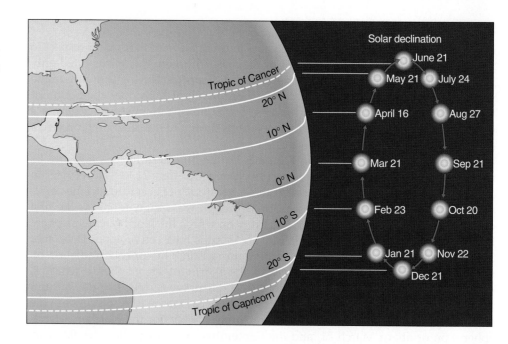

poleward of 23.5° N gives that latitude special significance, and we call it the **Tropic of Cancer**.

Likewise, on the December solstice, the sun is directly overhead at 23.5° S, the **Tropic of Capricorn**. On the two equinoxes, the subsolar point is on the equator. Thus, the subsolar point migrates 47° (that is, between 23.5° N and 23.5° S) over a 6-month period, and on any particular day it is located somewhere between the Tropics of Cancer and Capricorn. This seasonal movement of the subsolar point is similar to what would happen if instead of Earth orbiting the Sun, its axis were to slowly rock back and forth, toward and away from the Sun.

The latitudinal position of the subsolar point is the **solar declination**, which can be visualized as the latitude at which the noontime Sun appears directly overhead. Figure 2–14 plots the solar declination for several days of the year, while the arrows between the dates indicate the direction toward which the declination is moving at that time of year.

Now we are ready to see how the changing orientation of Earth with respect to the Sun directly affects the receipt of insolation through three mechanisms: (1) the length of the period of daylight during each 24-hour period, (2) the angle at which sunlight hits the surface, and (3) the amount of atmosphere that insolation must penetrate before it can reach Earth's surface.

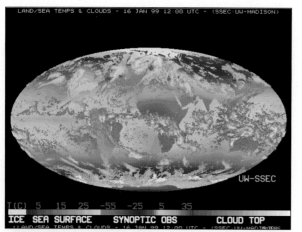

WEATHER IN MOTION
January and July Global Movies

PERIOD OF DAYLIGHT

One way the tilt of the axis influences energy receipt on Earth is by its effect on the lengths of day and night. We have already seen that if the axis were tilted 90° from the plane of Earth's orbit, there would be 1 day when the entire Northern Hemisphere underwent 24 hours of daylight and an equivalent period 6 months later of continuous darkness. But the axis is only tilted 23.5° from the plane of the orbit, so only the latitudes poleward of 66.5° (that is, 90° minus 23.5°) experience a 24-hour period of continuous daylight or night. These lines of latitude are the **Arctic Circle** (in the Northern Hemisphere) and the **Antarctic Circle** (in the Southern Hemisphere). This is illustrated in Figure 2–15. On the June solstice, any place north of the Arctic Circle has 24 hours of daylight. Just a short distance south of the Arctic Circle, there is almost (but not quite) 24 hours of daylight. Moving toward the equator, the period of daylight decreases until reaching the equator, where the day and night are both 12 hours long. Moving into the Southern Hemisphere, daylength shrinks until 66.5° S, where night is 24 hours long. The opposite pattern, of course, holds for the December solstice.

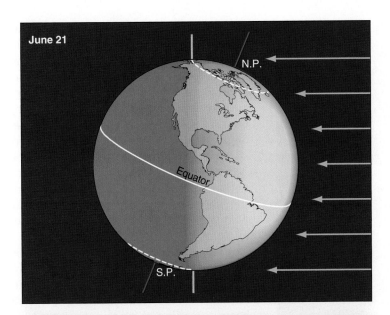

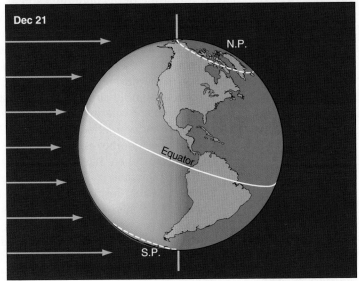

FIGURE 2–15
On the June solstice (a), every point north of 66.5° N has 24 hours of daylight and every point south of 66.5° S has continual night. During the December solstice (b), the situation is reversed. These latitudes are called the Arctic Circle and Antarctic Circle, respectively.

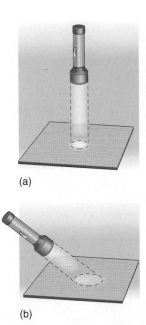

(a)

(b)

FIGURE 2–16
The intensity of radiant energy hitting a surface is affected by its angle of incidence. More direct illumination causes a more intense amount of heating.

SOLAR ANGLE

Think back to the last time you spent a few hours lying out in the Sun. If you went outside early in the day when the Sun was low above the horizon, you probably did not feel a great deal of warming from its rays. But as the Sun got higher in the sky, it became more effective at warming your body. This change was largely due to a decrease in beam spreading. **Beam spreading** is the increase in the surface area over which radiation is distributed in response to a decrease of solar angle, as illustrated in Figure 2–16. The greater the spreading, the less intense the radiation is. In part (a) of the figure, the incoming light is received at a 90° angle, which concentrates it to a small area and increases its ability to heat the surface. In (b), the rays hit the surface more obliquely and the energy is distributed over a greater area, leading to a less intense illumination (less energy per unit area). Thus, a beam of light is more effective at illuminating or warming a surface if it has a high angle of incidence (that is, the angle at which it hits the surface).

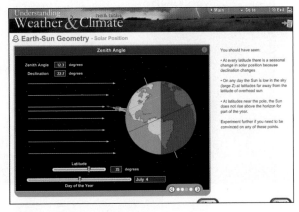

TUTORIAL
Earth-Sun Geometry

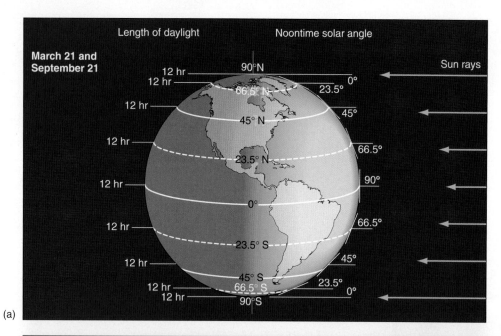

Length of daylight | Noontime solar angle

March 21 and September 21

(a)

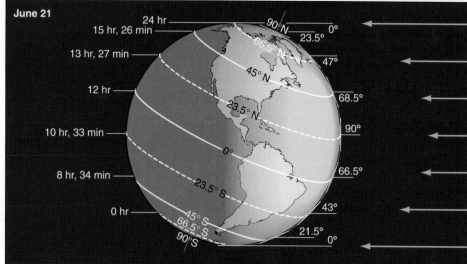

June 21

(b)

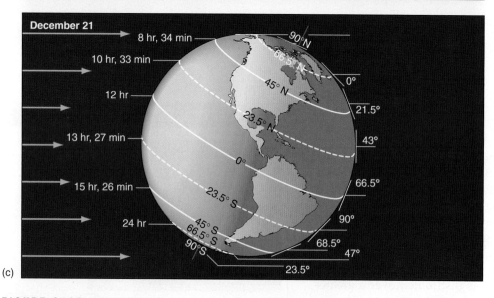

December 21

(c)

FIGURE 2-17
The length of day (left) and the noon solar angle (right) are shown for the equinoxes (a), June solstice (b), and December solstice (c).

Noontime sun angles for any given latitude can be easily determined if the solar declination is known. To do this, you simply subtract the latitude of a given location from 90° and then add the solar declination. Thus, on either of the equinoxes (solar declination = 0°) Toronto, Ontario (latitude 44° N), has a noontime solar angle of 90° − 44° = 46°. During the June solstice the noontime solar angle is 90° − 44° + 23.5° = 69.5°. Six months later the solar declination = −23.5°, with the negative sign indicating the sun is overhead in the Southern Hemisphere. At this time, the noontime solar angle = 90° − 44° − 23.5° = 22.5°.

The change in noontime solar angle over the course of a year can cause significant differences in the intensity of sunlight hitting the surface due to beam spreading. For example, beam spreading of the noontime sun at Toronto on the December solstice is almost two and a half times greater than on the June solstice. Thus, in the absence of other effects, the midday sun will be less than half as intense in the winter as in the summer.

Differences in the amount of beam spreading have a major role in causing the seasons. During the 6-month period between the March and September equinoxes, any latitude in the Northern Hemisphere has a more direct angle of incidence than does its Southern Hemisphere counterpart. Thus, insolation available to the Northern Hemisphere is subjected to less beam spreading, which promotes greater warming of the surface. During the following 6 months, the situation is reversed and the Southern Hemisphere has, on the whole, higher Sun angles.

Figure 2–17 shows the effects of noontime solar angle and length of daylight period for the solstices and equinoxes. On the June solstice, both factors work together to enhance warming in the Northern Hemisphere; on the December solstice, they combine for less effective heating in the north.

ATMOSPHERIC BEAM DEPLETION

The third way in which the tilt of the axis influences heating is in determining the amount of atmosphere that sunlight must penetrate before reaching the surface. As you can see in Figure 2–18a, insolation approaching the surface at a 90° angle passes through the atmosphere as directly as possible. Compare this to Figure 2–18b, in which sunlight approaches the surface at a low angle (as it does around sunrise or sunset). In this situation, a beam of sunlight must pass through a greater amount of atmosphere. Although the atmosphere is mostly transparent to incoming sunlight, some radiation is absorbed and even more is reflected back to space. The greater the thickness traveled, the more the beam is weakened. Because the solar altitude is lowest in the Northern Hemisphere near the December solstice, at that time more energy is lost due to atmospheric effects than at any other time of the year.

OVERALL EFFECTS OF PERIOD OF DAYLIGHT, SOLAR ANGLE, AND BEAM DEPLETION

We can now summarize some of the important points regarding controls on energy reaching Earth's surface. We must emphasize that we have not considered differences in atmospheric transparency or cloudiness at all; thus, the discussion assumes uniform optical properties. In other words, the patterns described below arise solely from geometrical considerations.

1. June solstice:
    a. Solar declination = 23.5° N.
    b. Every latitude in the Northern Hemisphere receives more energy than the corresponding latitude in the Southern Hemisphere.
    c. Every place north of the Arctic Circle receives 24 hours of daylight, and every place south of the Antarctic Circle has 24 hours of night.
    d. The equator receives 12 hours of daylight.
    e. For the Northern Hemisphere as a whole, sunlight travels through less atmosphere than it does in the Southern Hemisphere.

2. December solstice:
    a. Solar declination = 23.5° S.
    b. Every latitude in the Southern Hemisphere receives more energy than the corresponding latitude in the Northern Hemisphere.
    c. Every place south of the Antarctic Circle receives 24 hours of daylight, and every place north of the Arctic Circle has 24 hours of night.
    d. The equator receives 12 hours of daylight.
    e. For the Southern Hemisphere as a whole, sunlight travels through less atmosphere than it does in the Northern Hemisphere.

3. Equinoxes:
    a. Solar declination is 0°.
    b. Every place on Earth has 12 hours of daylight.
    c. Both hemispheres receive equal amounts of insolation.

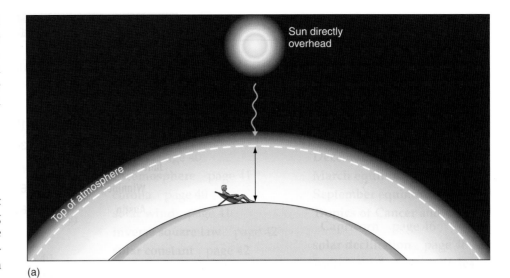
(a)

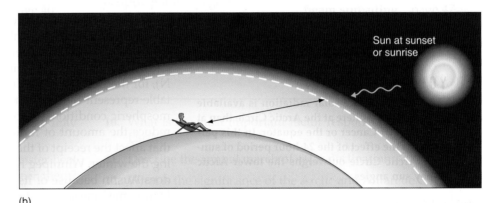
(b)

FIGURE 2–18
A high solar angle (a) allows sunlight to pass through the atmosphere with a relatively short path. Lower sun angles such as those near sunrise and sunset (b) require that the energy pass through more of the atmosphere. This increase in atmospheric mass results in a greater depletion of energy than in (a).

CHANGES IN ENERGY RECEIPT WITH LATITUDE

We have seen that seasonal changes in the orientation of Earth with respect to the Sun determine the availability of solar radiation across the globe. Obviously, a large supply of energy is favorable for warm temperatures and a lack of energy leads to cold conditions. But the impact of radiation availability does not end with temperature. As we will see in later chapters, energy receipts also affect the distribution of pressure, which in turn affects winds, cloudiness, precipitation, and other aspects of weather and climate.

3. Why is it not completely accurate to describe the energy coming from the Sun as visible radiation?

4. If Earth's speed of rotation were different from what it actually is, would there be a corresponding change in the amount of energy the planet as a whole receives?

5. At noon the solar angle will always be greater at Tucson, Arizona, than at Laramie, Wyoming. Is the same also true for 6 P.M.?

6. Locations near the equator typically have less seasonality than do locations farther away from the tropics. Explain why this is so.

7. Why is it that the solar angle cannot be considered a single influence on the amount of radiation reaching Earth's surface? Is the situation different for the Moon?

8. How might the temperature change in the course of a day differ on east-facing vs. west-facing slopes?

9. At noon at 45° N latitude, the solar angle is 45° above the southern horizon. What would the angle of incidence be on a north-facing slope of 45°? Would the slope of the surface affect both beam spreading and atmospheric path length?

10. Describe the apparent path of the Sun to a person standing at the North Pole on June 22.

11. On the equinoxes a person at the equator would see the sun rise exactly to the east, pass directly overhead at noon, and set exactly in the west—all over a 12-hour period. How will this change on the solstices?

Problems and Exercises

1. An instrument measures the radiation emitted from an ocean surface as 365 watts per square meter. What law would you apply to determine the ocean surface temperature? (More advanced question: What would the ocean surface temperature actually be? *Hint*: You will need to rearrange one of the equations given in this chapter.)

2. Assume that a body has an emissivity of 0.9 and a temperature of 300 K. Which would have a greater impact on the intensity of radiation emitted: a 50 percent reduction in the emissivity or a 5 percent reduction in the absolute temperature?

3. Saturn is about 1.42×10^{12} m from the Sun, or about 9.5 times as far from the Sun as Earth is. Calculate the solar constant for Saturn. Do you suppose the distribution of wavelengths of the sunlight received at that distance is different from the distribution Earth receives?

4. One factor that influences the amount of insolation available is the varying Earth–Sun distance. Using the distances for perihe-

lion and aphelion of 1.47×10^{11} m and 1.52×10^{11} m, respectively, determine the intensity of solar radiation at the top of Earth's atmosphere on those two days.

5. What is the difference in the noontime solar angle between the two solstices at a latitude of 10° N? How does this compare to the range of noontime solar angles at 30° N? Can you think of any significant outcomes of this difference?

6. Go to http://aa.usno.navy.mil/data/. Check the first of the two hypertext options under the category "Positions of the Sun and Moon." This site will allow you to determine the solar angle throughout the course of any date you select. Plot the solar angle of the Sun throughout the daylight period for the solstices and the equinoxes and notice how the pattern changes through the year. Next, for the solstices and equinoxes, plot and compare the differing solar elevations over the course of the day for East Lansing, Michigan; Knoxville, Tennessee; and Gainesville, Florida. What generalizations can you make?

Quantitative Problems

This chapter has introduced some important laws and concepts describing the type and amount of radiation received by Earth. You can enhance your understanding of these laws by solving some quantitative problems shown on this book's companion Web site, http://www.prenhall.com/aguado/. After entering

the site, go to the bottom of the page and select Chapter 2. Then highlight "Quantitative Examples" on the left hand panel. These brief problems should bolster your comprehension of the Stefan-Boltzmann and Wien's laws and the temperature scales discussed in this chapter.

Useful Web Sites

http://aa.usno.navy.mil/data/docs/AltAz.html
Some very useful astronomical data, especially with regard to Earth–Sun relationships. Interactive format allows you to request data for particular cities or latitude-longitude coordinates. Includes precise sunrise/sunset data as well as solar angle data.

http://aa.usno.navy.mil/data/docs/EarthSeasons.html
Dates and times of perihelion, aphelion, equinoxes, and solstices.

http://www.srrb.noaa.gov/highlights/sunrise/gen.html
Fully interactive site that allows you to get time of sunrise/sunset for a choice of world cities or for precise latitude-longitude coordinates. Also gives solar declination and several other items of Earth–Sun position information, complete with glossary to explain their meanings.

Media Enrichment

Tutorial
Radiation

The wave properties of radiation and blackbody emission, and the differences between solar and terrestrial radiation, are presented in this tutorial. Interactive controls are used to study the relationship between temperature and emitted radiation.

Tutorial
Earth-Sun Geometry

This tutorial covers the geometry of Earth's orbit and its effects on solar radiation. It uses three-dimensional diagrams to show how variations in daylength and solar position arise and includes animations depicting the processes of beam spreading and depletion.

Weather in Motion
January 1998 Ice Storm

This chapter opened with an account of the ice storm that hit the eastern United States and Canada in January 1998. This brief movie shows the movement of the storm as observed by satellite.

Weather in Motion
A Solar Eclipse

As Earth orbits the Sun, the Moon simultaneously revolves around Earth. Occasionally, the three bodies briefly line up in a straight line. If the Moon is between the Sun and Earth, the Moon's shadow passes across the Earth, forming a solar eclipse. In some instances the Earth–Sun distance is small enough that the center of the Moon's shadow (called the *penumbra*) crosses the Earth in a *total eclipse*. This movie depicts the path of the Moon's shadow across North America on February 26, 1998. Notice that only a small part of the planet experiences a total eclipse and that the shadow moves quickly. Because of this fast shadow movement, eclipses are short, lasting just a few minutes at most.

Weather in Motion
January and July Global Movies

We have seen in this chapter how the solar declination varies through the course of a year. The movies illustrate this annual progression by showing the January and July portions of the 1-year loop used in Chapter 1. The first thing you should look at is the differential movement of the zone of high land surface temperatures associated with the east-to-west migration of sunlight over the course of a day. In January the hottest regions are found in the daytime in the Southern Hemisphere desert regions of Australia, southern Africa, and South America. In July the maximum temperature zone shifts to southern Asia, North Africa, and the southern United States and northern Mexico. Notice also the very dramatic changes in temperature that take place in the higher latitudes, especially over Canada and northern Asia.

A couple of other interesting patterns can be observed in the distribution of clouds and the speed with which they migrate during January and July. During both months, a conspicuous zone of cloud cover extends east–west across the tropics. This zone, called the Intertropical Convergence Zone, or ITCZ (discussed in Chapter 8), is largely the result of surface heating near the equator. Because the zone of most intense heating migrates northward from January to July, this cloud band makes a similar, though much smaller, shift in location.

Another important feature exhibited by the two movies is the differential speed of movement of storm systems outside of the tropics. Notice that the cloud bands sweeping across North America during January move about twice as rapidly as those in July. The difference occurs because both daylength and solar angle give rise to a large gradient in heating across the winter hemisphere. As will be seen later in the book, this leads to vigorous movement of atmospheric systems. In summer, daylength and solar position combine to create small differences in heating, resulting in generally lower wind speeds.

It was early morning on October 24, 1998, and Tropical Storm Mitch had just intensified into a true hurricane in the western Atlantic. While forecasters were certain the storm would strengthen further and move westward, nobody knew that within a few days Mitch would become the worst hurricane to hit Central America since 1780 (Figure 3–1). Its ultimate devastation was hard to comprehend. The exact death toll from Hurricane Mitch will never be known but probably stands at somewhere between 9000 and 18,000 people. Wind gusts of more than 320 km/hr (200 mph) and rainfall amounts greater than 50 cm (20 in.) ravaged many villages. Worst hit was Honduras, where more than 20 percent of the population was suddenly homeless. The mountainous terrain of Nicaragua suffered devastating mudslides. A farmer, José Morales, recounted his experience with one of the worst of them, the Casitas mudslide in northwestern Nicaragua: "It was a ball of earth and trees, and suddenly I couldn't see houses anymore."

The misery caused by Hurricane Mitch still continues, as it may for many years to come. The battered nations of the region have had to contend with the threat of cholera, malaria, and dengue fever—not to mention the widespread hunger from the ruined crops. In Honduras about 70 percent of the crops were destroyed and a similar percentage of the transportation infrastructure demolished. Some relief experts have estimated that it will take 15 to 20 years before the effects of the storm will be fully undone.

Like other major hurricanes, Mitch covered an area greater than 250,000 square kilometers, lasted for about a week, dumped millions of tons of rainwater, and brought winds capable of wiping out whole villages. Such activity requires an enormous amount of energy and, as we saw in Chapter 2, solar radiation provides virtually all of that energy. But there is more to the story, because most of the energy contained in the atmosphere does not accrue by the *direct* absorption of solar radiation. Instead, the majority of the energy comes *indirectly* from the Sun after first having been absorbed by Earth's surface. From there, several processes combine to transfer this absorbed energy to the atmosphere. In this chapter, we examine this energy transfer, which provides the fuel for everyday weather and for catastrophic events such as Hurricane Mitch.

Atmospheric Influences on Insolation

Solar radiation reaching the top of the atmosphere does not pass unimpeded through the atmosphere, but rather is attenuated by a variety of processes. The atmosphere absorbs some radiation directly and thereby gains heat. Another portion of radiation disperses as weaker rays going out in many different directions through a process we call *scattering*. Some of the scattered radiation is directed back to space; the remainder is scattered forward as the light we see from the portion of the sky away from the solar disk. In either case, the energy that is scattered is not absorbed by the atmosphere and therefore does not contribute to its heating.

The remaining insolation is neither absorbed nor scattered and passes through the atmosphere without modification, reaching the surface as direct radiation. But not all the energy reaching the surface is absorbed. Instead, a fraction is scattered back to space and, like the radiation scattered by the atmosphere, it does not contribute to the heating of the planet.

These processes—the absorption, scattering, and transmission of solar radiation—directly affect the distribution of temperature throughout the atmosphere. They also explain a number of atmospheric phenomena of everyday interest, such as the blue sky on a clear day or the redness of a sunset. In this section, we explore the processes affecting incoming radiation.

ABSORPTION

Atmospheric gases, particulates, and droplets all reduce the intensity of insolation by **absorption**. Absorption represents an energy transfer to the absorber. This transfer has two effects: the absorber gains energy and warms, while the amount of energy delivered to Earth's surface is reduced.

The gases of the atmosphere are not equally effective at absorbing sunlight, and different wavelengths of radiation are not equally subject to absorption. Ultraviolet radiation, for example, is almost totally absorbed by ozone in the stratosphere. Visible radiation, in contrast, passes through the atmosphere with only a minimal amount of absorption. This is of no minor consequence, because if the atmosphere *were* able to absorb all the incoming solar energy, the sky would appear completely dark. Artificial lights would be useless, because their radiation would likewise be absorbed. The very fact that we can see great distances suggests that the atmosphere is not particularly good at absorbing visible radiation, an impression that turns out to be correct.

Near-infrared radiation, which represents nearly half the radiation emitted by the Sun, is absorbed mainly by two gases in the atmosphere—water vapor and (to a lesser extent) carbon dioxide. This is why direct sunlight in the desert feels so hot and shade is so welcome, whereas in humid regions the apparent temperature difference between standing in direct sunlight and standing in shade is relatively small. When the humidity is high, water vapor absorbs a significant portion of near-infrared radiation, thereby reducing the amount of energy available to warm your skin. On dry days, the lack of water vapor allows a greater amount of near-infrared radiation to penetrate the atmosphere and raise your skin temperature.

REFLECTION AND SCATTERING

The **reflection** of energy is a process whereby radiation making contact with some material is simply redirected away from the surface without being absorbed. The reason we are able to see is that the human eye can detect the receipt of visible radiation. Visible energy travels in all directions as it is reflected off objects in our field of view. Some of the reflected light comes into contact with our eyes, which in turn send signals to optical centers in our brains. All substances reflect visible light, but with vastly differing effectiveness. Objects do not reflect all wavelengths equally. A shirt, for example, will appear green if it most effectively reflects wavelengths in the green portion of the spectrum. A fresh patch of snow very effectively reflects visible light, while a piece of coal reflects only a small portion of the visible radiation hitting its surface. The percentage of visible light reflected by an object or substance is called its **albedo**.

Light can be reflected off a surface in a couple of different ways. When light strikes a mirror, it is reflected back as a beam of equal intensity, in a manner known as **specular reflection**. In contrast, when a beam is reflected from an object as a larger number of weaker rays traveling in many different directions, it is called **diffuse reflection**, or **scattering**. When scattering occurs, you cannot see an image of yourself on the reflecting surface as you can in a mirror. Consequently, although a surface of fresh snow might reflect back most of the visible light incident on it, you would not be able to check out your appearance by looking at it. The vast majority of natural surfaces are diffuse rather than specular reflectors.

In addition to large solid surfaces, gas molecules, particulates, and small droplets scatter radiation. Furthermore, although much is scattered back to space, much is also redirected forward to the surface. The scattered energy reaching Earth's surface is thus **diffuse radiation**, which is in contrast to unscattered **direct radiation**. Figure 3–2 illustrates the process of scattering and the transformation of direct radiation to diffuse radiation. You can think of it this way: The blocking of direct radiation is what creates shadows, but a surface in the shadow of the direct radiation is not completely dark because it is illuminated by diffuse radiation. Notice that whether accomplished by a gas molecule, particulate, or droplet, this result is still a scattering process in which the radiation is redirected but not absorbed.

The characteristics of radiation scattering by the atmosphere depend on the size of the scattering agents (the air molecules or suspended particles) relative to the wavelength of the incident electromagnetic energy. Three very general categories of scattering exist: Rayleigh scattering, Mie scattering, and nonselective scattering.

RAYLEIGH SCATTERING

Scattering agents smaller than about one-tenth the wavelength of incoming radiation disperse radiation in a manner known as **Rayleigh scattering**. Rayleigh scattering is performed by individual gas molecules in the atmosphere. It primarily affects shorter wavelengths. Rayleigh scattering is particularly effective for visible light, especially those colors with the shortest wavelengths, so blue light is more effectively scattered by air molecules than is longer-wavelength red light. Furthermore, Rayleigh scattering disperses radiation both forward and backward. Combined with its greater effectiveness in scattering shorter wavelengths, this characteristic leads to three interesting phenomena: the blue sky on a clear day, the blue tint of the atmosphere when viewed from space, and the redness of sunsets and sunrises.

Figure 3–3 illustrates how Rayleigh scattering produces a blue sky. As parallel beams of radiation enter the atmosphere, a portion of the light is redirected away from its original direction. A person looking upward, away from the direction of the Sun, can see some of the scattered light that has

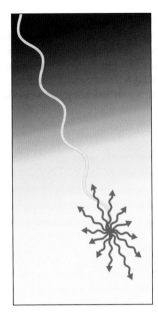

FIGURE 3–2
Scattering is a process whereby a beam of radiation is broken down into many weaker rays redirected in other directions.

FIGURE 3–3
The sky appears blue because the gases and particles in the atmosphere scatter some of the incoming solar radiation in all directions. Air molecules scatter shorter wavelengths most effectively. Someone at the surface looking skyward perceives blue light, the shortest wavelength of the visible portion of the spectrum.

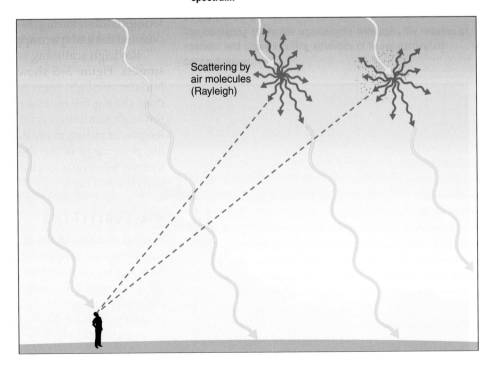

Scattering by air molecules (Rayleigh)

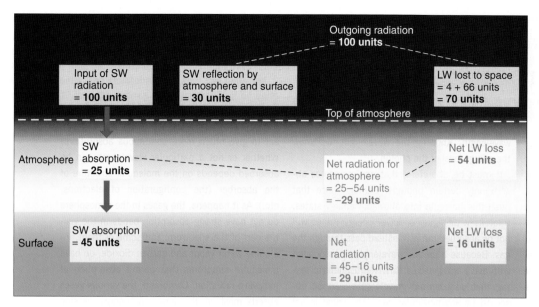

FIGURE 3–10

Net radiation is the end result of the absorption of insolation and the absorption and radiation of longwave radiation. The surface has a net radiation surplus of 29 units, while the atmosphere has a deficit of 29 units.

Figure 3–10 summarizes the net radiation balance for Earth. The atmosphere absorbs 25 units of solar radiation but undergoes a net loss of 54 units of thermal radiation, for a net deficit of 29 units. The surface absorbs 45 units of solar radiation but has a longwave deficit of 16, resulting in a net radiative surplus of 29 units. In other words, the atmosphere has a net deficit of radiative energy exactly equal to the surplus attained by the surface.

If radiation were the only means of exchanging energy, the surplus of radiative energy obtained by the surface would result in a perpetual warming, while the deficit of the atmosphere would lead to a continual cooling. Eventually our feet would be scorched by a terrifically hot ground while the rest of our bodies would freeze, surrounded by a bitterly cold atmosphere. This, of course, is not about to happen, because energy is transferred from the surface to the atmosphere and within the atmosphere by two other forms of heat transfer: conduction and convection. The net transfer of energy by these two processes allows the radiation surplus at the surface to be eliminated while at the same time offsetting the radiation deficit of the atmosphere.

CONDUCTION

Conduction, described in general terms in Chapter 2, helps transfer energy near the surface. As radiant energy is absorbed by a solid Earth surface during the middle of the day, a temperature gradient (a rate of change of temperature over distance) develops in the upper few centimeters of the ground. In other words, temperatures near the surface become greater than those a few centimeters below. As a result, conduction transfers energy downward. Warming of the ground during the day also sets up a temperature gradient within a very thin, adjacent sliver of air called the **laminar boundary layer**. Although air is usually highly mobile and capable of being easily mixed, very thin layers on the order of a few millimeters in thickness resist mixing. During the middle of the day, very strong temperature gradients can therefore develop in the laminar boundary layer, through which a substantial amount of conduction can occur. Energy conducted through the laminar boundary layer is then distributed through the rest of the atmosphere by a mixing process called *convection*.

CONVECTION

Convection is a process whereby heat is transferred by the bodily movement of a fluid—that is, a liquid or gas. In contrast to conduction, convection involves the actual displacement of molecules. Unlike conduction, which transfers energy

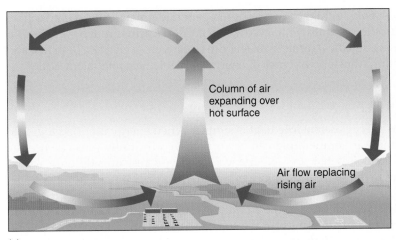

(a)   (b)

FIGURE 3–11
Convection (a) is a heat transfer mechanism involving the mixing of a fluid. In free convection, local heating can cause a parcel of air to rise and be replaced by adjacent air. Free convection can create updrafts able to keep a hawk airborne (b) without it having to flap its wings.

from the surface to the atmosphere, convection circulates this heat between the very lowest and the remaining portions of the atmosphere. The direction of heat transfer is upward when the surface temperature exceeds the air temperature (the normal situation in the middle of the day). At night the surface typically cools more rapidly than the air, and energy is transferred downward. Convection can be generated by two processes in fluids: local heating (free convection) and mechanical stirring (forced convection).

FREE CONVECTION

Free convection is the mixing process related to buoyancy, the tendency for a lighter fluid to float upward when surrounded by a denser fluid. Recall your days as a child, when you could annoy your parents by blowing bubbles through a straw in a glass of milk. As the air was injected into the milk, it would immediately rise upward because of its lesser density and cause turbulent mixing. This was free convection at work.

Free convection (shown in Figure 3–11a) often occurs when a localized parcel of air is heated more than the nearby air. Because warm air is less dense than cold, it is relatively buoyant and rises. On a warm summer day, we can see the effect of free convection by observing a circling hawk (Figure 3–11b) that stays airborne without flapping its wings. This flight is possible because the hawk's wings are designed to catch the rising parcels of buoyant air that carry it upward. Convection can have far more important impacts than helping to keep hawks in the air—it can lead to intense precipitation.

FORCED CONVECTION

Forced convection (also called **mechanical turbulence**) occurs when a fluid breaks into disorganized swirling motions as it undergoes a large-scale flow. When water flows through a river channel, for example, it does not flow uniformly, as would a very thick syrup. Instead, the flow breaks down into numerous *eddies*. Forced convection in the atmosphere is shown in Figure 3–12. Horizontally moving air undergoes the same type of turbulence. Instead of moving as a uniform mass, the air breaks up into numerous small parcels, each with its own speed and direction, that are superimposed on the larger-scale flow. Because there is a strong vertical component to the eddy motions, the forced convection helps transport energy from the top of the laminar boundary layer upward during the day.

WEATHER IN MOTION
Heavy Convection over Florida

FIGURE 3–12
Forced convection. Air is forced to mix vertically because of its low viscosity (ability to be held together) and the deflection of wind by surface features.

Generally speaking, higher wind speeds generate greater forced convection. Mechanical turbulence is also enhanced when air flows across rough surfaces (for example, forests and cities) rather than smooth ones such as glaciers. Both free and forced convection transfer two types of energy: sensible heat and latent heat.

SENSIBLE HEAT

The transfer of energy as **sensible heat** is simple. When energy is added to a substance, an increase in temperature can occur that we physically sense (hence the term *sensible*). This is what you experience when you rest outside on a warm, sunny day; the increase in your skin temperature results from a gain in sensible heat. The magnitude of temperature increase is related to two factors, the first of which is **specific heat**, defined as the amount of energy needed to produce a given temperature change per unit mass of the substance. In SI units, specific heat is expressed in joules per kilogram per kelvin. Everything else being equal, a substance with high specific heat warms slowly, because much energy is required to produce a given temperature change. Likewise, it also takes longer for a substance with a high specific heat to cool off, assuming the same rate of energy loss.

The temperature increase resulting from a surplus of energy receipt also depends on the *mass* of a substance. Not surprisingly, a given input of heat results in a greater rise in temperature if it is applied to only a small amount of mass. For example, compare the amount of energy needed to boil water for a cup of tea to that needed to take a warm bath. In just a few minutes, a single burner on a kitchen stove can have the water ready for the tea, but your hot water heater must supply considerably more energy for the large amount of bath water. These relationships are shown in Figure 3–13.

Sensible heat travels by conduction through the laminar boundary layer and is then dispersed upward by convection. Through these mechanisms, 8 of the 29 units of net radiation surplus for the surface are transferred to the atmosphere, where they help offset the net radiation deficit. The remaining 21 units are transferred to the atmosphere by the convection of heat in another form.

LATENT HEAT

Latent heat is the energy required to change the phase of a substance (that is, its state as a solid, liquid, or gas). In meteorology we are concerned almost exclusively with the heat involved in the phase changes of water.

Recall that all physical processes require energy. The evaporation of water and the melting of ice are no exceptions to this rule—for either process, energy must be supplied. In the case of melting ice, the energy is called the *latent heat of fusion*. For the change of phase from liquid to gas, the energy is called the *latent heat of evaporation*. It takes seven and a half times more energy (2,500,000 joules) to evaporate a kilogram of liquid water than it does to melt the same amount of ice (335,000 joules). Although both forms of latent heat can be important locally, on a global scale the latent heat of evaporation is far more influential.

When radiation is received at the surface, it can raise the temperature of the land or the water. If water happens to exist at the surface (or can be brought up from below the surface through the root systems of plants), some of the energy that might have been used to increase the surface temperature is instead used to evaporate some of the water. This results in a smaller temperature increase than would occur for a dry surface. You have probably experienced this while walking barefoot on a hot pavement. If the pavement is watered down, the surface cools as energy is taken from the ground and used for evaporation. The amount of energy consumed can be quite large, as much as 90 percent of absorbed solar radiation for a completely wet surface.

Let's use another common example to illustrate the concept of latent heat. We all know that perspiration is a mechanism that lowers our body temperatures and keeps us from overheating. But how does it work? Clearly, it's not a matter of

FIGURE 3–13
The heat content of a substance depends on several factors. In (a) the input of 4190 J of energy to a kilogram of water increases its temperature 1 °C, while a doubling of the energy input causes twice as much heating. The specific heat of a substance also influences the amount of temperature change resulting from an input of energy. In (b) the application of 4190 J to 1 kg of sandy soil produces more than five times the increase in temperature than it would for a kilogram of water. The amount of mass also affects the temperature change accruing for a given energy input. Note in (c) that the temperature increase for 2 kg of water is half as much as that for 1 kg.

sweat being cold—it's as warm as the body producing it. The reason sweat cools a person is latent heat. As you exercise, the heat produced as a byproduct causes your body temperature to rise. However, if your skin is covered with water and that water is free to evaporate, some of the energy produced by your body is used to evaporate the moisture rather than increase your body temperature.

The energy needed to evaporate water or melt ice is said to be latent because it does not disappear. It is held, "latent" in the atmosphere, to be released later when the reverse process occurs—the condensation of water vapor into liquid cloud or fog droplets. In effect, then, the evaporation of water makes energy available to the atmosphere that otherwise would have warmed the surface. It thus acts as an energy transfer mechanism, taking heat from the surface to the atmosphere. On a global average basis, the amount of energy transferred to the atmosphere as latent heat amounts to 21 units, which makes latent heat considerably more important as a mode of heat transfer than sensible heat (8 units). Perhaps this isn't surprising, given that the planet is mostly covered by ocean. The mean annual values of the net radiation components, latent and sensible heat, are depicted in Figure 3–14.

DID YOU KNOW?

When you drop an ice cube in a glass of water, the cooling that results is not primarily due to the lower temperature of the ice cube. It is mostly due to the latent heat of fusion involved in melting the ice. The energy used to melt the ice is taken from the water, thereby lowering its temperature.

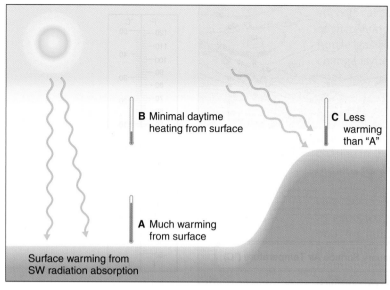

(a)

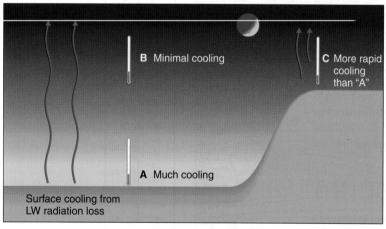

(b)

FIGURE 3–19
Effects of elevation and altitude on daily temperature patterns.

As was shown in Figure 1–9, temperatures in the troposphere typically decrease with altitude above sea level. This occurs because the surface is the primary source of direct heating for the troposphere, and increased altitude implies a greater distance from the energy source. Figure 3–19 contrasts day and night temperatures at three locations. Position A is located a couple of meters (about 6 ft) above the surface at sea level. Position B is 3000 m (about 10,000 ft) directly above A, and Position C is 3000 m above sea level but just a couple of meters above the mountain surface.

During the middle of the day, Position A responds to the absorption of solar radiation at the surface and warms as the surface transfers energy upward by convection and the emission of longwave energy. Position B, a considerable distance from the surface, undergoes virtually no warming. Position C, although at the same altitude as B, is nearer to the primary source of warming, and its daytime temperature rises appreciably.

At night, the surface below Position A cools by the emission of longwave radiation. The air 6 m above the surface also undergoes a lowering of temperature in response to the cooling of the underlying surface. At Position B, the air undergoes little cooling because of its distance from the surface. Position C chills rapidly because the sparse atmosphere above does not effectively absorb the outgoing radiation from the surface. Cooling is often enhanced by rapid evaporation into the drier high-altitude air. Perhaps while on a mountain camping trip you've noticed how rapidly a hot meal becomes stone cold. This is a consequence of strong evaporation and weak atmospheric counterradiation, compared to lower elevations. The overall effect of elevation above sea level and altitude above the ground, therefore, is that cooling and warming cycles are minimal high above the surface, and the air just above a high-elevation surface will undergo greater cycles of cooling and warming than will air at the same altitude but farther above the surface. It is not uncommon that the nighttime loss of longwave radiation at the surface can lead to lower temperatures near the surface than aloft. This reversal of the normal pattern in the troposphere (lower temperatures with increasing distance from the surface) is known as an **inversion**. The characteristics of inversions are described in Chapter 6.

ATMOSPHERIC CIRCULATION PATTERNS

As we will discuss in Chapter 8, an organized pattern of mean atmospheric pressure and air flow across the globe strongly influences the movement of warm and cold air, with a direct effect on temperature. These large-scale circulation patterns also influence the development of cloud cover, which has an indirect effect on temperature. Subtropical areas (latitudes 20° to 30° in both hemispheres), for example, tend to be regions of minimal cloud cover, and insolation passing through the atmosphere undergoes less attenuation on its way to the surface. In contrast, equatorial regions are often cloudy in the afternoon and experience a greater attenuation of incoming solar radiation. The result is that the highest temperatures on Earth tend to occur not at the equator but in the subtropics. Many other patterns of atmospheric circulation affect regional temperatures.

CONTRASTS BETWEEN LAND AND WATER

Because the atmosphere is heated primarily from below, it should be no surprise that the type of surface influences air temperature. The greatest influence arises because of contrasts between land and water. Water bodies are far more conservative

Earth's Equilibrium Temperature

3-2 PHYSICAL PRINCIPLES

If Earth had no atmosphere, and therefore no greenhouse effect (see page 69), the mean temperature of the planet would be much colder. Using the principles discussed thus far, we can easily estimate the magnitude of the greenhouse effect. To do so, we will compute the equilibrium temperature for a planet having no atmosphere. By comparing the computed and observed temperatures, we will see how the atmosphere influences Earth's temperature.

First, assume the planet acts as a blackbody with regard to longwave radiation, that the planetary albedo is 30 percent, and that the solar constant is 1367 watts per square meter. If Earth were a flat disk perpendicular to incoming radiation, each square meter would receive 1367 joules/second. But Earth is not a flat disk; it is a sphere, the surface area of which is four times larger than that of a disk of the same radius. Thus the intensity of radiation averaged over the sphere is one-fourth as large as for the imaginary disk. Because of this, each square meter of Earth receives 1367/4, or 342 watts/m². Given the planetary albedo of 30 percent, it must be

that 70 percent of this incoming radiation is absorbed. In other words, total absorbed radiation is

$$1367 \text{ watts/m}^2 \times 0.25 \times 0.7 = 239.2 \text{ watts/m}^2$$

The planet must lose exactly as much energy as it gains, and the intensity of radiation for a blackbody is determined by rearranging and applying the Stefan-Boltzmann law. Recall that the Stefan-Boltzmann law for a blackbody states that

$$I = \sigma T^4$$

We know, however, that the intensity of radiation for the planet without an atmosphere must be 239.2 watts per square meter. We then rearrange the equation to solve for T, rather than I, to get

$$T^4 = I/\sigma$$

which can be reduced to

$$T = (I/\sigma)^{0.25}$$

Using the values, $\sigma = 5.67 \times 10^{-8}$ watts/ $(\text{m}^2 \, \text{K}^4)$ and $I = 239.2 \text{ watts/m}^2$, the equilibrium temperature works out to 254.9 K (-18.3 °C, 0 °F). Thus, the mean temperature of Earth would be far colder without an atmosphere.

Note that our calculation is highly simplified and somewhat questionable. For example, we used 30 percent for the planetary albedo, but that value arises in part from the albedo of the atmosphere, which our imaginary planet lacks. Should we therefore have used present-day surface albedo in the computation? Perhaps, but surface albedo on the real Earth is in part the result of temperature, the very thing we are trying to compute! Ideally, we would treat albedo as a variable, allowing it to respond to changes in temperature.

We see that even a beginning question about the global effect of greenhouse gases raises complications that are not easy to address with a simple model. Given this, it is not surprising that realistic computer models of atmospheric behavior are enormously complex, requiring huge computer resources. Nonetheless, computer models have become indispensable for daily weather forecasting (Chapter 13) and tell us much of what we know about potential climatic changes due to human activity (Chapter 16).

than land surfaces with regard to their temperature, taking longer to warm and cool when subjected to comparable energy gains and losses.

San Francisco, California, and St. Louis, Missouri, together provide an interesting example of the effect of **continentality**—the effect of an inland location that favors greater temperature extremes. They are at similar latitudes and elevations and both are subject to a predominantly west-to-east airflow. Yet San Francisco, situated along the Pacific Coast, has more moderate temperatures than does its inland counterpart. During July, for example, its mean temperature is 15 °C (59 °F), while St. Louis averages 26 °C (79 °F). The temperature difference is similar but runs in the opposite direction in January when San Francisco has an average temperature of 10 °C (50 °F) and St. Louis a mean of 0 °C (32 °F).

Four reasons cause water bodies to be more conservative than landmasses with regard to temperature:

1. The specific heat of water is about five times as great as that of land.
2. Radiation received at the surface of a water body can penetrate to several tens of meters deep and distribute its energy throughout a very large mass. In contrast, the insolation absorbed by land heats only a very thin, opaque surface layer.
3. The warming of a water surface can be reduced considerably because of the vast supply of water available for evaporation. Because much energy is used in the evaporative process, less warming occurs.
4. Unlike solid land surfaces, water can be easily mixed both vertically and horizontally, allowing energy surpluses from one area to flow to regions of lower temperature.

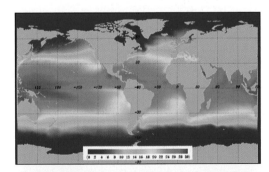

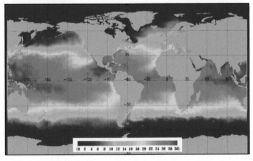

 WEATHER IN MOTION
Global Sea Surface Temperatures—Climatology
Global Sea Surface Temperatures—Actual

WARM AND COLD OCEAN CURRENTS

Figure 3–16 shows warm and cold ocean currents. The warm currents typically move poleward in the western portion of the ocean basins near the east coasts of continents in the middle latitudes, carrying large amounts of energy with them. Similarly, along the eastern margins of oceans, cold ocean currents dominate in the middle latitudes. Where the water temperatures are high, heat is transferred to the atmosphere and promotes higher air temperatures. Thus, the existence of a warm ocean current offshore can cause a location along the east coast of a continent to have higher temperatures than would a cold current offshore along a west coast.

Compare, for example, Los Angeles, California, and Charleston, South Carolina, two coastal locations at similar latitudes and elevations. Summers are considerably warmer at Charleston than at Los Angeles, largely (but not entirely) because the temperature of the ocean off Charleston is higher than the temperature of the ocean off Los Angeles. The high water temperatures of the Gulf of Mexico and the western margin of the Atlantic Ocean allow the transfer of an enormous amount of heat to the atmosphere, and the average July temperature in Charleston of 31 °C (88 °F) is substantially warmer than that of Los Angeles (23 °C; 73 °F). In winter, temperatures are lower at Charleston than at Los Angeles because the westerly winds blowing toward Los Angeles are subject to the moderating effects of the Pacific, whereas Charleston's winter temperatures are affected by colder prevailing winds passing over the continental interior. Thus, the influence of atmospheric circulation, which we noted previously, can interact with the position along the edge of a continent in influencing temperature patterns.

LOCAL CONDITIONS

A number of site-specific factors, such as slope orientation and steepness, can influence the temperature characteristics of an area. In the Northern Hemisphere, slopes that are south-facing receive mid-day sunlight at a more direct angle than do those oriented in other directions, thereby promoting greater energy receipt and higher surface temperatures. The heating of south-facing slopes often results in a greater amount of drying than on the opposite, north-facing slopes. Vegetation patterns often respond to the change in microclimate, with plants intolerant of dry conditions occupying north-facing slopes. Such a pattern is shown in Figure 3–20.

Ocean Surface Currents

WEATHER IMAGE
Ocean Currents

FIGURE 3–20
Slope aspect is one of the local factors affecting temperature. In the Northern Hemisphere, north-facing slopes typically receive less intense daytime heating and therefore exhibit lower temperatures. This retards the rate of surface evaporation, making more water available for plant life; thus, different types of vegetation are encountered on north- and south-facing slopes.

(a)

(b)

FIGURE 3–21
A dense vegetation cover lowers daytime temperatures because of its shadowing effect on incoming solar radiation (a). At night (b) the forest canopy retards the loss of longwave radiation to space, resulting in higher nighttime temperatures than in the open.

Densely wooded areas also have different temperature regimes than areas devoid of vegetation cover. In a region like that shown in Figure 3–21, a tall, dense vegetation cover reduces the amount of sunlight hitting the surface during the day; and considerable evaporation of water from leaf surfaces occurs. At night the plant canopy reduces the net longwave radiation losses. These factors lead to lower daytime temperatures but warmer evenings.

The effects of vegetation on local climate can often be put to use in a manner that increases human comfort. For example, it is often a good idea to plant deciduous trees on the equatorward side of houses. During the warm summer months, trees can cast shade on houses, thereby keeping interior temperatures down or reducing air conditioning costs. When winter arrives, the loss of the leaves from the deciduous trees minimizes the reduction in sunlight, and this can help keep the building warm.

Daily and Annual Temperature Patterns

The principles of heat transfer discussed in Chapter 2 and earlier in this chapter directly affect the daily (often called *diurnal*) and seasonal temperature changes that occur at any location. Let's first consider what happens over a 24-hour period

Because an instrument shelter is designed to reduce the influence of incoming radiation on the instruments, certain design criteria must be met. The shelter should be painted white so that its albedo will be maximized and reduce the absorption of radiation. It should also be paneled with slats rather than solid side walls to permit the free flow of air and the removal of any heat that might otherwise accumulate. The door must be mounted on the north side of the box (in the Northern Hemisphere) so that direct sunlight will not strike the instruments if the door is opened during the middle of the day. Finally, the shelter must conform to a standardized height, so that the thermometers will be mounted at 1.52 m (5 ft) above the ground.

In the United States, temperature is observed hourly by automated systems at National Weather Service offices and Federal Aviation Agency (FAA) facilities at airports across the country. The automated systems (described in Chapter 13) use resistance thermometers for temperature readings. This network is supplemented by observations made at a large number of cooperative agencies (such as U.S. Forest Service stations) and by individual volunteers. At the cooperative stations, observations are made once or twice daily, normally in the early morning and/or mid-afternoon, with maximum and minimum thermometers housed in instrument shelters. Environment Canada is responsible for temperature data acquisition in Canada.

Temperature Means and Ranges

In just about all aspects of daily life, we use descriptive statistics to talk about the things around us. Although the concept of an average, or mean, value of a property is fairly straightforward, applying the concept involves occasional complications. For example, trying to determine a daily mean temperature poses a dilemma—exactly how many times during the day must we measure it to obtain a true mean? We could make observations every hour, every minute, or even every second, with each method giving us a separate value.

The standard procedure is simple—the *daily mean* is defined as the average of the maximum and minimum temperature for a day. The advantage of this method is that the daily mean can be obtained even at weather stations having just the most basic instrumentation; a minimum and maximum thermometer are all we need to compute daily mean temperatures. The disadvantage of using just the maximum and minimum temperature is that it introduces a bias. Observe the daily temperature pattern for a particular day, shown in Figure 3–26. Notice that the nighttime temperatures remain nearly equal to the minimum throughout most of the night, while afternoon temperatures are near the maximum for only a few hours. If we were to obtain a daily mean by taking 24 hourly spaced observations and dividing by 24, the mean value would be lower than that obtained by using just the maximum and minimum temperatures. Nonetheless, averaging the maximum and minimum temperatures is the accepted method for obtaining daily mean temperatures even though it introduces a bias.

FIGURE 3–26

A continuous plot of temperature over a 24-hour period with clear skies. Note that the temperature is near that of the maximum for a relatively short time period. In contrast, the air temperature is near the daily minimum throughout most of the night.

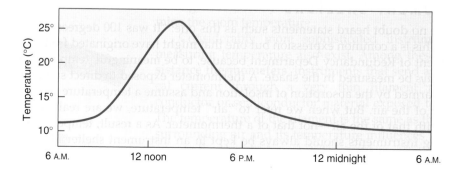

The *daily temperature range* is obtained simply by subtracting the minimum temperature from the maximum.

Having obtained daily mean temperatures for an entire month, we calculate the *monthly mean temperature* simply by summing the daily means and dividing by the number of days in the month. Similarly, an *annual mean temperature* is obtained by summing the monthly means for a year and dividing by 12. The *annual range* is the difference between the highest and lowest monthly mean temperatures.

GLOBAL EXTREMES

More impressive than mean temperatures are the maximum and minimum values ever recorded. Not surprisingly, they tend to occur at continental locations. The highest temperature ever recorded in North America was at Death Valley, California. Death Valley is located only a couple of hundred kilometers from the Pacific Ocean, but the Sierra Nevada mountain range presents a barrier that eliminates any moderating influence of the water. In addition, Death Valley's position below sea level and its sparse vegetation cover further promote high temperatures. On July 10, 1913, an all-time high temperature of 57 °C (134 °F) was recorded. The world's highest measured temperature was 58 °C (136 °F) at Azizia, Libya.

The lowest temperature ever recorded occurred at the Vostoc Research Station in Antarctica in 1983, with a reading of −89 °C (−129 °F). The research station is located atop thousands of meters of glacial ice, thereby combining the effects of high latitude, a continental locale, and a sparse atmosphere. In North America, the record low temperature of −63 °C (−81 °F) was observed at Shag, Yukon, in February 1947.

Undoubtedly, more extreme temperatures have occurred across the globe and North America at locations without temperature observation stations.

Some Useful Temperature Indices

There are several situations in which basic temperature readings can be modified to provide a better understanding of the effects of temperature. In some cases the temperature data can be combined with other variables, such as wind and humidity, to help assess the effects on human comfort. In other cases average temperatures might be adjusted for planning purposes. Several of these indices are described in the following three sections. An additional temperature index that incorporates the effect of humidity will be discussed in Chapter 4, which addresses atmospheric moisture.

WIND CHILL TEMPERATURES

Temperature by itself exerts a major impact on human comfort, but the discomfort caused by high or low temperatures can be compounded by other weather factors. If low temperatures are accompanied by windy conditions, a person's body loses heat much more rapidly than it would under calm conditions, due to an increase in sensible heat loss. Thus, a windy day with a temperature of −2 °C (28 °F) might feel colder than a calm day at −40 °C (−40 °F). As a result, when temperatures are low, it is common for weather reports to state both the actual temperature and how cold that temperature actually feels, the **wind chill temperature index** (or simply the *wind chill temperature*).

The earliest wind chill temperature index was based on low-temperature research conducted in Antarctica in 1945. Though the original research was not intended to quantify the combined effect of wind and low temperatures on people, the index derived from that work was widely used for more than a half-century. But meteorologists in Canada and the United States were aware of certain shortcomings in the index—the most notable of which was that it was inaccurate! People living in very cold environments were familiar enough with low temperatures to know that the wind chill values posted did not accurately describe how cold they really felt. Thus, researchers at the National Weather Service and the Meteorological Service of

DID YOU KNOW?

Of the 50 U.S. states, Montana holds the record for the greatest spread between its all-time maximum and minimum temperatures. That state recorded a high temperature of 47 °C (117 °F) at Medicine Lake in 1937 and low temperature of −57 °C (−70 °F) at Rogers Pass in 1954, for a range of 104 °C (187 °F). The all-time maximum temperatures for each state and their dates and locations of occurrence can be found at http://ggweather.com/climate/extremes_us.htm.

Recent Severe Heat Waves

Summer heat waves are certainly no rarity in the United States and Canada. Sooner or later everybody endures an episode of unpleasantly high temperatures. But heat waves can cause much more than a few days of discomfort—they can kill. One of the most notable heat waves of the last few decades was the relatively brief but severe event in mid-July 1995 in the north-central United States.

Although extremely high temperatures occurred from the Great Plains to the Atlantic Coast, nowhere was the problem more acute than in Chicago, Illinois, where 525 people died from the heat. The heavy mortality resulted from a combination of high temperatures (Midway Airport recorded an all-time high temperature of 41.1 °C, or 106 °F) and unusually high humidities. The heat and humidity combined to make the "apparent temperature" equivalent to 47 °C (117 °F). Though the searing daytime heat created plenty of misery on its own, it is believed that

the major factor leading to the many deaths is the fact that the extreme heat went uninterrupted, with the apparent temperature exceeding 31.5 °C (89 °F) for nearly 48 consecutive hours. (Recent research suggests that such conditions pose a greater danger than do brief periods of more extreme heat.)

Four years later, in 1999, another major July heat wave occurred in the eastern two-thirds of the United States. Once again, Illinois was in the center of the action, with more than half of the 232 fatalities across the Midwest occurring in Chicago. Missouri was the second hardest-hit state, with 61 fatalities. All across the region, power outages occurred from excessive demand, roads buckled, and crops wilted in the fields.

As July gave way to August, the heat moved eastward toward the Atlantic states, where it broke numerous weather records. Charleston, South Carolina, had an all-time high temperature of 40.5 °C (105 °F). Augusta, Georgia's

high temperature of 39.4 °C (103 °F) came on the sixth consecutive day in which the record for the daily maximum temperature was tied or exceeded. And on August 8, Raleigh-Durham, North Carolina, broke the 100 °F mark (37.8 °C) for the eleventh time that summer.

Though the death tolls were not as high as they were in 1995 and 1999, major heat waves in 2000 and 2001 resulted in numerous fatalities across the southern and southeastern United States. Such events serve as reminders that about one-third of all deaths directly attributed to weather events in the United States result from extremely high temperatures. The decade of the 1990s was remarkably warm relative to other periods in recorded history. This is particularly noteworthy because the topic of human-induced climatic warming has been a major issue for scientists and policymakers. This matter will be discussed further in later chapters of this book.

Canada joined forces and in November 2001 released a new wind chill index based on the results of tests on human volunteers. The six men and six women who provided their services took 90-minute walks on treadmills under controlled temperature and wind conditions. Four external temperature sensors were placed on each volunteer's face to determine surface heat loss. An additional temperature probe was placed inside each participant's cheek to determine the temperature gradient between the external skin and the inside of the mouth, and a small rectal probe was used to monitor body core temperatures. Observations of heat transfer from and within the subjects' bodies enabled the researchers to calculate cooling rates due to varying wind speeds. Tables 3–1 and 3–2 show new wind chill tables based on these experiments.

The new wind chill temperature index has values that differ markedly from those of the original index—usually indicating less severe wind chills than those obtained using the old formula. For example, at a temperature of −15 °C (5 °F) and a wind speed of 65 km/hr (40 mph), the old formula indicated a wind chill temperature of −43 °C (−45.4 °F). Under the same conditions the new index yields a wind chill temperature of −30 °C (−22 °F), a full 13 °C (23 °F) higher than that given by the old formula.

Despite the improvements in the new formula, it still has its shortcomings. For example, calculations do not take into account the potential warming effect of sunlight on a person's body. Nonetheless, the index provides people with guidance in the way they should dress and the types of activities they should undertake under cold and windy conditions. This is especially important when one considers the fact that extreme cold is the number one cause of fatalities directly attributable to weather.

Just as windy conditions can make low temperatures feel even colder, high humidities can cause warm days to feel oppressively hot. As a result, a heat index has been calculated that incorporates the effect of high atmospheric moisture at high temperatures. This index is discussed in Chapter 5, Atmospheric Moisture.

TABLE 3–1 Wind Chill Temperature (°C)

	TEMPERATURE (°C)									
WIND (km/hr)	5	0	−5	−10	−15	−20	−25	−30	−35	−40
5	4	−2	−7	−13	−19	−24	−30	−36	−41	−47
10	3	−3	−9	−15	−21	−27	−33	−39	−45	−51
15	2	−4	−11	−17	−23	−29	−35	−41	−48	−54
20	1	−5	−12	−18	−24	−31	−37	−43	−49	−56
25	1	−6	−12	−19	−25	−32	−38	−45	−51	−57
30	0	−7	−13	−20	−26	−33	−39	−46	−52	−59
35	0	−7	−14	−20	−27	−33	−40	−47	−53	−60
40	−1	−7	−14	−21	−27	−34	−41	−48	−54	−61
45	−1	−8	−15	−21	−28	−35	−42	−48	−55	−62
50	−1	−8	−15	−22	−29	−35	−42	−49	−56	−63
55	−2	−9	−15	−22	−29	−36	−43	−50	−57	−63
60	−2	−9	−16	−23	−30	−37	−43	−50	−57	−64
65	−2	−9	−16	−23	−30	−37	−44	−51	−58	−65
70	−2	−9	−16	−23	−30	−37	−44	−51	−59	−66
75	−3	−10	−17	−24	−31	−38	−45	−52	−59	−66
80	−3	−10	−17	−24	−31	−38	−45	−52	−60	−67

TABLE 3–2 Wind Chill Temperature (°F)

	TEMPERATURE (°F)																
WIND (mph)	40	35	30	25	20	15	10	5	0	−5	−10	−15	−20	−25	−30	−35	−40
5	36	31	25	19	13	7	1	−5	−11	−16	−22	−28	−34	−40	−46	−52	−57
10	34	27	21	15	9	3	−4	−10	−16	−22	−28	−35	−41	−47	−53	−59	−66
15	32	25	19	13	6	0	−7	−13	−19	−26	−32	−39	−45	−51	−58	−64	−71
20	30	24	17	11	4	−2	−9	−15	−22	−29	−35	−42	−48	−55	−61	−68	−74
25	29	23	16	9	3	−4	−11	−17	−24	−31	−37	−44	−51	−58	−64	−71	−78
30	28	22	15	8	1	−5	−12	−19	−26	−33	−39	−46	−53	−60	−67	−73	−80
35	28	21	14	7	0	−7	−14	−21	−27	−34	−41	−48	−55	−62	−69	−76	−82
40	27	20	13	6	−1	−8	−15	−22	−29	−36	−43	−50	−57	−64	−71	−78	−84
45	26	19	12	5	−2	−9	−16	−23	−30	−37	−44	−51	−58	−65	−72	−79	−86
50	26	19	12	4	−3	−10	−17	−24	−31	−38	−45	−52	−60	−67	−74	−81	−88
55	25	18	11	4	−3	−11	−18	−25	−32	−39	−46	−54	−61	−68	−75	−82	−89
60	25	17	10	3	−4	−11	−19	−26	−33	−40	−48	−55	−62	−69	−76	−84	−91

3. The latent heat for water is 2,500,000 joules per kilogram (kg), and the specific heat of water is 4190 joules per kg per degree Celsius of temperature change. Assume that a kg of water of water begins with a temperature of 20 °C (68 °F). Compare the amount of energy needed to bring the water to the boiling point to the amount of energy needed to evaporate the same amount of water.

4. View the maps of mean minimum temperatures in January and mean maximum temperatures in July at **http://www.climatesource.com/map_gallery.html**. Assess the relative importance of latitude, elevation, continentality, ocean current, and local conditions to these distributions.

Quantitative Problems

This chapter has described the energy balance of the atmosphere and the factors that influence global temperatures. The companion Web site for this book, **http://www.prenhall.com/aguado/**, offers a brief quiz in which you can test your knowledge of the material by answering several quantitative problems. The brief exercise should help you better understand equilibrium temperatures, sea-sonal effects of cloud cover on energy receipt, and relative expenditures of surface energy on latent and sensible heat transfer. After entering the site, go to the bottom of the page and select Chapter 3. Then highlight the Quantitative Examples line on the left hand panel.

Useful Web Sites

http://rredc.nrel.gov/solar/old_data/nsrdb/redbook/atlas/
Provides monthly and annual maps of solar radiation across the United States. For the material relevant to this chapter, be sure to check the box for *Horizontal Flat Plate* as instrument orientation mode.

http://earthobservatory.nasa.gov/Observatory/Datasets/netflux.erbe.html
Allows you to observe the shift in net radiation over multiple years. Maps are simultaneously matched with other biophysical distributions (such as vegetation) across the globe to show how they vary with net radiation. Monthly averages of net radiation can also be viewed individually or as an annual loop at http://itg1.meteor.wisc.edu/wxwise/museum/a2/a2net.html

Interesting maps depict mean energy budget characteristics for each month.

http://lwf.ncdc.noaa.gov/oa/climate/severeweather/temperatures.html
Features maps of temperature records for the United States. Also contains numerous special reports on temperature conditions.

http://www.ems.psu.edu/wx/usstats/uswxstats.html
Contains some very interesting maps, such as the current distribution of temperature across the 48 conterminous United States. Also provides the current temperature and other values averaged over the 48 conterminous United States.

http://www.crh.noaa.gov/dtx/New_Wind_Chill.htm
Provides wind chill and temperature index converters.

Media Enrichment

Weather in Motion

Heavy Convection over Florida

Earlier in this chapter, we discussed the role of convection in transferring heat between the surface and the atmosphere. Convection frequently leads to the formation of deep clouds that can yield heavy precipitation, lightning, hail—and even tornadoes.

In this movie, we see the formation of a major thunderstorm over southern Florida. The scene begins in the early morning with predominantly clear skies across the peninsula. (Subtract 4 hours from the UTC time at the bottom of the movie to convert to EDT.) Shortly after noon, a narrow band of small, isolated clouds emerges along the Atlantic Coast, along with a broad area of scattered clouds across the state. As the afternoon progresses, the cluster of small clouds along the coast gradually grows into a single larger thunderstorm. The situation changes rapidly at about 4 P.M., with an almost explosive growth of the thunderstorm along the coast. The first major cell of intense weather is followed by several larger ones, and by early evening the state has been fully obliterated from view.

This movie shows a large part of the spectrum of thunderstorm activity over just a brief time period. The small clouds that dominated the early part of the afternoon formed in response to localized heating that caused relatively small parcels of air to rise. Such clouds are commonplace, especially in the summer. But a number of conditions must be present for the small, fair-weather clouds to grow into major storms capable of violent weather. These conditions include a humid atmosphere near the surface and certain changes in temperature with height above the surface. These factors are discussed in Chapters 5–7.

Weather in Motion
Global Sea Surface Temperatures—Climatology

This movie shows the average distribution of sea surface temperatures across the world's oceans during the course of a year, beginning in June. Temperatures are warmest in the tropical regions and generally decrease with latitude. The distribution of temperatures does not remain fixed in time, however. Notice that for about 6 months prior to late August the warmer waters migrate northward, which is a lagged response to the varying solar declination. Also note the difference in temperatures off the west and east coasts of North America. At low latitudes the temperatures are warmer along the Atlantic coast than along the Pacific. In contrast, the east coast of Canada and New England tend to have cooler waters than the west coast of North America, because of the cool temperatures associated with the Labrador current.

Weather in Motion
Global Sea Surface Temperatures—Actual

A companion movie to the previous one, this movie shows actual sea surface temperatures observed from November 1998 to November 1999. The temperature distribution is very similar to the climatological average, although there is more fine-scale variation. Notice also that the temperatures along the east tropical Pacific are somewhat cooler than those for the climatological average. This is due to a La Niña situation, described in Chapter 8.

Weather Image
Ocean Currents

This image provides a unique depiction of ocean currents in the Pacific Ocean and a small portion of the western Atlantic, showing sea surface temperatures and the direction of flow. The cold California current is apparent along the extreme eastern Pacific, as is the warm, westward-moving current along a wide portion of the tropical Pacific. The North Pacific current is also evident, transporting warm water toward the Gulf of Alaska. Note that a small portion of the Gulf Stream can be seen off the east coast of North America.

ATMOSPHERIC PRESSURE AND WIND

CHAPTER OUTLINE

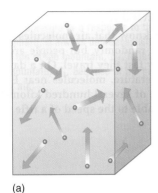

(a)

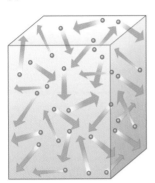

(b)

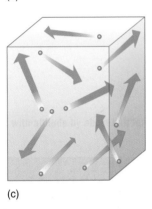

(c)

FIGURE 4–1
The movement of air molecules (indicated by the dots with the red arrows) within a sealed container exerts a pressure on the interior walls (a). The pressure can be increased by increasing the density of the molecules (b) or increasing the temperature (c). The speed of the molecules (and therefore the temperature) is indicated by the degree of redness and length of the arrows.

Wind can have a direct impact on how we feel, even under the mildest of weather conditions. A gentle breeze can make a hot afternoon more comfortable, or it can make a winter night bitterly cold. Occasionally, the movement of air can affect our lives in far more substantial ways. Residents of eastern Washington and Oregon were reminded of this on September 25, 1999, when winds up to 135 km/hr (85 mph) created a blinding dust storm that triggered a spate of accidents. Over the course of the day, 6 people were killed and 23 injured, and miles of highway were completely shut down for hours. The worst multiple-vehicle accident occurred along Interstate Highway 84 in northeastern Oregon. Fifty-eight-year-old Harold Fell described his experience: "An 18-wheeler passed us and the next thing I knew, it was stopped dead in the road. It just loomed up in front of me." Fell and his wife suffered only minor injuries, but they undoubtedly realized how lucky they were when they surveyed the accident scene and realized that 4 people were dead amid the wreckage of the 16 vehicles involved.

As concerned as we sometimes are with wind conditions, few of us pay much attention to a closely associated component of weather—*atmospheric pressure*. After all, how many times have you canceled a picnic because the pressure was too low? Or how many people do you know who have special clothes they wear only on days of high pressure?

Although seldom considered in everyday life, air pressure deeply affects other weather variables that have much more immediate impact. For example, horizontal variations in atmospheric pressure are directly responsible for the motion of the wind. And because air descends in areas of high surface pressure and rises in regions of low surface pressure, differences in air pressure strongly influence the likelihood of cloud formation and precipitation.

This chapter introduces the basic concepts of atmospheric pressure and its vertical and horizontal distributions. We discuss the relationship between pressure and other atmospheric variables and the processes that create horizontal and vertical variations in pressure. With this foundation, we can go on to discuss storm patterns in later chapters.

The Concept of Pressure

The atmosphere contains a tremendous number of gas molecules being pulled toward Earth by the force of gravity. These molecules exert a force on all surfaces with which they are in contact, and the amount of that force exerted per unit of surface area is **pressure** (see Box 4–1, *Physical Principles: Velocity, Acceleration, Force, and Pressure*). Of course, the concept of pressure is not confined to meteorology but rather is fundamental to all the physical sciences. In most physical science applications, the standard unit of pressure is the **pascal** (Pa), but in the United States meteorologists use the **millibar** (mb), which equals 100 Pa. Canadian meteorologists use yet another unit, the **kilopascal** (kPa), equal to 1000 Pa, or 10 mb. For purposes of comparison, air pressure at sea level is typically roughly 1000 mb (100 kPa)—or more precisely, 1013.2 mb.

To understand the characteristics of pressure, refer to Figure 4–1, which depicts a sealed container of air. The enclosed air molecules move about continually and exert a pressure on the interior walls of the container (a). The pressure of the air is proportional to the rate of collisions between the molecules and walls. We can increase the pressure two ways. The first way is by increasing the density of the air either by pumping more air into the container or by decreasing the volume of the container (b). The second is by increasing the air temperature, in which case

◀ The Moon rises over Mt. Everest, viewed from Annapurna base camp. At the top of Everest, air pressure is about one-third of that at sea level.

Velocity, Acceleration, Force, and Pressure

It is quite common in everyday conversation to hear the terms *force* and *pressure* used interchangeably, just as *velocity* and *speed* are often considered synonymous. In the language of science, however, intermixing these terms can lead to great confusion. Let us look briefly at how they differ.

VELOCITY AND ACCELERATION

Any object that moves has a particular **speed**, defined as the distance traveled per unit of time. Speed is related to, but not the same as, velocity. **Velocity** incorporates direction as well as speed. Think, for example, of two cars traveling at 20 meters per second (44 mph) but moving in opposite directions. Though they have the same speed, their velocities are not equal because of their different directions. This distinction is crucial for understanding our next quantity, **acceleration**, the change in velocity (not speed) with respect to time.

Because velocity includes both speed and direction, a change in either speed or direction is an acceleration. Consider a car that at one moment in time travels at 20 m/sec; one second later, the same car has a speed of 19 m/sec; one second later, the speed is 18 m/sec, and so on. As each second goes by, the car's speed decreases 1 m/sec (note that acceleration can be either positive or—as in this example—negative). An acceleration can also occur as a change in direction with respect to time, even for an object whose speed does not change. A car traveling at a con-

stant speed but gradually turning undergoes an acceleration, just like a car whose speed is changing.

In meteorology there is one particular acceleration of utmost importance—**gravity** (*g*). This acceleration, 9.8 m/sec/sec (32.1 ft/sec/sec), is nearly constant across the globe. There is a slight decrease in *g* from equator to pole, and also a very small difference in *g* from the surface to the upper atmosphere. For most applications, however, these variations in *g* are so slight they can be ignored.

FORCE AND PRESSURE

One of the most important tenets of physical science is Newton's Second Law, which relates the concept of **force** (denoted *F*) to mass *(m)* and acceleration *(a)*. Specifically, Newton's second law tells us that the acceleration of an object is proportional to the force acting on it and inversely proportional to its mass. Symbolically, this is expressed as

$$a = \frac{F}{m} \text{ or}$$
$$F = ma$$

Imagine that a fully loaded 18-wheeler truck is stopped at a traffic light next to a bicycle. As the light turns green, both begin to accelerate at the same rate. It's easy to see that if the two remain right next to each other, the much more massive truck requires a larger force (and more powerful "engine"). Likewise, if two bodies with equal mass are subjected to different forces, the one subjected to the greater force will undergo a greater acceleration.

Keep in mind that *F* in the equation above is the net force acting on the object. If various forces are acting simultaneously, they must all be considered together to determine the acceleration; we must account for both the magnitude and direction of each. As we will see with regard to falling raindrops (Chapter 7), forces acting in opposite directions reduce the net force and resulting acceleration, sometimes to zero.

Let's apply Newton's second law to our atmosphere. The atmosphere contains 5.14×10^{18} kg of mass. (To get an idea of what 5.14×10^{18} kg weighs, picture a million boxcars, each containing a billion elephants.) Multiplying the mass of the atmosphere by the acceleration of gravity, we determine that the force exerted on the atmosphere is about 5.0×10^{19} newtons (a newton [N] is the unit of force it takes to accelerate 1 kg one meter per second every second).

Force divided by the area on which it is exerted equals pressure. So dividing this force of 5.0×10^{19} newtons by the surface area of Earth gives us the average force per unit area, or average surface pressure of about 10.132 newtons per square centimeter. This is equivalent to 1013.2 mb, or about 14.7 pounds per square inch.

Having made the distinction between force and pressure, we now should address the question of how this distinction applies to the atmosphere. The answer is that despite the nearly constant total force of the atmosphere, its gases are not uniformly distributed across the planet. Consider

The Equation of State

Everyday experience indicates that gases tend to expand when heated and become denser when cooled. This suggests that temperature, density, and pressure are related to one another. As a matter of fact, their relationship is quite simple. It is described by the **equation of state** (also called the **ideal gas law**),

$$p = \rho RT$$

in which p is pressure expressed in pascals, ρ (the Greek letter rho) is density in kilograms per cubic meter, R is a constant equal to 287 joules per kilogram per kelvin, and T is temperature (in kelvins). To put the equation in words, it tells us a particular area at Earth's surface with an imaginary column extending upward to the top of the atmosphere, as shown in the top right of Figure 1. Greater surface pressures exist at the bases of atmospheric columns that contain a greater number of molecules, and lower surface pressures are found where less air occupies the column. Just how these differences in pressure arise is considered later in this chapter; for the time being, the important point is that surface pressure reflects the mass of atmosphere within the column, as shown in Figure 1.

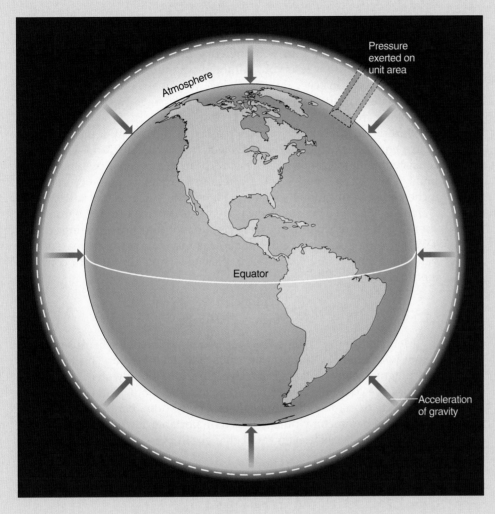

FIGURE 1

The downward force of the atmosphere is equal to the mass of the entire atmosphere times the acceleration of gravity. Because the amount of mass and the acceleration of gravity are constant through time, the force of the atmosphere does not change. Pressure is defined as the amount of force exerted per unit of area. Thus, the shaded area in the figure experiences a certain amount of pressure. Pressure varies because the mass of overlying air varies from place to place and time to time.

that if the air density increases while *temperature is held constant*, the pressure will increase. Similarly, at constant density, an increase in temperature leads to an increase in pressure.

This point leads us to a common confusion. Among the general public (and some students), there is often a temptation to look for a simple relationship between temperature and pressure. But the equation of state says that blanket statements such as these are not correct. A third variable, density, can overwhelm changes in the other two variables. For example, in the plains of North America, high pressure following a winter storm often brings bitterly cold temperatures, not soothing warmth. Although counterintuitive for some, such an occurrence is entirely consistent with the equation of state. (*Box 4–2, Physical Principles: Variations in Density* presents more information on the equation of state.)

DID YOU KNOW?

Throughout the troposphere the atmosphere consists of an extremely high concentration of molecules. At sea level, there are some 1 million billion molecules of air occupying each cubic millimeter of volume. But despite this large number of molecules, the atmosphere consists primarily of empty space. The air molecules are in fact spaced far apart relative to their size so that about only one-tenth of 1 percent of the volume is occupied by atoms. In contrast, liquids, which are much denser than gases, have much smaller intermolecular distances; about 70 percent of their volume is occupied by matter.

Measuring Pressure

Any instrument that measures pressure is called a *barometer.* Two types of barometers are most common for routine observations: one consisting of a tube partially filled with mercury, and another that uses collapsible chambers.

MERCURY BAROMETERS

The standard instrument for the measurement of pressure is the **mercury barometer** (Figure 4–4), invented by Evangelista Torricelli in 1643. It is a simple device made by filling a long tube with mercury and then inverting the tube so that the mercury spills into a reservoir. Although the tube is turned over completely, it does not empty. Instead, the air pushes downward on the pool of mercury and forces some of it up into the tube. The greater the air pressure, the higher the column of mercury.

   Barometric pressure is often expressed as the height of the column of mercury in a barometer, which at sea level averages 76 cm (29.92 in.). This measure is inconsistent with the concept of pressure, however, because pressure does not have units of length. In other words, expressing barometric pressure in centimeters or inches is as incongruous as stating somebody's age as "30 miles per hour" or weight as "$1.99." The length measurements obtained from a barometer are a response to the atmospheric pressure but are not direct observations of pressure itself. Meteorologists prefer a unit that measures force per unit area, such as pounds per square inch or millibars. The simple conversion formulas for converting barometric heights to millibars are

$$1 \text{ centimeter} = 13.32 \text{ mb and}$$
$$1 \text{ inch} = 33.865 \text{ mb}$$

   Mercury is an excellent fluid for use in a barometer because it is extremely heavy, with a density 13.6 times greater than that of water. This feature allows the instrument to be of a manageable size. Consider that if water was used instead of mercury, the column of water would need to be about 10 m (33 ft) tall to counterbalance the pressure of the atmosphere. On the other hand, although a three-story water-filled barometer would be less than portable, it would make for very precise measurements because even small changes in pressure would translate into large height changes.

CORRECTIONS TO MERCURY BAROMETER READINGS

One of the most important tools of a meteorologist is the weather map, which among other things plots the distribution of air pressure across the surface. Before barometer data can be used on the map, however, three corrections must be made to compensate for local factors that affect the readings.

   The first correction compensates for the influence of elevation (described earlier in this chapter). If surface pressure values were plotted on weather maps, they would give a false representation of the distribution of the atmosphere. This happens because high elevations have lower surface pressures than do low elevations, even if the sea level pressures are the same. To standardize the observations, we must convert surface pressure readings to sea level values. For a station situated 100 m (328 ft) above sea level, about a centimeter (0.4 in.) is added, corresponding

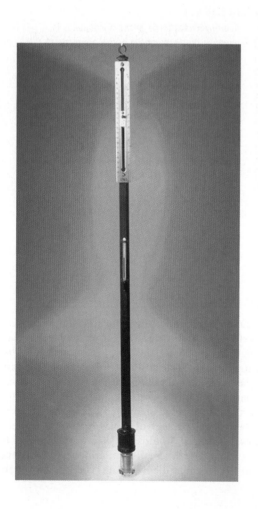

FIGURE 4–4
A mercury barometer.

Variations in Density

Perhaps you have wondered how much air weighs. The air around you has a particular density, and any volume of air contains a certain amount of mass. Changes in the density of air affect many everyday phenomena. For example, the density of the atmosphere influences how much lift a plane gets as it accelerates down a runway in preparation for takeoff. Likewise, automobile fuel injectors must account for variations in density to deliver the right mixture of gasoline and air into the car's engine. The density of air can even affect the amount of resistance the air exerts on a batted baseball, thereby influencing its distance traveled.

But are the variations in density really substantial? We can use the equation of state to see exactly how much variations in temperature affect the air's density. To do this, let's first rearrange the equation to the following:

$$\rho = p/RT,$$

Let's now also compare the air density for two situations: a warm day with a temperature of 308 K (35 °C or 95 °F) and a cold one with a temperature of 278 K (5 °C or 41 °F). For consistency we will assume that the pressure is 100,000 Pa (1000 mb; 100 kilopascals) in both instances.

Applying the equation of state for the warmer day, we find that the density of the air is

$$\rho = \frac{100{,}000 \text{ Pa}}{287 \text{ J kg}^{-1} \text{ K}^{-1} \times 308 \text{ K}} = 1.13 \text{ kg/m}^3{}^*$$

When we lower the air temperature to 278 K (5 °C or 41 °F), the equation yields an air density of

$$\rho = \frac{100{,}000 \text{ Pa}}{287 \text{ J kg}^{-1} \text{ K}^{-1} \times 278 \text{ K}} = 1.25 \text{ kg/m}^3$$

This is nearly 11 percent greater than the density the air had on the warmer day—a nontrivial amount.

In addition to temperature and pressure, the humidity of the air exerts an influence (although only a very minor one) on density. Let's see how. Molecular oxygen (O_2) and nitrogen (N_2) make up most of the mass of the atmosphere and exist in a constant proportion. Other, lesser constituents of the atmosphere are present in different amounts at different places and times, and because each has its own unique molecular weight (an expression of the relative amount of

---

*To get the units to balance, you must reduce the units of pascals (Pa) and joules (J) to their fundamental dimensions. Thus, Pa = kg m^{-1} sec^{-2} and J = kg m^2 sec^{-2}.

mass for molecules), their relative abundance can slightly affect the density of the atmosphere. Among these gases, water vapor usually accounts for about 1 percent of the atmospheric mass. Intuitively, we might assume that a greater humidity would favor a denser atmosphere. Actually, just the opposite is true.

Compare the amount of mass contained in individual molecules of water vapor and of the most abundant atmospheric gases. The molecular weights of nitrogen and oxygen are 28.01 and 32.00, respectively, and the mean molecular weight of the dry atmosphere is 28.5. Water vapor, on the other hand, has a molecular weight of only 18.01. Thus, as the proportion of the air occupied by water vapor increases, an accompanying reduction in the mean molecular weight of the atmosphere must occur. All other things being equal, humid air is less dense than dry air.

Incorporating the effect of varying moisture content requires only a small modification to the equation of state. Calculations using the revised formula show that at 15 °C (59 °F), air density declines by only 0.6 percent for a 1 percent increase in water vapor (from dry air to 1 percent water vapor).

---

to about 13 mb. At higher elevations, a much greater adjustment might be needed. At Denver, Colorado (the "Mile High City"), for example, the correction is about 16 cm (6.24 in.), or 213 mb.

The second correction deals with the similarity between a mercury barometer and a thermometer. Just as the mercury in a thermometer expands with increasing temperature, so does the mercury in a barometer. The expansion reduces the density of the fluid and requires that it attain a greater height to offset the pressure of the atmosphere. In other words, on a hot day the height of the mercury column is greater than on a cold day, even if the atmospheric pressure is the same. For this reason, mercury barometers always have a thermometer attached to determine the temperature of the instrument, and a correction table tells us what height the mercury column would be if the temperature were at the standard value of 0 °C (32 °F). At normal room temperature, this correction is small, requiring the subtraction of about 0.25 mm (0.01 in.).

The third correction accounts for the slightly greater acceleration of gravity at higher latitudes. To standardize the readings from all latitudes, we convert them to what they would be if the local gravity were equal to that at 45° north or south, or midway between the equator and poles. The latitudinal changes in gravity are small, however, and corrections are usually on the order of 0.25 mm (0.01 in.).

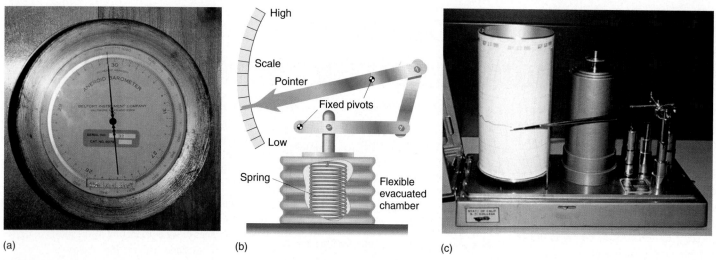

(a)　　　　　　　　　　　(b)　　　　　　　　　　　(c)

FIGURE 4–5
An aneroid barometer (a) and its workings (b). A barograph (c).

ANEROID BAROMETERS

Mercury barometers are precise instruments, but they are also expensive and inconvenient to relocate. An alternative instrument for measuring pressure is the **aneroid** (meaning "without liquid") **barometer** (Figure 4–5a). Aneroid barometers are relatively inexpensive and can be quite accurate. They contain a collapsible chamber from which some of the air has been removed (b). The atmosphere presses on the chamber and compresses it by an amount proportional to the air pressure. A pointing device connected to a lever mechanism indicates the air pressure.

Aneroid barometers, which are often found in homes, must be calibrated when first installed. The user simply finds out the current sea level pressure and sets the instrument by turning a small screw on the back of the casing. Because there is no expandable fluid in an aneroid barometer, the instrument requires no temperature correction. Furthermore, the effects of altitude and latitude are already accounted for when the instrument is first calibrated. Thus, once calibrated, an aneroid barometer gives the sea level pressure without corrections or adjustments.

Sometimes it is useful to have a continual record of pressure through time. Aneroid devices that plot continuous values of pressure are called **barographs** (Figure 4–5c). A rotating drum (usually set to one rotation per week) turns a chart so that a pen traces a permanent record of the changing pressure.

The Distribution of Pressure

The distribution of sea level pressure across the globe is a highly variable characteristic of the atmosphere. To visualize this distribution, meteorologists plot lines called isobars on weather maps.

Each **isobar** is drawn so that it connects points having exactly the same sea level pressure, and locations between any two isobars have pressures between those represented by the two lines. Isobars are drawn at intervals of 4 mb on U.S. surface weather maps, so the pressure difference between adjacent isobars is the same everywhere on the map. The advantage of this is that the distance between adjacent isobars provides information about how rapidly pressure changes from one place to another. In other words, the spacing of the isobars indicates the strength of the **pressure gradient**, or rate of change in pressure, in the same way that spacing of isotherms reveals temperature gradients. A dense clustering of

TUTORIAL
Pressure Gradients

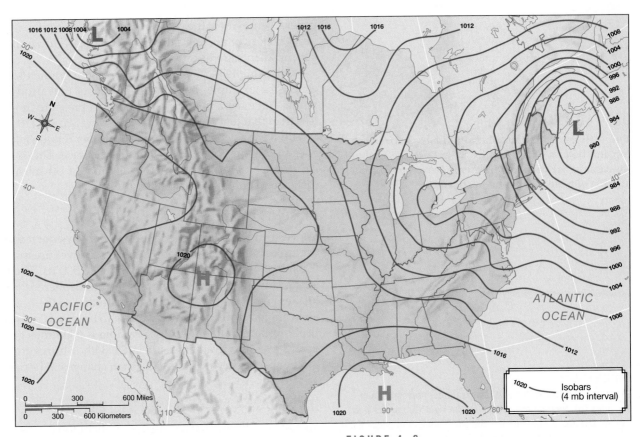

FIGURE 4–6

A weather map showing the distribution of sea level air pressure on March 4, 1994. Note that the pressure is relatively low over the northeastern United States and eastern Canada. Also note that the highest and lowest pressure on the map are only within about 4 percent of each other.

isobars indicates a steep pressure gradient (a rapid change in pressure with distance), while widely spaced isobars indicate a weak gradient.

By way of example, Figure 4–6 maps the sea level pressure distribution as it existed on March 4, 1994. The pressure over New England and southeast Canada was lower than over most of the West, and the strongest pressure gradient was over eastern North America.

PRESSURE GRADIENTS

Pressure gradients provide the impetus for the movement of air we call *wind*. Imagine two people pushing against each other. The person who exerts the greater force pushes the other one back, and the greater the difference in force applied, the faster the pushed person will move. The same concept applies to air. If the air over one region exerts a greater pressure than the air over an adjacent region, the higher-pressure air will spread out toward the zone of lower pressure as wind. The pressure gradient gives rise to a force, called the **pressure gradient force**, that sets the air in motion. For pressure gradients measured at constant altitude, we use the term *horizontal* pressure gradient force and call the resulting motion *wind*. Everything else being equal, the greater the pressure gradient force, the greater the wind speed.

HORIZONTAL PRESSURE GRADIENTS

The map of sea level pressure shown in Figure 4–6 is fairly typical, having low- and high-pressure areas of average magnitude. Notice that the changes in pressure across the map are small. The lowest pressure observed is about 977 mb, while the highest is about 1021 mb. This 44 mb difference represents a mere 4 percent or so of the average pressure. Note also that the physical distance separating the areas of highest and lowest pressure is about 3000 km (1800 mi). In the most general sense, then, the pressure gradients across the map are on the order of 40 mb per 3000 km, or about 1 mb per 75 km. Clearly, on a continental scale at least, pressure gradients are usually small.

On a smaller scale, horizontal pressure gradients can be much greater. Hurricanes, for example, have steep gradients that produce violent and destructive winds. Yet even a hurricane may have a pressure in its interior only about 50 mb less than that just outside the storm, some 300 km (180 mi) away. Such a hurricane would have a pressure change of 1 mb per 6 km, yielding only a 5 percent difference in pressure over a considerable distance. This is in marked contrast to vertical pressure gradients, wherein a drop of 50 mb can occur within a vertical distance of only half a kilometer (0.3 mi).

VERTICAL PRESSURE GRADIENTS

We've seen that atmospheric pressure always decreases with altitude. Notice, for example, in Figure 4–3 that the mean sea level pressure of 1013.2 mb decreases to 500 mb at an altitude of 5640 m (about 18,000 ft). Thus, the average vertical pressure gradient in the lower half of the atmosphere is about 500 mb per 5640 m, or just less than 1 mb per 10 m. Compare that to the horizontal pressure gradient of an average hurricane, which we saw to be about 1 mb per 6000 m. The *average* vertical pressure gradient in this example is 600 times greater than the *extreme* horizontal pressure gradient associated with a hurricane! In sum, vertical pressure gradients are very much greater than changes in horizontal pressure.

HYDROSTATIC EQUILIBRIUM

You already know that a pressure gradient force causes wind to flow from high to low pressure and that air pressure rapidly decreases with altitude. Given these two facts, you might infer that the wind must always blow upward. If this were the case, it would have troublesome implications for humans on the surface, who would suffocate as all the air around us literally exploded out to space in response to the vertical pressure gradient force.

Before you panic, however, consider a second relevant fact: Gravity pulls all mass, including the atmosphere, downward. Then why doesn't the atmosphere collapse all the way down to the point where we would be able to breathe only by getting on our hands and knees and sucking up the air that has fallen to the surface? Because the vertical pressure gradient force and the force of gravity are normally of nearly equal value and operate in opposite directions, a situation called **hydrostatic equilibrium**.

When the gravitational force exactly equals the vertical pressure gradient force in magnitude, no vertical acceleration occurs. When the gravitational force slightly exceeds the vertical pressure gradient force, downward motions result. Such downward motions are always very slow. On the other hand, the upward-directed pressure gradient force sometimes greatly exceeds the gravitational force, and updrafts in excess of 160 km/hr (100 mph) can develop. Such updrafts are associated with powerful thunderstorms. Although the gravitational and vertical pressure gradient forces are normally almost in balance, the exact value of each varies from place to place and time to time. The downward gravitational force on a volume of air is proportional to its mass (remember that force = mass × acceleration), so a dense atmo-

FIGURE 4–7

Two columns of air with equal temperatures, pressures, and densities (a). Heating the column on the right (b) causes it to expand upward. It still contains the same amount of mass, but it has a lower density to compensate for its greater height. The pressure drops 500 mb over 5700 m within the warm air; it only take 5640 m of ascent for the pressure to drop the same amount in the cool air. Thus, the cool air has the greater vertical pressure gradient.

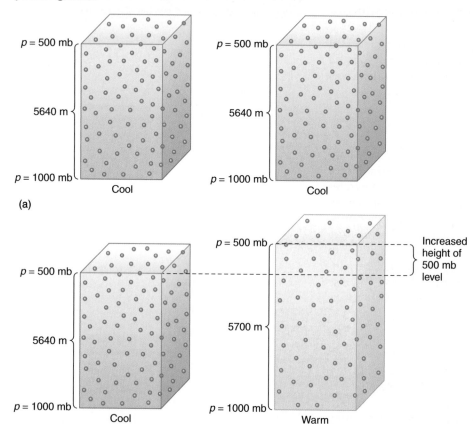

(a)

(b)

13. Describe the roles (if any) that wind speed, latitude, and direction of motion have in determining the magnitude of the Coriolis force.

14. What are geostrophic and gradient flows? Why don't they occur near the surface?

15. What are supergeostrophic and subgeostrophic flows?

16. Define the terms *cyclone, anticyclone, trough,* and *ridge.*

17. Briefly describe the movement of air around cyclones and anticyclones in the Northern and Southern Hemispheres.

18. What do anemometers and aerovanes measure?

Critical Thinking

1. Pressurized cans of shaving cream advise users not to expose the product to excessive heat. What might happen if the advice is not followed? Will this potential problem remain throughout the life of the product?

2. On a particular day, the vertical pressure gradient at the surface is −11 pascals per meter. What is the vertical pressure gradient in units of millibars per kilometer? Would you be able to use this gradient to exactly determine the pressure at the top of a building 200 m tall?

3. If a low-pressure region were to instantaneously replace a high-pressure system (assuming normally encountered values of high and low pressure), do you think you would be able to notice the difference by the pressure in your ears? Why or why not?

4. Would a particular pressure gradient produce the same exact wind speed over an Arizona desert that it would over a dense forest of tall trees? Why or why not?

5. The pilot of a small plane wants to fly at a constant height above the surface. Can the pilot fly at a constant pressure level (such as 500 mb) to assure the constant height above the ground? Why or why not?

6. A rule of thumb is that the 850 mb level often marks the boundary between the free atmosphere and the boundary layer. Are there parts of North America where this relationship is likely not to be valid? If so, where?

7. The Coriolis force applies equally to objects moving in any horizontal direction. Do you think the Coriolis force also affects objects moving directly up or down? If so, how would latitude affect the magnitude of the force?

8. Consider a 90-story skyscraper with high-speed elevators. Would a person ascending from the 46th to the 90th floor undergo the same degree of ear popping as a person ascending from the 1st floor to the 45th? Why or why not?

Problems and Exercises

1. Refer to any Web site below that produces surface and 500 mb maps. Examine the current surface map and identify the major cyclones, anticyclones, troughs, and ridges at the surface. Then look at the 500 mb map. Does the same general pattern emerge? Do the troughs and ridges at the 500 mb level occur directly over the corresponding features on the surface map? (Relationships between the surface and 500 mb levels will be discussed further in Chapter 10.)

2. On a daily basis, go to the Weather Channel's Web site at **www.weather.com**. Read the narrative describing the general weather pattern across the United States and identify the most notable weather events occurring across the country. Then look at the surface and 500 mb weather maps. This process will help you to become more familiar with normal pressure distributions and the type of weather often associated with

them. As you proceed through this text, the pressure patterns and their association with daily weather will become more meaningful to you.

3. Examine today's 500 mb weather map. You will probably find a trend toward decreasing 500 mb heights with increasing latitude. Are there any exceptions on the map to that general pattern? If so, observe the surface temperatures across North America. Do the temperature patterns have any association with the 500 mb pattern? If so, describe them.

4. Observe a surface weather map that plots isobars and station models (**http://www.atmo.arizona.edu/products/** is a good source). Do the air flow patterns around cyclones and anticyclones shown by the station models completely correspond with the generalizations made in this chapter? If not, why not?

Quantitative Problems

Differences in atmospheric pressure across the globe affect all the other elements of weather. Several quantitative problems are presented in this book's Web site to help you understand the concept of pressure, how it decreases with height, and its sensitivity to changes in moisture. It also provides problems to help illustrate the effect of latitude on the Coriolis force. To get to the problems, enter the Web site at **www.prenticehall.com/aguado** and click on Chapter 4. Then click on the "Quantitative Problems" section on the bar at the left.

Useful Web Sites

http://www.atmo.arizona.edu/products/
Offers numerous images, maps, and animations. Scroll down to the last two thumbnails to view the current surface and 500 mb weather maps for North America. These maps are plotted with the same conventions that weather forecasters have used for many decades. You can observe an animation of the 500 mb surface for North America (also showing the water vapor distribution) by clicking on the appropriate thumbnail under the heading *GOES Water Vapor Images*.

http://weather.uwyo.edu/upperair/uamap.html
Allows the user to produce upper air maps at different levels for any portion of the globe. Maps are archived for approximately 2 weeks.

http://weather.unisys.com/eta/index.html
Includes maps of the forecasted upper air patterns for a number of upper-level pressure surfaces, based on one of the primary forecast models. Output from other models is also available from this site, as are forecast maps for sea level pressure and expected precipitation.

http://www.princeton.edu/%7Eoa/safety/altitude.html
Provides interesting information on how low pressure associated with altitude affects humans.

Media Enrichment

Tutorial

Pressure Gradients

This tutorial covers the essentials of pressure gradients—both at the surface and aloft. See how pressure maps depict underlying pressure patterns and learn the relationship between isobaric surfaces and pressure gradients.

Tutorial

The Coriolis Force

This tutorial will help you visualize Coriolis accelerations—why they appear in Newton's law and how they vary with latitude and wind speed. This is a "must-do" if you doubt any of the four Coriolis properties described earlier in the chapter.

Tutorial

Atmospheric Forces and Wind

This is an extensive review of the equation of motion, showing the joint effects of the various forces involved in horizontal motion. It covers the development of geostrophic flow (and its close relative, gradient flow) and shows wind flow around cyclones and anticyclones. In addition, an interactive exercise lets you experiment with various combinations of friction and pressure gradient to see the resulting effects on wind speed and direction. Animations are used throughout to present many issues difficult to depict with static diagrams.

Weather in Motion

Changing Wind Patterns at Five Levels of the Atmosphere

This movie shows wind patterns as they evolve over a 5-day period at five different levels. You can view each of the five "slices" of the atmosphere by clicking the appropriate circle. Notice that the wind field changes its overall shape and migrates in the downwind direction, generally from west to east. These patterns will be discussed further in Chapter 8.

PART TWO
Water in the Atmosphere

Snowcapped mountains above the foggy Elwah Valley in Olympic National Park in Washington at sunset.

Despite the snowstorm on the afternoon of January, 9, 1997, air traffic into and out of Detroit Metropolitan Airport continued as normal, just as it had throughout the Midwest. Shortly before 4 P.M., Comair Flight 3272 was approaching for landing. The flight, which had taken off about an hour earlier from Cincinnati/Northern Kentucky International Airport, had proceeded normally until the pilot suddenly lost control of the aircraft. According to witnesses such as Ted Rath, who watched the horrifying scene, the twin-engine plane rolled three times and then crashed into a field 18 miles short of Runway 3. There were no survivors among the 29 passengers and crew.

Air safety experts immediately suspected icing as the cause of the accident. Further evidence, such as the fact that the aircraft had encountered two previous failures of its deicing system, added further support to the suspicion. The National Transportation Safety Board ultimately concluded that a coating of ice on the plane's wings and a low approach speed caused the plane to lose the necessary aerodynamic lift.

Although other crashes have been caused by aircraft icing, such incidents are extremely rare. The fact remains that weather conditions can pose significant problems to aviation. Dangerous weather most often takes the form of extreme turbulence or rapidly changing wind conditions. Sometimes even relatively mild conditions can present significant risks, as was the case near Detroit. A mere fog bank or layer of overcast clouds can reduce visibility and pose a threat to safe ground and air travel. Despite the fact that clouds and fog are common to our everyday lives, many of us have a weak understanding of how they form.

This chapter opens Part Two, "Water in the Atmosphere," by describing the fundamentals of atmospheric moisture. It lays out the processes by which water can change from one phase to another and describes the common measures by which we express humidity. The chapter also describes the fundamentals involved in fog and cloud formation. Chapter 6 describes the processes of cloud development and the resultant cloud forms. Chapter 7 discusses how cloud droplets grow large enough to fall as precipitation. The topics discussed here are vital to understanding some of the most common weather phenomena, as well as those that sometimes have major human impacts.

The Hydrologic Cycle

Precipitation in all its forms—rain, snow, hail, and so on—is for many people the most notable feature of the atmosphere. Though the amount and timing of precipitation vary markedly from one region of Earth to another, the total amount of precipitation for the entire globe is relatively constant from one year to the next—at about 104 cm (41 in.) per year. And yet the atmosphere doesn't run out of water! Clearly, there is continual replenishment of water lost through precipitation. In other words, just as we have described for other gases, water vapor is constantly added to and removed from the atmosphere. As long as these balance, the atmospheric store will remain constant. The movement of water between and within the atmosphere and Earth is referred to as the **hydrologic cycle**. As it happens, the hydrologic cycle, depicted in Figure 5–1, is among the fastest of all geochemical cycles; atmospheric residence time for water vapor is only 10 days or so. The hydrologic cycle is a continuous series of processes that occur simultaneously. As with any cycle, the entire process has no real end or beginning.

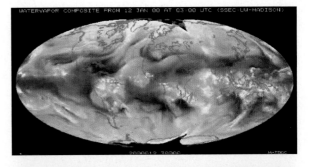

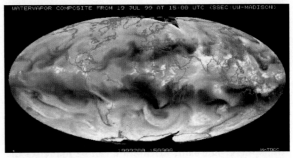

WEATHER IN MOTION
January and July Water Vapor Movies

◀ Liquid water secreted as sweat is converted to water vapor using energy supplied by the player's warm scalp. But the vapor condenses quickly in the cold environment, forming a visible "cloud" around his head. Yet another change in phase occurs in the left side of the photo where the cloud evaporates into drier air downwind.

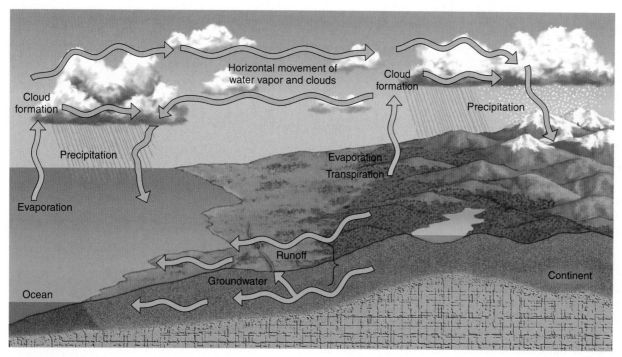

FIGURE 5-1
The hydrologic cycle.

Let's begin this discussion with the evaporation of water from Earth's surface into the atmosphere. Evaporation can occur directly from the oceans or from water bodies on land surfaces (such as lakes and rivers). It can also occur indirectly through plants via a process called *transpiration*. The water vapor that goes into the air eventually becomes water droplets or ice crystals in the form of clouds or fog. Many fog and cloud droplets or crystals will evaporate back into the air, but others precipitate down to the surface.

The shortest route the cycle can take is from the ocean to the atmosphere and back to the ocean. The situation is a bit more involved if water precipitates onto land. There, some precipitation might not reach the surface directly and may instead fall onto vegetation, accumulating as a coating of water or ice. The water that undergoes this process, called *interception*, might then drip or trickle down (after melting, if the precipitation fell as snow) the plant to the surface or evaporate back into the air. In some environments a sizable percentage—as much as 40 percent in some situations—of the precipitation that has fallen can be evaporated back into the atmosphere after interception.

Rainfall that does reach the land surface, either directly or after interception, might then flow above the surface into rivers, which then transport the water into a lake or ocean, or it can evaporate directly back into the atmosphere. If the precipitation falls as ice (the most obvious example being snow) it might temporarily remain on the ground before melting, or it might be locked away for eons as part of a glacier.

Liquid water at the ground does not always flow along the land surface but instead penetrates into the ground in the process called *infiltration*. Such water is pulled downward by gravity and can collect in the pores of underlying soil or rock as *groundwater*. Much of this groundwater eventually seeps into rivers for eventual transport toward the ocean, where the cycle continues. But much is almost immediately withdrawn by plants and transpired to the atmosphere. Still smaller quantities enter the animal kingdom as plants are consumed by browsers of all sizes, from elephants to microorganisms. And an ever-increasing fraction is drawn off by humans for agriculture (from which most is transpired), industry, and residential uses.

This chapter deals with water in its vapor state and its transformation to and from the liquid and solid phases in the atmosphere. The two chapters following

DID YOU KNOW?

Just like the planet on which you live, you have your own hydrologic cycle. The human body is 50 percent to 60 percent water, on average, and none of the water molecules in your body reside there permanently. In Chapter 1 we saw that the residence time of water vapor in the atmosphere can be calculated by dividing its mass by the rate at which it enters and leaves the atmosphere. The same calculation can be made for the water in your body, and that calculation yields a residence time of about 14 days—not all that different from the residence time of water in the atmosphere (10 days). Also, most of the water molecules currently in your body are ancient, because water is a very stable molecule. Thus, those molecules in your body are well traveled and have probably been a part of the lives of many a famous person—at least temporarily.

will examine clouds, the processes that form them, and precipitation. In reading those sections, keep in mind that the processes discussed represent individual components of the hydrologic cycle.

Water Vapor and Liquid Water

Although matter in the gaseous phase is highly compressible, the density of a gas cannot be increased to an arbitrarily high level. At some point a limit is reached, forcing a change to liquid or solid state. For one atmospheric gas, water vapor, that limit is routinely achieved at temperatures and pressures found on Earth. (Other gases, such as nitrogen and oxygen, can be liquefied only at very low temperatures.) Air that contains as much water as possible is said to be *saturated,* and the introduction of additional water vapor results in formation of water droplets or ice crystals. The concept of saturation is fundamental to understanding the processes that form clouds and fogs. We begin our discussion with a hypothetical laboratory experiment that describes the general principles of evaporation and condensation. We then apply those principles to the processes that take place in the real atmosphere.

EVAPORATION AND CONDENSATION

Figure 5–2 depicts a hypothetical experiment, in which a tightly sealed container is partially filled with pure water (H_2O). Although it may seem obvious at this juncture, let's stipulate that the water in the jar has a perfectly flat surface. Furthermore, assume that at the onset of the experiment the water surface is covered by an impermeable coating, so no water vapor exists in the volume of the container above the water surface. Whether the volume above the water surface contains any air is entirely irrelevant to this experiment. The volume can contain normal air, pure hydrogen, methane, or fumes from French perfume—it can even be a complete vacuum. All that matters with respect to the evaporation/condensation process is that no water vapor be present initially.

Figure 5–2b shows what happens when we remove the covering on the liquid water surface. Without the covering, some of the molecules at the surface can escape into the overlying volume as water vapor. The process whereby molecules break free of the liquid volume is known as **evaporation**. The opposite process is **condensation**, wherein water vapor molecules randomly collide with the water surface and bond with adjacent molecules. At the beginning of our hypothetical experiment, no condensation could occur because no water vapor was present. As evaporation begins, however, water vapor starts to accumulate above the surface of the liquid.

At the early stages of evaporation, the low water vapor content prevents much condensation from occurring, and the rate of evaporation exceeds that of condensation. This leads to an increase in the amount of water vapor present. With increasing water vapor content, however, the condensation rate likewise increases. Eventually, the amount of water vapor above the surface is enough for the rates of condensation and evaporation to become equal, as shown in Figure 5–2c. A constant amount of water vapor now exists in the volume above the water surface due to offsetting gains and losses by evaporation and condensation. The resulting equilibrium state is called **saturation**.

The state of saturation described here can occur whether or not air (or other gases, for that matter) exists in the container. In other words, the water vapor is not "held" by the air (although this erroneous

FIGURE 5–2

A hypothetical jar containing pure water with a flat surface, and an overlying volume that initially contains no water vapor (a). When evaporation begins (b), water vapor accumulates in the volume above the liquid water. Initially, no condensation can occur because of the absence of water vapor above the liquid. But as evaporation contributes moisture to the overlying volume, some condensation can occur. Evaporation exceeds condensation for a while and thereby increases the water vapor content. Eventually enough water vapor is above the liquid for condensation to equal evaporation (c). At this point, saturation occurs.

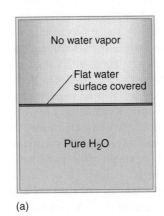

No water vapor

Flat water surface covered

Pure H_2O

Evaporation rate   Condensation rate

(a)

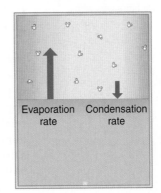

Evaporation rate   Condensation rate

(b)

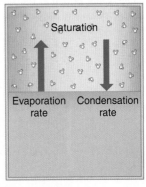

Saturation

Evaporation rate   Condensation rate

(c)

statement is frequently made). Water vapor is a gas, just like the other components of the air. Thus, it does not need to be "held" by air any more than the oxygen, nitrogen, argon, and other gases of the atmosphere need to be held by water vapor! When the air is saturated, there is simply an equilibrium between evaporation and condensation; the dry air plays no role in achieving this state. It is also important to realize that the exchange of water vapor and liquid described here applies as well to the change of phase between water vapor and ice. The change of phase directly from ice to water vapor, without passing into the liquid phase, is called **sublimation**. The reverse process (from water vapor to ice) is called **deposition**. (Meteorologists sometimes use the word *sublimation* to apply to vapor-to-solid phase changes as well as solid-to-vapor. Because opposite processes should not have the same name, the use of the term *deposition* is preferred for vapor-to-ice changes.)

Indices of Water Vapor Content

Humidity refers to the amount of water vapor in the air. Humidity can be expressed in a number of ways—in terms of the density of water vapor, the pressure exerted by the water vapor, the percentage of the amount of water vapor that can actually exist, or several other methods. There is no single "correct" measure, but, rather, each has its own advantages and disadvantages, depending on the intended use. All measures of humidity have one thing in common, however—they apply exclusively to water vapor, and not to liquid droplets or ice crystals suspended in or falling through the air. Let's now take a look at these measures.

VAPOR PRESSURE

In Chapters 1 and 4, we saw that the air exerts pressure on all surfaces. Each gas that makes up the atmosphere contributes to the total air pressure, with the most abundant permanent gases accounting for most of the pressure. Because water vapor seldom accounts for more the 4 percent of the total atmospheric mass, it exerts only a small percentage of the total air pressure. The part of the total atmospheric pressure due to water vapor is referred to as the **vapor pressure**. Like the atmospheric pressure, vapor pressure is commonly expressed in units of millibars (mb) by U.S. meteorologists, and as kilopascals (kPa) by their Canadian counterparts, though in most scientific applications the pascal (Pa) is the preferred unit (100 Pa = 1 mb = 0.1 kPa).

The vapor pressure of a volume of air depends on both the temperature and the density of water vapor molecules (Figure 5–3). If the air temperature is high, water vapor molecules (along with all the other gaseous constituents of the atmosphere) move more rapidly and exert a greater pressure. Similarly, a greater concentration of water vapor molecules means that a greater amount of mass is available to exert pressure. In practice, temperature influences are small compared to density changes, so vapor pressure closely follows changes in the density or abundance of water molecules.

Because there is a maximum amount of water vapor that can exist, there is a corresponding maximum vapor pressure, called the **saturation vapor pressure**. The saturation vapor pressure does not represent the current amount of moisture in the air; rather, it is an expression of the maximum that *can* exist. The saturation vapor pressure depends on only one variable—temperature. Figure 5–4 shows the relationship between saturation vapor pressure and temperature, with higher temperatures having higher saturation vapor pressures. For example, at 40 °C the saturation vapor pressure is 73.8 mb, while at 0 °C it is only 6.1 mb, less than one-tenth as much.

The increase in saturation vapor pressure with temperature is not linear. At low temperatures there is only a modest increase in saturation vapor pressure, but

FIGURE 5–3

The movement of molecules exerts a pressure on surfaces, called *vapor pressure*. The vapor pressure increases with concentration and temperature.

at high temperatures saturation vapor pressure grows rapidly. For example, a 2 °C increase in temperature, from 0 °C to 2 °C, increases the saturation vapor pressure from 6.1 mb to 7.1 mb, only a 1 mb difference. Raising the temperature the same amount from a higher starting point, from 40 °C to 42 °C, raises the saturation vapor pressure by 7.7 mb, from 73.8 to 81.5 mb. This nonlinear behavior is captured in a simple statement: At temperatures normally encountered near Earth's surface the saturation vapor pressure approximately doubles for every 10 °C (18 °F) increase in temperature.

ABSOLUTE HUMIDITY

Another measure of water vapor content is the **absolute humidity**, which is simply the density of water vapor, expressed as the number of grams of water vapor contained in a cubic meter of air. Because absolute humidity represents the amount of moisture contained in a volume of air, its value changes whenever air expands or contracts. Thus, for example, if an air parcel expands (as it does when it is heated or lifted upward), its absolute humidity will fall, even though no water vapor is removed from the parcel. Because absolute humidity suffers from this drawback and has no strong advantage over any other index, it is not widely used.

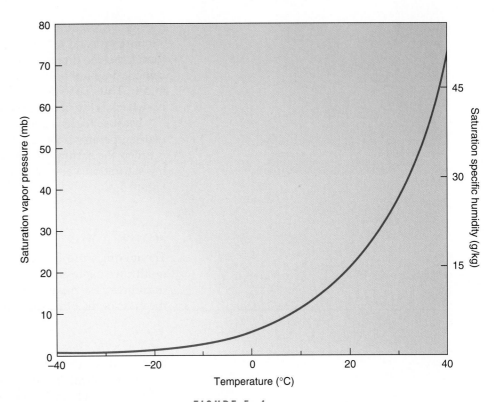

FIGURE 5–4

Saturation vapor pressure and saturation specific humidity as a function of temperature. The curve is steeper at higher temperatures, meaning that saturation vapor pressure is more sensitive to temperature changes when the air is warm.

SPECIFIC HUMIDITY

Although not normally encountered outside scientific applications, **specific humidity** is a useful index for representing atmospheric moisture. Specific humidity expresses the mass of water vapor existing in a given mass of air. Consider, for example, a volume containing exactly 1 kg of air (at sea level such a volume would be about 0.8 cubic meters, or about 27 cu. ft). Of that kilogram, some number of grams would be water vapor. The proportion of the atmospheric mass accounted for by water vapor is the specific humidity. Most often, specific humidity is expressed as the number of grams of water vapor per kilogram of air. Because the water vapor outside the tropics usually is less than 2 percent of the mass of the air near the surface, specific humidities are normally less than 20 grams of water vapor per kilogram of air. Specific humidity, q, is expressed mathematically as

$$q = \frac{m_v}{m} = \frac{m_v}{m_v + m_d}$$

where m_v = the mass of water vapor, m = the mass of atmosphere, and m_d = the mass of dry air (all the atmospheric gases other than water vapor). Unlike vapor pressure, specific humidity is affected to a small degree by atmospheric pressure, because it depends in part on the total mass of the atmosphere, m.

Unlike absolute humidity, specific humidity has the advantage of not changing as air expands or contracts. When a kilogram of air expands, its mass is unchanged (it is still 1 kg), and the proportion that is water vapor is unchanged. As a result, the specific humidity is unaffected. Likewise, specific humidity is not temperature dependent. If a kilogram of air contains 1 g of water vapor, it still contains 1 g after heating. For this reason, specific humidity is a good indicator for comparing water vapor in the air at different locations whose air temperatures might be different from each other.

For example, if Toronto, Ontario, has a specific humidity of 10 grams of water vapor per kilogram of air on a given day, and Albuquerque, New Mexico, has

5 g/kg, we can infer that Toronto has twice as much water vapor in the air as does Albuquerque, no matter what their temperatures are. This may not seem very profound, but the direct correspondence between specific humidity and water vapor content does not hold for the more frequently used index of moisture, relative humidity. Thus, specific humidity is a useful measure of water vapor whose only real drawback is the general public's unfamiliarity with the term.

Because there is a maximum amount of water vapor that can exist at a particular temperature, there is likewise a maximum specific humidity. This maximum is called the **saturation specific humidity**. This property is directly analogous to the saturation vapor pressure and increases in the nonlinear manner shown in Figure 5–4.

MIXING RATIO

The **mixing ratio** is very similar to specific humidity. In the case of specific humidity, we express the mass of water vapor in the air as a proportion of *all* the air. In contrast, the mixing ratio, r, is a measure of the mass of water vapor relative to the mass of the other gases of the atmosphere, or

$$r = m_v/m_d$$

(Note that the denominator denotes a mass of *dry* air as opposed to *all* air.) Numerically, the mixing ratio and specific humidity will always have nearly equal values. This is because the amount of water vapor in the air is always small, so that whether or not it is counted in the denominator hardly changes the ratio.

A simple example should clarify the similarity between specific humidity and mixing ratio. If the specific humidity is 10 grams of water vapor per kilogram of air, the mixing ratio is 10 grams of water vapor per 990 grams of dry air. Note that 10 divided by 990 equals 10.011. In other words, if the specific humidity is 10.0 g/kg, the mixing ratio is only 1.1 percent higher, or 10.011 g/kg.

Using the mixing ratio as an index of moisture content offers the same advantages as using specific humidity. The maximum possible mixing ratio is called the **saturation mixing ratio**.

RELATIVE HUMIDITY

The most familiar measure of water vapor content is **relative humidity**, RH, which relates the amount of water vapor in the air to the maximum possible at the current temperature. Equivalently, it can be thought of as

RH = (specific humidity/saturation specific humidity) × 100%

To see how this works, let's refer to the top example in Figure 5–5, in which the actual specific humidity is 6 g/kg of air, and the temperature of 14 °C (57 °F) yields a saturation specific humidity of 10 g/kg of air. The relative humidity would thus be

$$\text{RH} = \frac{6}{10} \times 100\% = 0.6 \times 100\% = 60\%$$

The relative humidity is not uniquely determined by the amount of water vapor present. Because more water vapor can exist in warm air than in cold air, the relative humidity depends on both the actual moisture content and the air temperature. If the temperature of the air increases, more water vapor can exist and the ratio of the amount of water vapor in the air relative to saturation decreases. Thus, the relative humidity declines even if the moisture content is unchanged. Again referring to the example in Figure 5–5, let's consider what would happen if the amount of water vapor remains constant but the temperature increases from its original 14 °C to 25 °C (77 °F). At the new temperature,

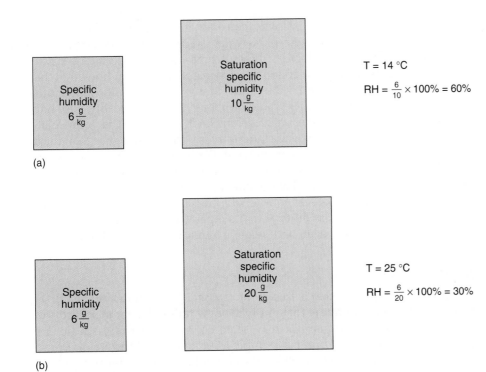

(a)

(b)

FIGURE 5-5

The relationship of relative humidity to temperature. In (a), the temperature of 14 °C has a saturation specific humidity of 10 grams of water vapor per kilogram of air. If the actual specific humidity is 6 g/kg, the relative humidity is 60%. In (b), the specific humidity is still 6 g/kg, but the higher temperature results in a greater saturation specific humidity. As a result, the relative humidity is less than in (a), even though the density of water vapor is the same.

the saturation specific humidity increases to 20 grams of water vapor per kilogram of air, and the relative humidity becomes

$$RH = \frac{6}{20} \times 100\% = 0.3 \times 100\% = 30\%$$

The relative humidity decreased even though the amount of water vapor remained constant! This is a significant drawback to any index that is supposed to be a measure of humidity.

Because of its dependence on temperature, the relative humidity will change throughout the course of the day even if the amount of moisture in the air is unchanged. Relative humidity is usually highest in the early morning—not because of abundant water vapor, but simply because the temperature is lower. As the day warms up, the relative humidity typically declines because the saturation specific humidity increases. Figure 5–6 shows a typical pattern of daily temperature and relative humidity values. Notice how relative humidity varies by a factor of 3 over the course of the day. Almost all of that variation arises from the change in air temperature.

The influence of temperature on relative humidity creates another problem—it confounds direct comparisons of moisture contents at different places with unequal temperatures. Consider, for example, a cold morning in Montreal, Quebec, where the temperature is –20 °C (–4 °F) and the specific humidity is 0.7 g/kg. At −20 °C, the saturation specific humidity is 0.78 grams of water vapor per kilogram of air, and the resultant relative humidity is [(0.70 ÷ 0.78) × 100%], or 89.7%. Now compare that to the warmer situation at Atlanta, where the temperature is 10 °C (50 °F) and the specific humidity 6.2 g/kg (nearly nine times greater than at Montreal!). At 10 °C, the saturation specific humidity is 7.7 grams of water vapor per kilogram of air, so the relative humidity is [(6.2 ÷ 7.7) × 100%], or 79.9%. Notice that the relative humidity is lower at Atlanta than at Montreal despite the fact that it contains much more water vapor. This illustrates why relative humidity is a poor choice for comparing the amount of water vapor in the air at one place to that at another.

Some people are confused about the true meaning of relative humidity. Some believe the term represents the percentage of the air that is water

FIGURE 5-6

A typical plot of temperature and relative humidity over a 24–hour period. (The temperature scale is shown on the left vertical axis and the relative humidity on the right vertical axis.) Observe that as the temperature increases, the relative humidity decreases, and vice versa. In this example there is a substantial change in the relative humidity even though the actual water vapor content over the course of the day underwent only minimal changes. This shows the strong dependence of relative humidity on air temperature (and thus one of its serious limitations as an indicator of moisture content).

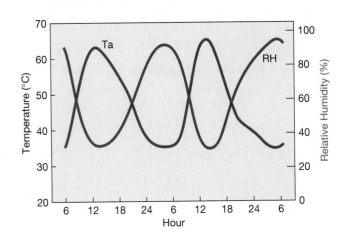

vapor. This is not correct. To see why, consider an instance in which the relative humidity is 100%. If the air were 100% water vapor, it would include no nitrogen or oxygen, and we would have a difficult time breathing, let alone discussing water vapor content! Another common misperception is about how high the relative humidity can be on a hot, humid day. Many people would estimate that on such a day the relative humidity would be about 99%. But in reality, very hot days never have relative humidities approaching that value. That is because at high temperatures the saturation specific humidity is very much higher than the actual specific humidities likely to be encountered. For example, if the temperature is 35 °C (95 °F), the saturation specific humidity is 36.8 grams of water vapor per kilogram of air. But we have seen that outside the tropics it is unusual for the specific humidity to exceed 20 g/kg—even when the air is humid. Thus, a 99% relative humidity is not a realistic possibility at that temperature. Indeed, warm days can be extremely uncomfortable even with relative humidities of only about 50%.

DEW POINT

A useful moisture index that is free of the temperature relationship just described is the **dew point temperature** (or simply the **dew point**), the temperature at which saturation occurs. This quantity may seem confusing at first because it is expressed as a temperature, but it is a simple index to use and easy to interpret. And it is dependent almost exclusively on the amount of water vapor present.

Consider the parcel of unsaturated air in Figure 5–7. Initially the air temperature was 14 °C (58 °F), yielding a saturation specific humidity of 10 grams of water vapor per kilogram of air. The initial specific humidity was 8 g/kg. The relative humidity was therefore 80%. As the air cools, its relative humidity increases, and if the air is cooled sufficiently its relative humidity reaches 100% and it becomes saturated. Any further cooling leads to the removal of water vapor by condensation. In this example, the dew point is 10 °C (50 °F) because that is the temperature at which the saturation specific humidity is 8 g/kg. Notice that even though the relative humidity increased as the temperature decreased, the dew point remained constant at 10 °C.

FIGURE 5–7

The dew point is an expression of water vapor content, although it is expressed as a temperature. In (a), the temperature exceeds the dew point and the air is unsaturated. When the air temperature is lowered so that the saturation specific humidity is the same as the actual specific humidity (b), the air temperature and dew point are equal. Further cooling (c) leads to an equal reduction in the air temperature and dew point so that they remain equal to each other.

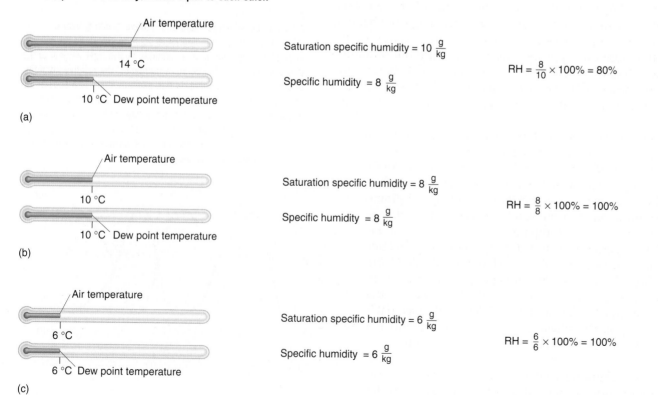

Dew Point and Nighttime Minimum Temperatures

5–1 FORECASTING

Knowledge of the current dew point temperature is a useful tool to the forecaster for the prediction of the following morning's low temperature. If no major wind shifts or other weather changes are anticipated, the minimum temperature will often approximate the dew point. Consider a hypothetical evening with an air temperature of 15 °C (59 °F) and a dew point of 5 °C (41 °F). The spread between the air temperature and the dew point temperature is not very large, and a 10 °C (18 °F) lowering of the air temperature is feasible. If the air temperature does indeed drop to the dew point and there is little or no wind, a radiation fog has a good chance of forming. The fog would then inhibit further cooling, partly because latent heat is released, and partly because water droplets are extremely effective at absorbing longwave radiation from the surface. Without the loss of radiation, the surface temperature would remain almost constant, and the overnight low would equal the dew point temperature.

The relationship between dew point and minimum temperature will not hold under certain conditions. The first has to do with the changes in the big weather picture. Imagine, for example, that a mass of warmer air is moving into the forecast region. This large body of air can replace the one present at the time of the forecast and bring with it higher nighttime temperatures. Similarly, the passage of an advancing cold front (briefly described in Chapter 1 and discussed in more detail in Chapter 9) can lead to significant drops in temperature below the current dew point.

Both heavy cloud cover and strong winds inhibit a drop in air temperature, and their presence may keep minimum air temperatures above the dew point. Cloud cover achieves this effect because of its absorption and downward reradiation of longwave energy. Strong winds prevent large temperature decreases at the surface by vertical mixing. A shallow layer of cold air that would otherwise develop is easily disrupted, leading to higher surface temperatures and more uniform temperatures with height.

Minimum temperatures won't go down to the dew point if the difference between the air temperature and the dew point temperature is very large. One can readily see how this might occur if a desert has a high temperature of 45 °C (113 °F) and a dew point of 0 °C (32 °F). Even with calm winds and no cloud cover, a cooling of 45 °C is unlikely over the course of a short summer night. Though the temperature won't always drop down as low as the dew point, it is always certain that unless a front passes through or the wind direction changes significantly, the minimum temperature will not fall much below the evening dew point.

What would have happened if the specific humidity had remained constant and the temperature had increased from its initial 14 °C? The relative humidity would have decreased, yet the dew point would have remained constant. The dew point would not have changed because eventual cooling of the air to 8 °C still would have led to saturation.

The dew point is a valuable indicator of the moisture content; when the dew point is high, abundant water vapor is in the air. Moreover, when combined with air temperature, it is an indicator of the relative humidity. When the dew point is much lower than the air temperature, the relative humidity is very low. When the dew point is nearly equal to the air temperature, the relative humidity is high. Furthermore, when the air temperature and the dew point are equal, the air is saturated and the relative humidity is 100 percent.

Unlike relative humidity, the dew point does not change simply because air temperature changes. Moreover, if one location has a higher dew point than another, it will also have a greater amount of water vapor in the air, assuming the same air pressure. Once you are familiar with dew point, it is probably the most effective index of water vapor content. Dew points on very humid, hot days are typically in the low 20s on the Celsius scale (low 70s Fahrenheit). (When you see a dew point of 70 °F or higher, you can plan on a sleepless night unless you have air conditioning.) On comfortable days that are neither humid nor dry, dew points may be in the low teens Celsius (mid-50s Fahrenheit); very dry days can have dew points in the minus 20s or lower on the Celsius scale (0 °F). The dew point temperature can sometimes serve as a predictor of overnight cooling, as explained in Box 5–1, *Forecasting: Dew Point and Nighttime Minimum Temperatures*.

The dew point is always equal to or less than the air temperature; under no circumstances does it ever exceed the temperature. So what happens if the air temperature is lowered to the dew point and then cooled further? In that case, the amount of water vapor exceeds the amount that can now exist, and the surplus is

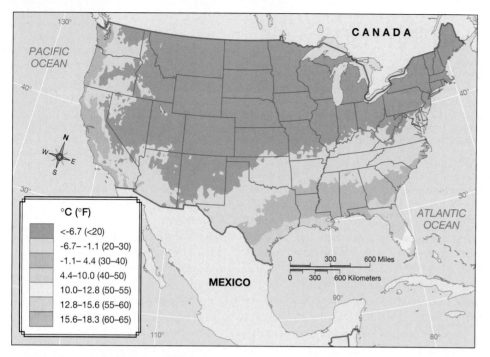

°C (°F)
<-6.7 (<20)
-6.7– -1.1 (20–30)
-1.1– 4.4 (30–40)
4.4–10.0 (40–50)
10.0–12.8 (50–55)
12.8–15.6 (55–60)
15.6–18.3 (60–65)

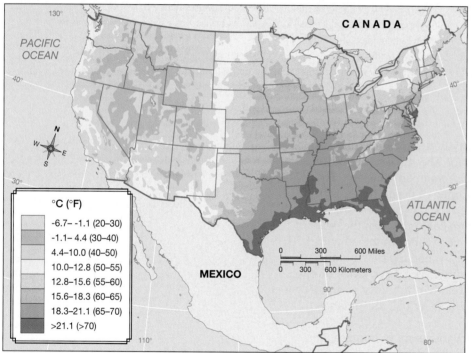

°C (°F)
-6.7– -1.1 (20–30)
-1.1– 4.4 (30–40)
4.4–10.0 (40–50)
10.0–12.8 (50–55)
12.8–15.6 (55–60)
15.6–18.3 (60–65)
18.3–21.1 (65–70)
>21.1 (>70)

FIGURE 5–8

The average distribution of dew points across the United States in January (top) and July (bottom).

removed from the air. This happens by condensation of the water vapor to form a liquid or by the formation of ice crystals. In either case, the dew point decreases at the same rate as the air temperature, and the two remain equal to each other. This is illustrated in Figure 5–7b and c. When the temperature was lowered to 10 °C in Figure 5–7b, the air became saturated with 8 grams of water vapor per kilogram of air. As the air cooled further to 6 °C (Figure 5–7c), the saturation specific humidity decreased to 6 g/kg. Because the specific humidity, by definition, cannot exceed the saturation specific humidity, 2 grams of water vapor (8 grams minus 6 grams) had to be removed from each kilogram of air by condensation. The removal of water vapor kept the specific humidity equal to that of the saturation specific humidity and also lowered the dew point. Note that when the temperature at which saturation would occur is below 0 °C (32 °F), we use the term **frost point** instead of dew point.

Distribution of Water Vapor

Water vapor gets into the atmosphere either from local evaporation or from the horizontal transport (advection) of moisture from other locations. The affect of advection on the distribution of water vapor is clearly evident in Figure 5–8, which shows the spatial distribution of mean dew points (in °F and in °C) across the United States in January (top) and July (bottom). Looking at the eastern two-thirds of the country first, it is clear that for both months the amount of water vapor generally decreases with distance from the Gulf of Mexico. Because of the Gulf's high water temperatures, moisture is readily evaporated into the atmosphere year round, and this moisture can be transported northward. The decline in water vapor content is seen in a north–south direction and also moving westward from about the Mississippi River toward the Rocky Mountains during the summer. During the winter months, the amount of moisture extending into the Great Plains is low and only a minimal amount of east–west variation exists.

The effect of distance from the source of moisture is also evident in the West, with water vapor generally decreasing from the Pacific Coast to the Rocky Mountains. The most rapid drop occurs very near the coast because local mountains block off substantial amounts of moisture from inland areas.

Comparing the two maps, you will note a substantial increase in the amount of water vapor in the air in July over that in January. This should not be surprising because lower January temperatures preclude the existence of high water vapor contents. Thus, for example, along the Ohio River Valley the average dew point increases from about –7 °C (20 °F) in January to perhaps 17 °C (63 °F) in July. This is why residents of much of the country, especially those east of the Rockies, are subject to uncomfortable dry skin in the winter—only to find themselves sweating profusely during the summer.

The general patterns described for the United States also apply to most of Canada. *Box 5–2, Forecasting: Vertical Profiles of Moisture,* provides further information on the analysis of water vapor distributions.

Methods of Achieving Saturation

Air can become saturated by any one of three general processes: adding water vapor to the air; mixing cold air with warm, moist air; and lowering the temperature to the dew point. The first of these processes can be seen in your bathroom when you take a warm shower. The warm water from a showerhead evaporates moisture into the air in the room and brings it to the saturation point. Condensation first forms on your mirrors and other surfaces, and then a general fog develops. In the natural environment, the evaporation of water from falling raindrops can

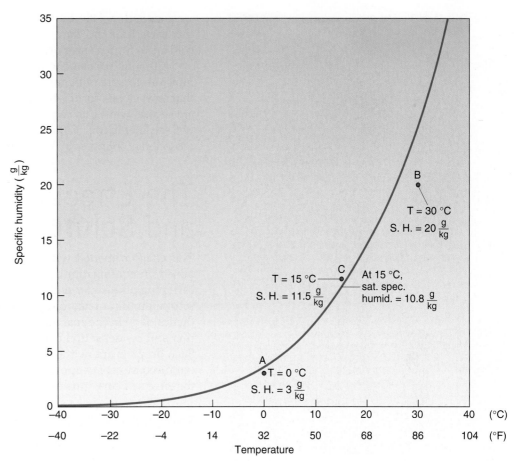

FIGURE 5–9

Saturation by the mixing of warm, moist air with cold air. Parcel A has a low temperature and can therefore contain only a small amount of water vapor. The warm parcel (Parcel B) has a higher moisture content. Parcel C results when Parcels A and B are mixed together and has a temperature in between those of the original parcels. The amount of moisture in the air is greater than that which can exist at the new temperature, so the excess water vapor condenses.

raise the dew point in the air beneath the cloud from which the rain falls. If enough vapor is added to the air to saturate it, a **precipitation fog** forms beneath the cloud.

Condensation from the second process, the mixing of cold with warm moist air, is illustrated in Figure 5–9. Consider parcels A and B of unsaturated air. Parcel A has a temperature of 0 °C (32 °F) and a specific humidity of 3 grams of water vapor per kilogram of air; Parcel B has a temperature of 30 °C (86 °F) and a specific humidity of 20 g/kg. If equal amounts of the two parcels are mixed together, the new parcel has a temperature of 15 °C (59 °F), exactly midway between the temperatures of the original parcels. However, such is not the case for the specific humidity.

Although the total H_2O content in the mixed parcel is exactly between the amounts of the original parcels, some of the moisture occurs in the form of a liquid rather than as water vapor. Here is why: The midpoint between the original specific humidity values is 11.5 g/kg, but at 15 °C the saturation specific humidity is only 10.8 g/kg. In other words, the air contains 0.7 g more water than can exist in the vapor form. The surplus therefore condenses to form fog droplets.

The preceding process is what causes contrails to form behind aircraft traveling at high altitudes. As the jet engines burn fuel, they put out a large amount of heat as well as water vapor. The air is extremely turbulent in the wake of the aircraft, so the hot, moist (but unsaturated) exhaust from the engines rapidly mixes with the cold surrounding air. At the subfreezing temperatures of the upper troposphere, the vapor directly forms into ice crystals or into liquid droplets that eventually freeze to form the contrail.

A similar but naturally occurring phenomenon is known as **steam fog**. As we learned in Chapter 3, water bodies are rather slow to change temperature. As a result, lakes can remain relatively warm well into the fall or early winter, even as air temperatures become low. Because of evaporation and the upward transfer of sensible heat, a thin, transitional layer of air exists just above the water surface that is

FIGURE 5–10
Steam fog.

warmer and moister than the air above. If a mass of cold air abruptly passes over the warm lake, the warm, moist transitional air mixes with the overlying layer of cold air to form a layer of fog a meter or two thick (Figure 5–10).

Although we can cite several examples of how clouds form by the increase of moisture content or by mixing warm, moist air with cold, dry air, experience shows that most clouds form when the air temperature is lowered to the dew point. There are several ways this cooling can occur, requiring considerable explanation, as we will see later in this chapter. For now we simply note that this third mechanism, atmospheric cooling, is by far the most common process for cloud formation.

The Effects of Curvature and Solution

This chapter opened with a hypothetical experiment in which water in a jar attained an equilibrium between condensation and evaporation. The experiment, assuming pure water with a flat surface, provided a foundation for understanding saturation. But meteorology studies the atmosphere—not what goes on in hypothetical jars. In the real atmosphere, we are concerned with the rates of evaporation and condensation across the surfaces of suspended cloud and fog droplets. Such droplets are neither flat nor made up of pure H_2O. We therefore expand on our discussion of evaporation, condensation, and equilibrium to take into account the effects of curvature and impurity of cloud and fog droplets.

EFFECT OF CURVATURE

Water droplets exist in nature not as tiny cubes with flat sides, but rather as microscopically small spheres with considerable curvature. Compare the two droplets shown in Figure 5–11. The one on the left is much larger than the one on the right and therefore has less curvature. We could even consider a more extreme example—Earth itself. Most of us are pretty certain by now that the planet is not

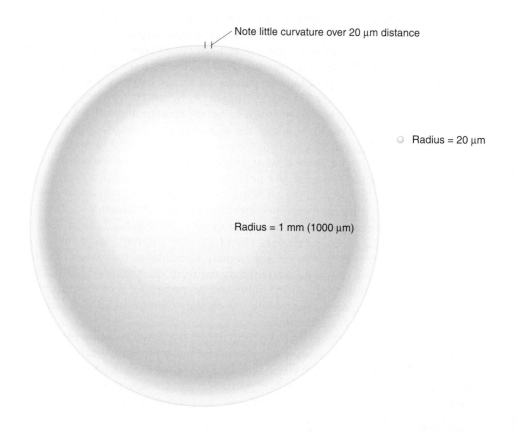

Note little curvature over 20 μm distance

Radius = 20 μm

Radius = 1 mm (1000 μm)

FIGURE 5–11
Large droplets have less curvature than small droplets.

Vertical Profiles of Moisture

In Chapter 3, you saw how simplified thermodynamic diagrams can be used to plot the vertical profile of temperature. Because dew point values are likewise expressed as temperatures, they too can be plotted on thermodynamic charts. In fact, by plotting temperatures and dew points simultaneously, you can obtain considerable information on cloud conditions. Refer to the sounding of temperature and dew point in Figure 1, taken from Stapleton Airport, near Denver, Colorado, at midnight, Greenwich mean time, on April 12, 2002.

The example in Figure 1 plots temperature (the curve on the right) and dew point (left). If you contrast this to the profile shown in Chapter 3 (page 87), taken at Slidell, Louisiana, you will notice that this sounding begins at a much lower pressure—at about the 840 mb level. The reason for this is very simple. Denver's elevation of 1625 m (5330 ft) causes its surface pressure to be much lower than that at Slidell's nearly sea level location (remember, pressure *always* decreases with elevation).

In Figure 1, the temperature at the surface is 17 °C (63 °F), and the dew point is −4 °C(25 °F). This large difference between the two values indicates that the relative humidity is low (calculated to be 23%). But as distance from the surface increases, temperature decreases more rapidly than dew point. At about the 560 mb level (at a height of about 4850 m, or 15,900 ft, above the surface), the two values become equal to each other and the air is saturated (a slight measurement error accounts for the plotted temperature and dew point values not being exactly equal). The dew point and the air temperature then decrease at the same rate up to about the 460 mb level (about 6500 m, or 21,300 ft, above the surface), above which temperature once again exceeds the dew point. Because the air is saturated in the layer of air between 4850 m and 6500 m above the surface, we can infer that a cloud occupies that 1650–m thick layer.

The thermodynamic diagram shown here is slightly more complex than the one in Chapter 3, because it includes an additional set of lines that provides one more type of moisture information. The brown lines that slope gently to the left as they extend upward indicate the specific humidity and the saturation specific humidity at any level.

Let's first see how the plot of the air temperature profile can be used to determine the saturation specific humidity at a given pressure level. Notice that at the 700 mb level the air temperature is 2 °C. It so happens that one of the sloping lines (labeled at the bottom of the diagram with the number 7) nearly intersects the temperature profile at the 700 mb level (highlighted by a red dot). This indicates that at the air at the 700 mb level has a saturation specific humidity of just under 7 grams of water vapor per kilogram of air.

We can follow a similar procedure to find out what the actual specific humidity is at the 700 mb level. To do this, we follow the dew point profile up to the 700 mb level. The point where the profile crosses the 700 mb line occurs right between the two sloping lines labeled 2 and 4. In other words, the actual specific humidity at the 700 mb level is right between 2 grams of water vapor per kilogram of air and 4 g/kg—with 3 g/kg a very good approximation.

After estimating the actual specific humidity and saturation specific humidity at the 700 mb level, it is easy to use the values to obtain the relative humidity at the 700 mb level:

$$RH = (\text{specific humidity}/ \text{saturation specific humidity}) \times 100\%$$

$$= (3/7) \times 100\% = 43\%$$

This procedure can be performed at the surface or any other level of the atmosphere.

Later in this chapter, we will see how thermodynamic diagrams can give us information in forecasting the likelihood of cloud development.

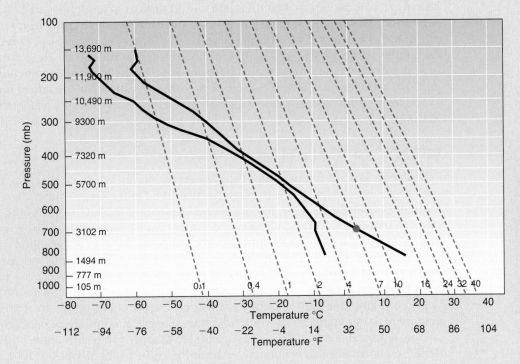

FIGURE 1

A sounding of temperature and dew point. This chart plots temperature and dew points throughout the troposphere and much of the stratosphere. The slightly sloping brown lines depict values of specific humidity in grams of water vapor per kilogram of air. Meteorologists use these lines along with the plots of temperature and dew point to determine the saturation specific humidity and the actual specific humidity, respectively.

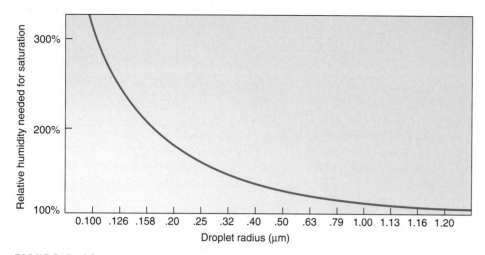

FIGURE 5–12

Small droplets of pure water require relative humidities above 100% to remain in equilibrium.

flat but spherical. However, because of Earth's large size, its curvature is inconspicuous, and a centimeter of distance across its surface essentially forms a straight line. Moreover, a straight-edge ruler can lie flat against its surface. But hold a ruler over a tennis ball and only a small part of it is in contact with the ball's surface; it is more strongly curved than Earth.

Cloud droplets are much smaller than tennis balls, of course, and thus exhibit an even greater curvature. But what does curvature have to do with saturation? The answer is that curvature has an effect on evaporation from cloud droplet surfaces and therefore on the vapor pressure necessary for saturation. Effects arising from surface tension lead to differences in the saturation point, as described shortly.

The graph of saturation vapor pressure versus temperature, shown as Figure 5–4, applies only to flat surfaces of pure H_2O. For curved water surfaces, the evaporation rate is greater. The enhanced rate of evaporation requires that condensation also be increased for the two to remain in balance. Thus, a highly curved droplet of pure water at any given temperature has a higher saturation vapor pressure than indicated in Figure 5–4. Stated another way, highly curved droplets of pure water require relative humidities in excess of 100% to keep them from evaporating away.

Figure 5–12 illustrates the effect of droplet size on the relative humidity needed to maintain an existing droplet of pure H_2O. For very small droplets (those with radii of about 0.1 μm), relative humidities in excess of 300% are needed to achieve an equilibrium between evaporation and condensation. In other words, those droplets require a *supersaturation* of 200%. The degree of supersaturation necessary to maintain a droplet rapidly decreases with increasing droplet size. Droplets with radii larger than 10 μm require supersaturations of only about 10%.

If the atmosphere were devoid of any aerosols, condensation would occur only by **homogeneous nucleation**, in which droplets form by the chance collision and bonding of water vapor molecules under supersaturated conditions. Such droplets would necessarily have only a small number of molecules and a high degree of curvature, and therefore would only exist at high levels of supersaturation. This process seldom if ever occurs, because certain **hygroscopic** (water-attracting) aerosols in the atmosphere assist the formation of droplets at relative humidities far below those necessary for homogeneous nucleation. The formation of water droplets onto hygroscopic particles is called **heterogeneous nucleation**, and the particles onto which the droplets form are called **condensation nuclei**. When condensation occurs, the condensation nuclei dissolve into the water to form a *solution*.

EFFECT OF SOLUTION

When materials are dissolved in water, a certain number of the molecules at the surface are those of the *solute* (the material dissolved in the water) rather than H_2O molecules. With fewer H_2O molecules at the surface, the rate of evaporation is lower than for pure water. As a result, solutions require less water vapor above the surface to maintain an equilibrium between evaporation and condensation.

If condensation in the atmosphere occurred by homogeneous nucleation, newly formed droplets would consist of pure water. But in fact droplets actually form by heterogeneous nucleation and the incorporation of the nuclei into the water solution. This has the opposite effect of droplet curvature by lowering the amount of water vapor necessary to keep the droplets in equilibrium. Under most circumstances, this *solute effect* is about equal to that of droplet curvature, and condensation normally occurs at relative humidities near or slightly below 100%.

Although the proportion of aerosols in the air that are hygroscopic is small, the atmosphere contains so many suspended particles that condensation nuclei are always abundant. Some materials are more hygroscopic than others, and large aerosols are generally more effective than smaller ones. Some are even capable of attracting water at relative humidities below 90% and forming extremely small droplets. We observe such droplets as **haze**.

Until very recently, scientists believed that most condensation nuclei in the atmosphere were natural, consisting mostly of continental dust, sea salt, and aerosols derived from volcanic eruptions, natural fires, and gases given off by marine phytoplankton. Research undertaken during the 1990s now indicates that the role of human activity was previously understated and that anthropogenic sources may account for the majority of condensation nuclei over industrialized areas in the Northern Hemisphere. The effect of human activity on aerosol concentration even extends to oceanic areas in the Northern Hemisphere. In the Southern Hemisphere, human influences are much smaller and are confined mainly to regions of heavy industrial activity and frequent burning.

Although salt particles are very hygroscopic, they are less abundant over land than over their oceanic source. Over land, salt particles account mainly for the very largest of cloud condensation nuclei, but they are relatively rare compared to other materials.

ICE NUCLEI

So far we have examined the formation of liquid water droplets when air becomes saturated. But saturation can occur at very low temperatures, which suggests that rather than liquid water droplets, ice crystals may form. On the other hand, many of us have walked through fog composed of liquid droplets even though the temperature was below 0 °C. So what really happens when saturation occurs at temperatures below the freezing point? Does this lead to the condensation of liquid droplets or to the deposition of ice crystals? The answer is that it depends. Strangely, although ice always melts at 0 °C, water in the atmosphere does not normally freeze at 0 °C.

If saturation occurs at temperatures between 0 °C and −4 °C, the surplus water vapor invariably condenses to form **supercooled water** (water having a temperature below the melting point of ice but nonetheless existing in a liquid state). Ice does not form within this range of temperatures. Just as the formation of liquid droplets at relative humidities near 100% requires the presence of condensation nuclei, the formation of ice crystals at temperatures near 0 °C requires **ice nuclei**. Unlike condensation nuclei, which are always abundant, ice nuclei are rare in the atmosphere. This is because an ice nucleus must have a six-sided structure that mimics the alignment of molecules in an ice crystal (although exceptions to this rule exist).

A material's ability to act as an ice nucleus is temperature dependent, and no materials are effective ice nuclei at temperatures above −4 °C. Though ice can exist at temperatures between −4 °C and 0 °C, it does not form spontaneously in the atmosphere in this range. (In fact there is little ice in the atmosphere at temperatures above −10 °C.) Thus, between −4 °C and 0 °C, the removal of water vapor occurs only by the condensation of supercooled water.

As temperature decreases, the likelihood of ice formation increases, and at temperatures between about −10 °C (14 °F) and −40 °C (−40 °F) saturation can lead to the nucleation of ice crystals, supercooled droplets, or both. In clouds having temperatures within this range, liquid droplets and solid crystals will usually coexist, but with a greater proportion of ice at lower temperatures. At temperatures below −30 °C (−22 °F), the cloud will be mostly ice crystals. For temperatures below −40 °C, saturation leads to the formation of ice crystals only, with or without the presence of ice nuclei. As we will see in Chapter 7, the coexistence of ice crystals and liquid droplets in clouds is extremely important for the development of precipitation outside the tropics.

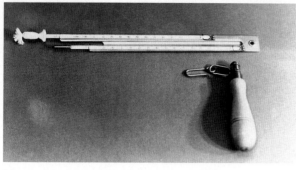

(a)

(b)

FIGURE 5–13
(a) Sling psychrometer and (b) hygrothermograph.

Among the materials that serve as ice nuclei are components of natural soils called *clays.* Clay materials have a platy structure, microscopic sizes, and strong electrical attractions. These characteristics make them very difficult to dislodge from the surface and incorporate into the atmosphere, which largely explains their scarcity. Clays occur in a variety of compositions, of which the most effective seems to be *kaolinite* (aluminum silicate). Other types of clay serve as ice nuclei but are active only at lower temperatures.

Unfortunately, it is very difficult for cloud physicists to obtain much knowledge about the sources of ice nuclei because they tend to be small in size and rare in number. Observations do seem to show, however, that a few ice crystals contain a foreign ice nucleus. This has led to the idea that ice crystals themselves may be very important ice nuclei. In this model, existing ice crystals break into small six-sided fragments, which then act as nuclei for new crystals. Other possible ice nuclei include fragments of decaying leaves, volcanic material, and certain bacteria.

Measuring Humidity

Considering the fact that water vapor is an invisible gas mixed with all the other gases of the atmosphere, you might suspect that its measurement would entail some highly sophisticated instrumentation. Such is not the case. The simplest and most widely used instrument for measuring humidity, the **sling psychrometer** (Figure 5–13a), consists of a pair of thermometers, one of which has a cotton wick around the bulb that is saturated with water. The other thermometer has no such covering and simply measures the air temperature. The two thermometers, called the **wet bulb** and **dry bulb thermometers**, respectively, are mounted to a pivoting device that allows them to be circulated ("slung") through the surrounding air. If the air is unsaturated, water evaporates from the wet bulb, whose temperature falls as latent heat is consumed. After about a minute or so of circulating, the amount of heat lost by evaporation is offset by the input of sensible heat from the surrounding, warmer air, and the cooling ceases. Thereafter, the wet bulb maintains a constant temperature no matter how long the instrument is swung around.

The difference between the dry and wet bulb temperatures, called the **wet bulb depression**, depends on the moisture content of the air. If the air is completely saturated, no net evaporation occurs from the wet bulb thermometer, no latent heat is lost, and the wet bulb temperature equals the dry bulb temperature. On the other hand, if the humidity is low, plenty of evaporation will take place from the wet bulb, and its temperature will drop considerably before reaching an equilibrium value. To determine the moisture content, first you note the difference between the dry and wet bulb temperatures. Then, with the use of tables such as Tables 5–1 and 5–2, you obtain the dew point, relative humidity, or any other humidity measure by finding the value corresponding to the row for the air temperature and the column for the wet bulb depression.

Some psychrometers are equipped with fans that circulate air across the bulbs of the two thermometers. These **aspirated psychrometers** save the user the effort needed to sling the thermometers through the air (as well as the aggravation of cleaning up the mess after accidentally striking nearby objects). Another alternative to the sling psychrometer is the **hair hygrometer**, whose basic part is a band of human hair. Hair expands and contracts in response to the relative humidity. By connecting a hair to a lever mechanism, we can easily determine its water vapor content. Often, the hygrometer is coupled with a bimetallic strip and rotating drum to give a continuous record of temperature and humidity. Such a **hygrothermograph** is shown in Figure 5–13b. (The distribution of water vapor across the globe is routinely observed by water vapor imagery. This is described in *Box 5–3, Forecasting: Water Vapor and Infrared Satellite Imagery.*)

(a)

(b)

FIGURE 5–17
Dew (a) and frost (b).

lack of wind precludes the mixing of warmer air from above. Together these conditions promote rapid cooling within a shallow layer of air immediately adjacent to the surface.

FROST

The formation of **frost** is similar to that of dew, except that saturation occurs when the temperature is below 0 °C. When the air temperature is lowered to the frost point, very small ice crystals are deposited onto solid surfaces, giving them a bright white appearance, as shown in Figure 5–17b. This type of deposition, sometimes referred to as *white frost* or *hoar frost*, occurs by the transformation of water vapor directly into ice, without going through the liquid phase.

Because it consists of a huge number of separate ice crystals rather than a solid, continuous coating of ice, frost on the windshield of a car is often easy to remove by a swift brushing with a credit card or window scraper. This is in contrast to a more troublesome type of condensation called *frozen dew*.

FROZEN DEW

Frozen dew differs from frost in both its structure and its manner of formation. It begins when saturation forms liquid dew at temperatures slightly above 0 °C. When further cooling brings its temperature below the freezing point, the liquid solidifies into a thin, continuous layer of ice. In contrast to frost, frozen dew is neither milky white nor easy to remove but instead bonds tightly to any surface on which it forms.

Because it is a continuous coating of solid ice, a mere brushing does not come close to removing frozen dew. In addition to coating your car's windshield, the ice can make it difficult or impossible to get your key into the lock mechanism of your door. And even if you are lucky enough to be able to turn the key, the door can become frozen to the car frame as if it were welded shut. On the other hand, it may be just as well for you that you are not able to get into your car because of *black ice*, the smooth coating of frozen dew that forms on road surfaces. This is especially likely to form over bridges, causing dangerous, slippery driving conditions.

FIGURE 5–18
Fog.

FOG

Fog is essentially a cloud whose base is at or near ground level (Figure 5–18). It can be extremely shallow, on the order of a meter or so in depth, or it can extend for tens of meters above the surface. Like any other form of condensation, fog can form by the lowering of the air temperature to the dew point, an increase in the water vapor content, or the mixing of cold air with warm, moist air.

PRECIPITATION AND STEAM FOGS

We have already described one example of condensation resulting from the addition of water vapor to the air, *precipitation fog*, which results from the evaporation of falling raindrops. Another, although very localized, type of fog occurs from adding water vapor to the air—the type you see right in front of you when you exhale on a cold day. We have also discussed the formation of *steam fogs*, which occur when cold, dry air mixes with warm, moist air above a water surface. All other types of fog result from a cooling of the air to the dew point. They include radiation fogs, advection fogs, and up-slope fogs.

RADIATION FOG

Radiation fogs (sometimes called *ground fogs*) develop when the nighttime loss of longwave radiation causes cooling to the dew point. Like dew, a radiation fog is most likely to form on cloudless nights when longwave radiation from the surface easily escapes to space. Unlike dew it is most likely to form with light winds of about 5 km/hr (3 mph) rather than in perfectly still air. Light breezes promote a gentle stirring of the lower atmosphere, which permits condensation to form throughout a layer of air. When the wind speed is much greater than 5 km/hr, the excess turbulence brings warm air down toward the surface and thereby inhibits cooling to the dew point.

Most radiation fogs begin to dissipate within a few hours of sunrise. Although we sometimes talk about a fog "lifting," that is not what really happens. It is probably better (although still somewhat imprecise) to describe it as "burning off." When sunlight penetrates the fog, it warms the surface, which in turn warms the overlying air. As the air temperature increases, the fog droplets gradually evaporate. Because the evaporation of droplets is most rapid near the surface, the fog appears to lift, although it really undergoes no vertical displacement.

When radiation fogs are especially well developed, they can scatter backward the greater part of incoming solar radiation. Because the amount of energy reaching the surface is now reduced, the fog can persist throughout the day, especially in the winter when the days are short and the sun angles are low. Under the most extreme circumstances, fogs can persist for days on end.

A prime location for a persistent radiation fog is the Central Valley of California (Figure 5–19). To the west of the valley, the Coast Ranges block the moderating effects of the Pacific Ocean, while the Sierra Nevada isolates the valley from the east. In addition to the clear skies and light winds that often predominate during the winter, this heavily agricultural region has copious amounts of moisture evaporated into the air from the irrigated farmland. When the radiation fog (locally referred to as a *Tule fog*) covers the valley, visibility along the two major north–south highways can be dangerously reduced to near zero. This happened with fatal consequences on February 5, 2002, on California State Highway 99, south of Fresno. As the fog reduced visibility to as little as 15 m (50 ft), California Highway Patrol "pace cars" were dispatched to lead traffic at safe speeds. But, as has happened many times in the past, the effort was unsuccessful. Eighty-seven cars were involved in two chain reaction pile-ups that left two persons dead and many others injured. Just one month earlier, a similar set of accidents caused two other fatalities along Interstate 5 near the Sacramento airport.

On March 5, 2002—one month after the major pile-up near Fresno—a thick, morning fog led to a major crash involving 125 vehicles on Interstate 75 in northwest Georgia, just south of Chattanooga, Tennessee. Four people died in the accident. This is another region where radiation fogs are common, and that morning the National Weather Service issued numerous advisories about hazardous driving conditions. Despite the warning, the near-zero visibility made any travel along the highway a risky endeavor. According to Sheriff Phil Summers, "It did not appear that there was any fault other than fog."

Radiation fogs—and the risks they impose on travelers—are frequent occurrences wherever air has the opportunity to cool at night with gentle stirring. Various state transportation agencies have tried to alleviate the threat of such accidents by painting reflective "fog lines" on at-risk highways and installing automated weather stations to provide real-time weather information, but measures such as these cannot overcome the inherent danger imposed by dense fog.

Radiation fogs are formed by diabatic cooling and are therefore associated with cold air. Because cold air is denser than warm air with otherwise similar characteristics, radiation fogs often settle into local areas of low elevation, where they are called *valley fogs*. It is tempting to attribute the fog's high density to the heavy droplets of liquid water, but they are not the cause. It is the low temperature—not the existence of water droplets—that causes the fog to be dense and settle into valleys. Because all the water droplets replace an equal mass of water vapor from which they formed, their presence does not make the air any heavier.

ADVECTION FOG

Advection fogs (Figure 5–20) form when relatively warm, moist air moves horizontally over a cooler surface (the term *advection* refers to horizontal movement). As the air passes over the cooler surface, it transfers heat downward; this causes it to cool diabatically. If sufficient cooling occurs, a fog forms. Such fogs can be advected for considerable distances and persist well downwind of the area over which they form. One of the most famous examples of this phenomenon occurs during the summer months over the San Francisco Bay area. As relatively warm Pacific air drifts eastward, it passes over the narrow, cold, southward-flowing California ocean current. The cooling of the air offshore forms the fog, which drifts eastward toward San Francisco. It is quite common for the fog to fully engulf San Francisco while Oakland and Berkeley, across the bay, remain warm and sunny.

Advection fogs can also form over water when warm and cold ocean currents are in proximity to each other. Off the coast of New England and the Maritime Provinces, the cold Labrador current flows just to the north of the Gulf Stream. When the moist air from the Gulf Stream region drifts

FIGURE 5–19
The Central Valley of California often gets radiation fogs in winter that last for days on end. These persistent fogs are visible from space.

FIGURE 5–20
An advection fog at San Francisco Bay.

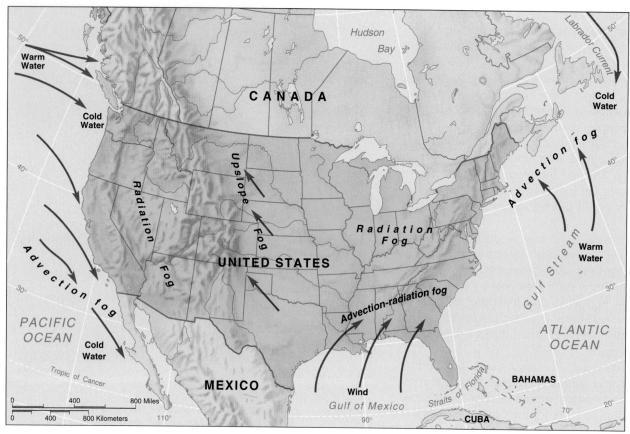

FIGURE 5–21
The different types of fog commonly found in North America.

over the Labrador current, it can be cooled to the dew point to form a persistent, dense fog that can last for weeks. Although most common in summer, it can occur any time during the year.

Another example of an advection fog is the fog that often covers London during the winter. Warm air passing over the warm Gulf Stream is advected over England, where it is chilled by the surface to form a thick fog.

The winds associated with advection fogs are often greater than those of radiation fogs. This promotes greater turbulence and allows the droplets to circulate to greater heights. Advection fogs can have thicknesses up to about a half kilometer (1500 ft).

UPSLOPE FOG

Of the three types of fog caused by the cooling of air, only **upslope fog** is formed by adiabatic cooling. When air flows along a gently sloping surface, it expands and cools as it moves upward. The western slope of the Great Plains of the United States provides an excellent setting for this type of condensation. Westward from the Mississippi River valley, the elevation gradually increases toward the foothills of the Rocky Mountains. Moist air from the Gulf of Mexico cools adiabatically as it ascends the slope of the plains to create widespread fog. Figure 5–21 presents a highly generalized description of the types of fog most prevalent over regions of the United States and Canada.

Distribution of Fog

Figure 5–22 depicts the number of heavy fog days (defined as limiting visibility to a quarter of a mile or less) across the 48 conterminous United States. The three significant centers of heavy fog are along the Pacific Northwest, New England, and the middle Appalachians. The Pacific Northwest and the coast of British Columbia experience numerous advection fogs, as westerly winds advect moist air over the cold California current. But it is not correct to attribute all of the fog formation to the cooling effect of the cold surface waters. As damp, cool air reaches the shore, fog formation is abetted by the *orographic* effect (the lifting effect caused as air crosses a mountain or similar barrier) as air approaches the steep Coast Ranges. Cape Disappointment, Washington, wins the prize for the foggiest U.S. location, being shrouded in heavy fog nearly one-third of the time. The zone of most persistent fog is confined primarily to the coastal region because the Coast Ranges block the flow of moisture inland.

New England and the Maritime Provinces experience a large number of heavy fog days. Along the coast, advection fogs dominate, with the coast of Maine having the highest incidence of dense cover. The advection fogs are most prevalent in summer. Inland, radiation fogs are very common at some of the higher elevation areas of New Hampshire and Vermont.

The upper reaches of the Appalachians, the third major focus of fog, undergo a large number of radiation fogs, particularly in late summer and fall. Not surprisingly, fog is at a minimum in the desert Southwest.

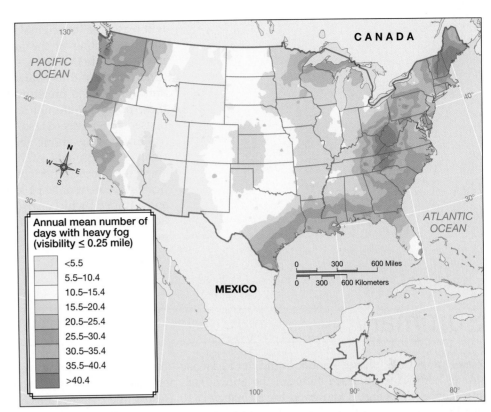

FIGURE 5–22
Average annual number of days with heavy fog in United States.

Formation and Dissipation of Cloud Droplets

Farther from the surface than fog, dew, or frost, clouds are usually the result of the adiabatic cooling associated with rising air parcels. In this section we take a closer look at such parcels as they rise, become saturated, and continue upward above the level at which condensation first occurs.

For reasons we will not discuss here, the dew point decreases as the air rises, at the rate of about 0.2 °C/100 m (1.1 °F/1000 ft). This decrease is called the **dew point lapse rate**. As unsaturated air is lifted, its temperature therefore approaches the dew point by 0.8 °C for every 100 m of ascent (i.e., 1.0 °C minus 0.2 °C). Thus, if the air temperature and dew point start out at 18 °C and 10 °C, respectively, an ascent of 1000 m is necessary to cause the air to be saturated.

Raising an air parcel above the lifting condensation level initially leads to the formation of small cloud droplets. But at about 50 m or so above the LCL, all the condensation nuclei in the air will have attracted water, and further uplift leads only to the growth of existing water droplets. In other words, no new droplets form as the air continues to rise; instead, the existing droplets grow larger.

The processes that lead to the formation of a cloud do not continue forever. At some point in time, lifting will cease and there will be no further condensation. When this happens, further cloud development comes to an end.

Now consider what happens if a rising parcel of air reverses its movement and begins to subside. The parcel now warms at the *saturated* adiabatic lapse rate, because the warming by compression is partially offset by the gradual evaporation of the droplets. The evaporation continues until the parcel has descended back to the original level at which condensation occurred. If the air then sinks below the original LCL, it warms at the DALR, because all the droplets will have evaporated. If brought down to the initial level at which uplift first began, the air will reassume its original temperature and dew point. The net effect of all this is that the cooling of the air and the condensation of liquid water are *reversible processes.* Note that we assume all the condensation products remain in the atmosphere and are thus available for evaporation during descent. Of course, the real atmosphere *loses* moisture by precipitation of rain and snow, which means that these processes are not strictly reversible. But a relatively small portion of condensed water falls out, so the concept remains generally valid.

In thinking about vertical motions and moisture, it's interesting that very small displacements can have such large consequences. After all, we hardly think about the effects of horizontal movements covering 100 km (62 mi). But in the vertical, movement of just 1 km (0.62 mi) can make the difference between a fine day and a ruined picnic, as rising air leads to clouds, and clouds give rise to precipitation. Like the formation of clouds, precipitation is no simple matter, as we will see in Chapter 7.

Summary

Water in its three phases—solid, liquid, and gaseous—constantly moves across the interface between the atmosphere and Earth's surface through what is known as the *hydrologic cycle.* Despite the fact that water accounts for only a small proportion of the mass of the air, in its three phases it is extremely important to the atmosphere—and to all life on Earth. To fully understand its role in the atmosphere, we need to comprehend the concept of saturation. At the surface of liquid water, molecules constantly move about. As they do so, some randomly break free of the surface to become water vapor. In a partially filled jar, this adds vapor to the volume above the water surface and thereby increases the water vapor content. As the water vapor concentration increases, so does the opposite process—condensation. As long as the rate of evaporation exceeds condensation, the amount of water vapor increases. The condensation rate in the jar eventually becomes equal to the evaporation rate, and saturation occurs.

The foregoing principle applies to the real atmosphere. In meteorology, we are particularly concerned with the exchange of water between a suspended cloud droplet and the air that surrounds it, because the equilibrium between evaporation and condensation is a prerequisite to the persistence of clouds and fog. When evaporation and condensation are in equilibrium, the air is saturated and small droplets can remain in the air without evaporating away.

There is no single best expression of humidity. One way of expressing the amount of water vapor is by the pressure it exerts (vapor pressure). The specific humidity and mixing ratio are similar ratios that relate the mass of water vapor to the air in which it is contained and to the mass of the other gases, respectively. Another measure, absolute humidity, is not widely used but is simply the density of water vapor. In contrast, relative humidity is a quantity we hear about almost daily, even though its partial dependence on temperature presents a serious drawback. Perhaps the most useful index is dew point, expressed as the temperature to which the air must be lowered for saturation to occur.

Although it might seem that condensation should occur at exactly 100 percent relative humidity, the situation in reality is slightly more complex. Condensation is controlled by two factors having opposite effects: curvature and the abundance of hygroscopic nuclei. Curvature retards condensation; without the presence of condensation nuclei, condensation would occur only at very high relative humidities. Natural and anthropogenic nuclei are sufficiently numerous that condensation usually occurs at relative humidities between 98% and 100.1%.

Humidity is an easy property to measure. It can be determined using paired thermometers (a psychrometer) that provide dry and wet bulb temperatures. The difference between the two temperatures is the wet bulb depression which, when combined with dry bulb temperature, permits the use of simple tables to determine relative humidity and dew point. Humidity affects a human's susceptibility to heat-related dangers. This susceptibility is reflected in the heat index.

For any kind of condensation (such as dew, fog, or clouds) to form, the dew point must equal the air temperature. This can result from raising the vapor content of the air to the saturation level; mixing warm, moist air with cooler, dry air; or by lowering the air temperature to the dew point. The latter process is most important for cloud formation. Lowering the air temperature does not require that heat be removed from the air. In fact, most clouds form by adiabatic cooling—the lowering of the air temperature without the removal of heat. Adiabatic cooling results from the expansion of air that occurs when it is lifted, and is a direct application of the first law of thermodynamics.

Non-cloud forms of condensation—dew, frost, frozen dew, and fog—are distinguished from clouds by their proximity to or direct contact with the planet's surface. Though one type of fog (upslope fog) results from adiabatic processes, the others result from diabatic cooling of the air, involving the loss of energy.

We have discussed how rising air leads to adiabatic cooling and to saturation and cloud formation, but we have yet to provide any explanation for these vertical motions. As it happens, several mechanisms are capable of generating uplift, some of which are quite intricate and closely tied to the cloud form that results. The topic deserves extended discussion, which we provide in the next chapter.

Key Terms

hydrologic cycle page 123	relative humidity page 128	wet/dry bulb thermometer page 138	dry adiabatic lapse rate page 144
evaporation page 125	dew point (temperature) page 130	wet bulb depression page 138	lifting condensation level page 144
condensation page 125	frost point page 132	aspirated psychrometer page 138	saturated (wet) adiabatic lapse rate page 144
saturation page 125	precipitation fog page 133	hair hygrometer page 138	environmental (ambient) lapse rate page 144
sublimation page 126	steam fog page 133	hygrothermograph page 138	dew page 146
deposition page 126	homogeneous nucleation page 136	heat index page 141	frost page 147
humidity page 126	hygroscopic page 136	apparent temperature page 141	frozen dew page 147
vapor pressure page 126	heterogeneous nucleation page 136	diabatic process page 143	radiation fog page 148
saturation vapor pressure page 126	condensation nucleus page 136	second law of thermodynamics page 143	advection fog page 149
absolute humidity page 127	haze page 137	adiabatic process page 143	upslope fog page 150
specific humidity page 127	supercooled water page 137	first law of thermodynamics page 143	dew point lapse rate page 151
saturation specific humidity page 128	ice nucleus page 137		
mixing ratio page 128	sling psychrometer page 138		
saturation mixing ratio page 128			

Review Questions

1. What is the hydrologic cycle?

2. Why is it incorrect to refer to the air as "holding" water vapor?

3. What are deposition and sublimation?

4. What is vapor pressure? In what units of measure is it expressed?

5. Explain the concepts of equilibrium and saturation.

6. What units of measurement are used to describe mixing ratio and specific humidity? Why are the two values nearly equal?

7. Why is absolute humidity seldom used?

8. Define relative humidity.

9. Why is relative humidity a poor indicator of the amount of water vapor in the air?

10. Define dew point. What characteristics make this measure superior to relative humidity?

11. Why can't the dew point temperature exceed the air temperature? What happens if the air temperature is lowered to a value below the initial dew point?

12. Describe the distribution of average dew point across the United States in summer and winter.

13. What are the three general methods by which the air can become saturated?

14. What are the effects of droplet curvature and solution on the amount of water vapor needed for saturation?

15. Why doesn't homogeneous nucleation form water droplets in the atmosphere?

16. What are condensation nuclei and ice nuclei? Are they typically made of the same materials? Which is more abundant in the atmosphere?

17. What is supercooled water?

18. What are psychrometers? How do they work?

19. Define dry bulb temperature, wet bulb temperature, and wet bulb depression.

20. What is the heat index?

21. What is the first law of thermodynamics and how does it apply to cloud development?

22. Explain the difference between diabatic and adiabatic processes.

23. What are the numerical values of the dry and saturated adiabatic lapse rates? Under what circumstances are they applicable?

24. What does the environmental lapse rate refer to?

25. Describe the various processes that can lead to the formation of dew.

26. What is the difference between frozen dew and frost?

27. Describe the various processes that can lead to the formation of fog.

Critical Thinking

1. When rubbing alcohol is applied to a person's skin, it feels colder than the application of water would. Why?

2. A person sleeps through the night without waking up, but awakes in the morning weighing slightly less than the night before. What happened?

3. A crowded classroom is filled with students. In what way might the presence of the students affect the dew point and relative humidity in the room?

4. A person parks her car in the driveway on a warm afternoon and notices a small puddle of water beneath the car a few minutes later. Explain how using the car's air conditioning can account for the puddle.

5. At Wheeling, West Virginia, the evening temperature is 55 °F and the dew point is 48 °F. How would you assess the likelihood of fog forming overnight?

6. A map of North America shows the average distribution of vapor pressure across the continent. Will the distribution on the map be *only* a function of the amount of water vapor in the air, or will the distribution be affected by another factor as well? Explain.

7. The temperature within a forest is −2 °C (28 °F) and there is frost on the trees but no fog. Outside the woods there is a fog. Why wasn't this fog in the woods?

8. Is fog more likely to occur downwind or upwind of an oil refinery? Why?

9. All fogs are made of water droplets or ice crystals. Despite the fact that they have the same composition, how would you know if a particular fog is a radiation, advection, or upslope fog?

10. Diesel engines, like four-stroke engines, work because of burning fuel, but they do not require a spark plug or similar device for initiating the burning. Apply your knowledge of the first law of thermodynamics to explain how the fuel can be forced to burn.

Problems and Exercises

1. Assume that a kilogram of air consists of 995 g of dry air and 5 g of water vapor. Show that the specific humidity and mixing ratio are very nearly equal.

2. Assume that a kitchen measures 4 meters by 5 meters by 3 meters. If the air density is 1 kg/m³ and the specific humidity is 10 g water vapor per kg of air, how much water vapor is in the room? If the doors and windows were sealed shut, would boiling 1 kg of water into the air make a substantial change in the humidity of the room?

3. The dry and wet bulb readings in Honolulu are 80 °F and 69 °F. At Charlottesville, Virginia, the readings are 50 °F and 45 °F.
   a. Use Tables 5–1 and 5–2 to determine the relative humidity and the dew point for both locations.
   b. Which of the two locations is more humid?

4. The dry and wet bulb temperatures are 70 °F and 54 °F. Use Tables 5–1 and 5–2 to answer the following questions:
   a. What are the dew point and the relative humidity?
   b. What will the dew point and relative humidity be if the air temperature increases to 80 °F? (*Hint:* Do not assume the same wet bulb depression.)
   c. What will the dew point and relative humidity be if the air temperature drops to 39 °F? (*Hint:* You don't need the tables for this one.)
   d. What will the dew point and relative humidity be if the air temperature drops further, to 35 °F?

5. The numerical value of the specific heat for air in the first law of thermodynamics, c_v, is strictly valid only for air with no water vapor. Water vapor has a specific heat approximately twice as great as that of c_v. Will a humid mass of air undergoing expansion therefore undergo more or less cooling than would a dry mass of air?

6. Assume a parcel of air starts out at the surface with a temperature of 12 °C and a dew point of 9.6 °C. Then it is lifted.
   a. What will the air temperature be at the 100 m level?
   b. What will the dew point be at the 100 m level? (*Hint:* Don't forget the dew point lapse rate.)
   c. At what height will condensation occur?
   d. What will the temperature be when condensation occurs?
   e. What will the dew point be when condensation occurs?
   f. What will the temperature be 500 m above the surface?
   g. If the parcel of air is lowered back to the surface (assuming none of the condensed moisture was removed as rain), what will the temperature and dew point be?

Quantitative Problems

As this chapter has shown, several indices are used to express the moisture content of the air, with each having some advantages and disadvantages. In meteorology we are particularly concerned with how these indices change in response to the lifting of air adiabatically. One excellent way to reinforce your understanding of these measures is by solving the simple problem set given on the Chapter 5 page of this book's Web site (**http://www.prenhall.com/aguado/**). These questions are particularly valuable as an aid to understanding how saturation is achieved in rising parcels.

Useful Web Sites

http://www.usatoday.com/weather/whumcalc.htm
Much useful information on humidity, with links to charts and calculators.

http://www.ems.psu.edu/wx/usstats/dewpstats.html
Map of current dew point distribution across the United States.

http://www.wunderground.com/US/Region/US/HeatIndex.html
Map of the current heat index across the United States.

Media Enrichment

Weather in Motion
January and July Water Vapor Movies

These two movies show primarily the flow of water *vapor*, based on observations at 3-hour intervals. As we will see in the rest of this chapter, there is a very big distinction between water vapor and water in its other two phases. Dark portions of the screen indicate dry air, whereas brighter areas have a higher water vapor content. Very bright areas indicate high cloud cover, often the tops of thunderstorm clouds.

In January we see a broad movement of moisture, especially in the middle latitudes. Areas of strong convection occur primarily in the tropics. Moreover, we see bands of moisture at times flowing into the southern United States as parts of air flows called *subtropical jets* (which we describe in Chapter 8). In July the zone of tropical convection shifts northward, in response to the shift in solar declination. Also, thunderstorms are more common in the middle latitudes of the Northern Hemisphere in this movie than they are in the January movie.

6 | CLOUD DEVELOPMENT AND FORMS

All of us pay special attention to clouds at one time or another. Some are particularly beautiful, while others portend severe weather. But certain people do a lot more than passively observe clouds. *Storm chasers* set out in teams during the severe weather season with the goal of observing the formation and movement of tornadoes. Some storm chasers are nonprofessional weather enthusiasts, while others are trained meteorologists with advanced degrees or students working toward their credentials. The professionals and students go out not just for the thrill of witnessing severe weather firsthand (though the excitement accounts for much of the pursuit) but also to acquire information that helps us understand these phenomena. Using a combination of tools ranging from advanced technology to sheer intuition, professional storm chasers pursue potentially severe storms to understand the mechanisms within and near clouds that lead to tornado development. These efforts have helped meteorologists understand what parts of active thunderstorm clouds are most likely to produce tornadoes. These efforts have also revealed that large areas of rotation within storm clouds often precede tornado formation by about 20 minutes to half an hour—thus providing advanced warning to forecasters and the public.

Storm chasing came of age in the spring of 1979, when the Severe Environmental Storms and Mesoscale Experiment (SESAME) began in the central United States. Project scientists assembled information from radar units, weather balloons, and field observations. By a remarkable coincidence, the spring of 1979 saw an unusually devastating tornado hit the region. During the afternoon of April 10, several teams of storm chasers outside of Seymore, Texas, observed a *wall cloud*, a large protrusion extending below the main base of a thunderstorm where tornadoes commonly develop. Howard Bluestein, now one of the leading experts on severe storms, photographed the wall cloud in Figure 6–1. Soon afterward, a tornado occurred that the storm chasers were able to observe from its inception to its demise, about 15 minutes later. But the main event was soon to follow. Off to the northeast, another tornado had just formed. Though the observation teams tried to keep pace, the storm outran them as it tracked toward Wichita Falls. When the chasers arrived in town, they witnessed the horrible destruction that had occurred minutes before: 3000 homes were destroyed and 42 people killed. The tornado was still swirling in full force and moving northeast. As it continued out of Wichita Falls, the chasers saw the back side of the storm, which Bluestein describes as looking like an atomic bomb explosion.

Most clouds are far less interesting than those that spawn tornadoes. Yet even clouds associated with the most violent weather result from the same processes that cause condensation and deposition in fair-weather clouds. From Chapter 5 we know that condensation or deposition can occur by (1) adding water vapor to the air; (2) mixing warm, moist air with cold air; or (3) lowering the air temperature to the dew point by adiabatic cooling of rising air. Although the first two processes can lead to cloud formation in many situations, lowering of air temperature to the dew point is the most common mechanism of cloud formation (especially for clouds that cause precipitation). This chapter discusses the processes and conditions associated with the formation of clouds due to upward motions. It also describes the cloud types that form as a result of those processes.

FIGURE 6–1
Wall cloud associated with severe thunderstorm near Seymore, Texas.

Mechanisms That Lift Air

Four mechanisms lift air so that condensation and cloud formation can occur:

1. Orographic lifting, forcing air above a mountain barrier
2. Frontal lifting, displacement of one air mass over another
3. Convergence, horizontal movement of air into an area at low levels
4. Localized convective lifting due to buoyancy

◀ Storm chasers risk life and limb to track violent storms.

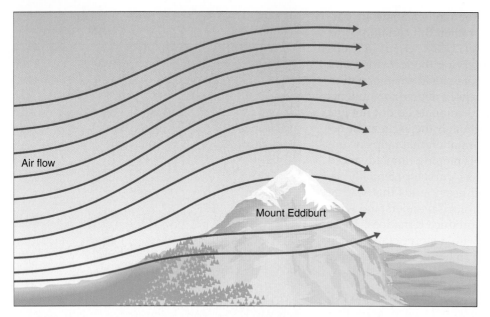

FIGURE 6–2

Orographic uplift. When air approaches a topographic barrier, it can be lifted upward or deflected around the barrier.

OROGRAPHIC UPLIFT

As shown in Figure 6–2, air flowing toward a hill or mountain will be deflected around and over the barrier. The upward displacement of air that leads to adiabatic cooling and possible cloud formation is called **orographic uplift** (or the **orographic effect**). The height to which these *orographic clouds* can rise is not limited to the height of the hill or mountain; their tops can extend many hundreds of meters higher and even into the lower stratosphere. The height of cloud tops is strongly related to characteristics of the air that vary from day to day, an issue we will describe in more detail later in this chapter. Figure 6–3 shows the development of orographic clouds. Notice that the thickness of the cloud greatly exceeds the height of the orographic barrier that helped create it.

Downwind of a mountain ridge, on its leeward side, air descends the slope and warms by compression to create a **rain shadow** effect, an area of lower precipitation. The Sierra Nevada mountain range (Figure 6–4) provides a dramatic illustration of this effect. The ridge crest of the Sierra runs north-to-south and is essentially perpendicular to the predominantly west-to-east air flow. With much of the range being higher than 3500 m (11,500 ft), precipitation on the western, windward side is greatly enhanced because of orographic lifting; in places, the mean annual precipitation exceeds 250 cm (100 in.). The eastern slope of the range is extremely steep and the valley floor is low, sometimes below sea level. Thus, the descending air on

(a)

(b)

FIGURE 6–3

The development of cumulus clouds over the mountains east of San Diego, California. Notice the increased vertical development between (a) and (b). In (c) the cloud has decreased in depth due to precipitation. Photos were taken at about 10-minute intervals.

(c)

(a)

(b)

FIGURE 6-4
The Sierra Nevada forms a major barrier to winds that generally blow from the west. This promotes enhanced precipitation on the windward side (a), and a rain shadow on the leeward side (b).

the leeward slope creates one of the strongest rain shadow effects on Earth. Death Valley, one of the driest places in North America, is just east of the range whose windward slopes accumulate the majority of California's usable water. A comparable rain shadow effect exists in South America, where the Andes Mountains form an abrupt barrier to the westerly winds.

FRONTAL LIFTING

Although temperature normally changes from place to place, experience tells us that such changes are usually quite gradual. In other words, if the temperature is 10 °C (50 °F) in Toronto, Ontario, chances are that the temperature in Buffalo, New York, about 100 km (60 mi) away, won't be very much different. Sometimes, however, transition zones exist in which great temperature differences occur across relatively short distances. These transition zones, called *fronts*, are not like vertical walls separating warm and cold air but rather slope gently, as we will discuss in Chapter 9.

Air flow along frontal boundaries results in the widespread development of clouds in either of two ways. When cold air advances toward warmer air (a situation called a **cold front**), the denser cold air displaces the lighter warm air ahead of it, as shown in Figure 6–5a. When warm air flows toward a wedge of cold air (a **warm front**), the warm air is forced upward in much the same way that the orographic effect causes air to rise above a mountain barrier (Figure 6–5b).

(a)

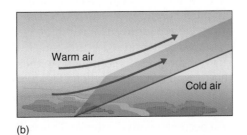

(b)

FIGURE 6-5
Frontal boundaries. Cold fronts (a) cause uplift as cold air advances on warmer, less dense air. Uplift occurs along a warm front (b) when warm air overruns the cold wedge of air ahead of it.

CONVERGENCE

Because the mass of the atmosphere is not uniformly distributed across Earth's surface, large areas of high and low surface pressure exist. These pressure differences set the air in motion in the familiar effect we call the wind. Not surprisingly, the pattern of wind that results is very much related to the pattern of pressure.

In particular, when a low-pressure cell is near the surface, winds in the lower atmosphere tend to converge on the center of the low from all directions. Horizontal movement toward a common location implies an accumulation of mass called **horizontal convergence**, or just **convergence** for short. Does convergence lead to increasing density near the center of the low, with imported air confined to its original altitude? No—instead, vertical motions near the center of the low carry away about as much mass as is carried in. Air will rise. This will be explained in more detail later, but for now we can just make the connection between low-level convergence and rising air with adiabatic cooling.

LOCALIZED CONVECTION

We saw in Chapter 3 that free convection is lifting that results from heating the air near the surface. It is often accompanied by updrafts strong enough to form clouds and precipitation. During the warm season, heating of Earth's surface can produce free convection over a fairly limited area and create the brief afternoon thunderstorms that disrupt summer picnics. In Canada and the United States east of the Rocky Mountains, high moisture content of the air sometimes allows for tall clouds with bases at relatively low altitudes. Such conditions favor vigorous precipitation over small regions (free convection by its nature does not create updrafts more than several tens of meters in diameter). Even in the deserts of the Southwest, which are usually low in water vapor, intense heating can lead to localized convection intense enough to cause thundershowers.

Free convection arises from buoyancy, the tendency for a lighter fluid to float upward through a denser one. By itself, buoyancy can initiate uplift. But buoyancy can also speed or slow the uplift begun by the orographic effect, frontal lifting, and convergence. As we'll see next, meteorology uses the concept of static stability to summarize the effect of buoyancy on uplift.

Static Stability and the Environmental Lapse Rate

Sometimes the atmosphere is easily displaced and an air parcel given an initial boost upward continues to rise, even after the original lifting process ceases. At other times, the atmosphere resists such lifting. The air's susceptibility to uplift is called its **static stability**. *Statically unstable* air becomes buoyant when lifted and continues to rise if given an initial upward push; *statically stable* air resists upward displacement and sinks back to its original level when the lifting mechanism ceases. *Statically neutral air* neither rises on its own following an initial lift nor sinks back to its original level; it simply comes to rest at the height to which it was displaced.

Static stability is closely related to buoyancy. When a parcel of air is less dense than the air around it, it has a positive buoyancy and floats upward. (In fact, buoyant parcels of air increase in speed as they move upward—even to the point of creating violent updrafts.) Air that is denser than its surroundings sinks if not subjected to continued lifting forces. Density differences between a parcel and its surroundings are determined by their temperatures. If the parcel is warmer than the surrounding air, it will be less dense and experience a lifting force. If it is colder, it will be more dense and have "negative" buoyancy.

If a rising parcel cools at a rate that makes it colder than the surrounding air, it will become relatively dense. This will tend to suppress uplift. If the lifted air cools

TUTORIAL
Atmospheric Stability

TABLE 6–1	Ten Principal Cloud Types	
HIGH CLOUDS (HEIGHTS GREATER THAN 6000 m, OR 19,000 ft)		
Cirrus (Ci)	(Figure 6–16)	Thin, white, whispy clouds resembling mares' tails.
Cirrostratus (Cs)	(Figure 6–19)	Extensive, shallow clouds somewhat transparent to sunlight, producing a halo around the Sun or Moon.
Cirrocumulus (Cc)	(Figure 6–20)	High, layered cloud with billows or parallel rolls.
MIDDLE CLOUDS (HEIGHTS BETWEEN 2000 m AND 6000 m, OR 6000 TO 19,000 ft)		
Altostratus (As)	(Figure 6–21)	Extensive, watery, layered cloud. Allows some penetration of sunlight but Moon or Sun appears as bright spot within cloud.
Altocumulus (Ac)	(Figure 6–22)	Shallow, mid-level cloud containing patches or rolls. Generally more opaque and having less distinct margins than cirrocumulus.
LOW CLOUDS (BELOW 2000 m, OR 6000 ft)		
Stratus (St)	(Figure 6–23)	Uniform layer of low cloud ranging from whitish to gray.
Nimbostratus (Ns)	(Figure 6–24)	Low cloud prouducing light rain. Produces darker skies than altostratus.
Stratocumulus (Sc)	(Figure 6–25)	Low-level equivalent to atltocumulus.
CLOUDS WITH VERTICAL DEVELOPMENT (MAY EXTEND THROUGH MUCH OF ATMOSPHERE)		
Cumulus (Cu)	(Figures 6–26 and 6–28)	Detached billowy clouds with flat bases and moderate vertical development. Sharply defined boundaries.
Cumulonimbus (Cb)	(Figure 6–29)	Clouds with intense vertical development with characteristic anvil. May be tens of thousands of meters thick. Appear very dark when viewed from below.

HIGH CLOUDS

High clouds are generally above 6000 m (19,000 ft). They are almost universally composed of ice crystals instead of liquid droplets. Recall that within the troposphere the average temperature decreases from 15 °C (59 °F) at sea level at a rate of 6.5 °C/1000 m (3.6 °F/1000 ft). As a result, the average temperature for high clouds is usually no higher than –35 °C (–31 °F); cooling to the frost point causes the formation of ice crystals instead of supercooled droplets.

FIGURE 6–15
Generalized cloud chart.

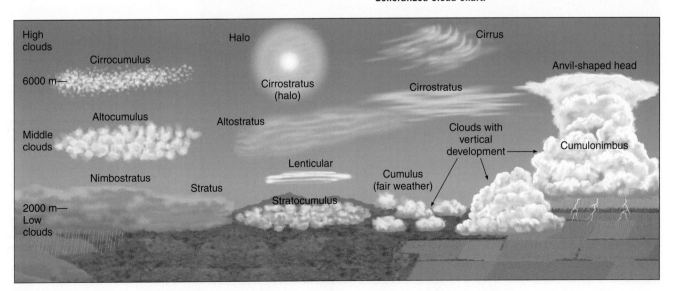

NAME	TYPES OF CLOUD	DESCRIPTION
TABLE 6–2	Cloud Species	
Calvus	Cumulonimbus	Upper portion of cumulonimbus loses distinct outline as ice crystals form in lieu of water droplets.
Capillatus	Cumulonimbus	Fibrous upper portion consists of ice crystals. Further developed than cumulonimbus calvus.
Castellanus	Cirrus, cirrocumulus, altostratus, and stratocumulus	Towerlike vertical development on portion of cloud.
Congestus	Cumulus	Undertaking rapid and significant vertical development.
Fractus	Stratus and cumulus	Broken or ragged.
Humilis	Cumulus	Having slight vertical extent.
Incus	Cumulonimbus	Having anvil-shaped top.
Lentiucularis	Stratocumulus, altocumulus, cirrocumulus	Lens shaped.
Mammatus	Cumulonimbus	Pouchlike.
Pileus	Cumulus, cumulonimbus, stratocumulus	Caplike cloud immediately above larger cloud.
Translucidus	Altocumulus, altostratus, stratocumulus, and stratus	Somewhat transparent so that the Sun, Moon, or higher clouds might be discernible.

Where surface temperatures are very low, clouds composed exclusively of ice can occur at altitudes as low as 3000 m (10,000 ft). Thus, the definition of a high cloud (above 6000 m), is somewhat temperature dependent.

The simplest of the high clouds are cirrus (abbreviated Ci), which are wispy aggregations of ice crystals (Figure 6–16). The average thicknesses of cirrus clouds is about 1.5 km (1 mi), but they can be as thick as 8 km (5 mi). Given the very low temperatures at which they exist, they have little water vapor from which to form ice. So although they may be easily visible, the water content of these clouds is extremely low. In fact, the ice content of cirrus clouds is typically only about 0.025 grams per cubic meter, or about one-thousandth of an ounce per cubic yard.

Although the entire mass of ice contained in a cirrus cloud is small, the individual crystals can be as long as 8 mm (0.3 in.). These crystals fall at a speed of about 0.5 meters per second (about 1 mile per hour), which is sufficient for them to overcome updrafts and descend as *fall streaks* (Figure 6–17a).

Fall streaks gradually sublimate (that is, they change to water vapor without first melting) as they fall through the warmer, dry air below and dissipate high above the surface. As shown in Figure 6–17b, the higher, faster-moving crystals (as indicated by the length of the arrows on the figure) are blown horizontally faster than those below, which gives the fall streaks their characteristic curved appearance. Horizontal swirls are also evident in cirrus clouds because of curving winds, leading to mares' tails.

FIGURE 6–16

Cirrus clouds are wispy clouds of ice crystals.

(a)

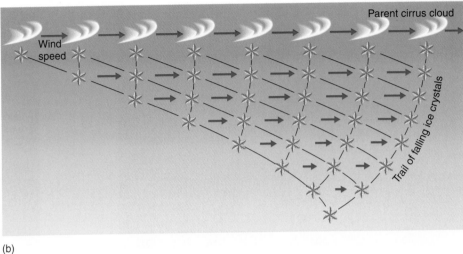

(b)

FIGURE 6–17
Fall streaks (a) result from falling ice crystals. The ice crystals fall through air (b) with gradually decreasing wind speeds and become comma shaped.

Contrails (Figure 6–18), are a type of ice cloud frequently caused by jet aircraft. The very hot engine exhaust contains considerable water vapor as a result of fuel combustion, and turbulence in the wake of the aircraft rapidly mixes the exhaust with the cold, ambient air. As was explained in Chapter 5, the mixing of warm moist air with cold air can lead to saturation and, in this case, the rapid formation of ice crystals.

Cirrostratus (Cs) clouds (Figure 6–19), like cirrus, are composed entirely of ice but tend to be more extensive horizontally and have a lower concentration of crystals than cirrus. Though cirrostratus clouds reduce the amount of solar radiation reaching the surface, enough direct sunlight penetrates to allow objects at the surface to cast shadows. Furthermore, they do not fully obscure the Moon or Sun behind them. Instead, when viewed through a layer of cirrostratus, the Moon or Sun has a whitish, milky appearance but a clear outline.

FIGURE 6–18
Aircraft contrails.

FIGURE 6–19
Cirrostratus clouds. Such clouds often create a halo around the Sun or Moon.

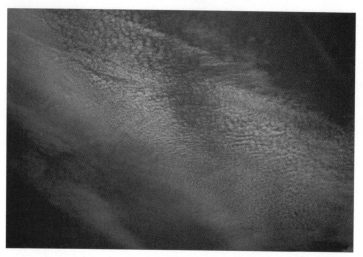

FIGURE 6–20
Cirrocumulus clouds. These frequently occur in rows of individual, puffy clouds.

FIGURE 6–21
Altostratus clouds. These middle-level, layered clouds are composed of water droplets.

DID YOU KNOW?

For three days following the tragedy of September 11, 2001, all commercial aircraft were grounded in the United States. Formation of contrails stopped immediately; according to one study, this increased the daily temperature range across the country during that period. The evidence suggests that the temporary disappearance of contrails increased the amount of sunlight reaching the surface during the daytime (thus raising daytime surface temperatures), while increasing longwave radiation losses at night (thereby lowering nighttime temperatures). Researchers believe that the daily temperature ranges across the country averaged about 1.8 °C (3.2 °F) more than they normally would over the period. It is likely that over the Midwest, where contrails are most likely to form, the increase in the daily range may have been even greater.

FIGURE 6–22
Altocumulus clouds. Layered, mid-level altocumulus clouds are often arranged in bands.

A characteristic feature of cirrostratus clouds is the *halo*, a circular arc around the Sun or Moon formed by the refraction (bending) of light as it passes through the ice crystals. Ice crystals bend much of the passing sunlight or moonlight 22° away from its initial direction. So if you face toward the Sun or Moon, refracted sunlight will be directed toward you from a ring that surrounds the Sun at a 22° angle.

Cirrocumulus (Cc) are often among the most beautiful of clouds. They are composed of ice crystals that arrange themselves into long rows of individual, puffy clouds (Figure 6–20). Cirrocumulus form during episodes of *wind shear*, a condition in which the wind speed and/or direction changes with height. Wind shear often occurs ahead of advancing storm systems, so cirrocumulus clouds are often a precursor to precipitation. Because of their resemblance to fish scales, cirrocumulus clouds are associated with the term "mackerel sky."

MIDDLE CLOUDS

Middle clouds occur between 2000 and 6000 m (6500 and 19,000 ft) above the surface and are usually composed of liquid droplets. The two major categories in this group are both prefixed by *alto*, which means "middle."

Altostratus (As) clouds (Figure 6–21) are the middle-level counterparts to cirrostratus. They differ from cirrostratus in that they are more extensive and composed primarily of liquid water. Altostratus scatter a large proportion of incoming sunlight back to space, thereby reducing the amount of sunlight that reaches the surface. The insolation that does make its way to the surface consists primarily or exclusively of diffuse radiation, so one way to distinguish altostratus from cirrostratus is by the absence of shadows. Furthermore, when viewing the Sun or Moon behind altostratus, one sees a bright spot behind the clouds instead of a halo.

Altocumulus (Ac) (Figure 6–22) are layered clouds that form long bands or contain a series of puffy clouds arranged in rows. They are often gray in color, although one part of the cloud may be darker than the rest. Consisting mainly of liquid droplets rather than ice crystals, they usually lack the beauty of cirrocumulus.

LOW CLOUDS

Low clouds have bases below 2000 m. Stratus (St) (Figure 6–23) are layered clouds that form when extensive areas of stable air are lifted. They are normally between 0.5 and 1 km (0.3 to 0.5 mi) thick, in

A New Orleans neighborhood destroyed by Hurricane Katrina.

CHAPTER

EIGHT   **Atmospheric Circulation and Pressure Distributions**

NINE   **Air Masses and Fronts**

CHAPTER OUTLINE

FIGURE 8–1
Crops wither from prolonged drought.

During the summer of 1999, vast portions of the eastern United States and Canada experienced several weeks of searing heat, with temperatures persistently well above 38 °C (100 °F). These high temperatures, combined with minimal rainfall, exacerbated a water shortage that had begun during the dry summer of 1998 (Figure 8–1). By mid-1999, much of the region, especially in the mid-Atlantic states, was in the midst of its worst drought in decades. River flows and reservoir levels were critically low, forcing several eastern governors to declare drought emergencies outlawing such nonessential water use as watering lawns and washing cars.

Agricultural losses were estimated to be $800 million. The heat and drought severely reduced crop harvests for many farmers. Others endured a different problem: unusually early ripening of produce that caused an early glut and reduced prices. These low prices kept farmers such as Dale Benson of Delaware from selling his produce at local auctions.

The drought also brought about some unexpected animal behavior. Black bears, for example, wandered into suburban areas outside major eastern cities in search of food. One bear was spotted a mere 15 km (10 mi) from downtown Baltimore, and another broke into a family's kitchen near Stillwater, New Jersey.

Events such as these typically result from pressure patterns that, once established, persist for unusually long periods of time. Relief comes only when the pressure pattern evolves to permit wetter conditions. In Chapter 4 we saw that atmospheric pressure varies from one place to another, but its distribution is not haphazard. Instead, well-defined patterns dominate the distribution of pressure and winds across the global surface. The largest-scale patterns, called the **general circulation,** can be considered the background against which unusual events occur, such as the drought described above. Likewise, even mundane daily wind and pressure variations can be thought of as departures from the general circulation.

Our first goal in this chapter is to describe dominant planetary wind motions and look at the processes that generate them. In particular, we examine the interrelationships between the winds of the upper and lower atmosphere and the connections that occur at the boundary between the surface of the oceans and the lower

◀ A blizzard complicates life on Wall Street, New York City.

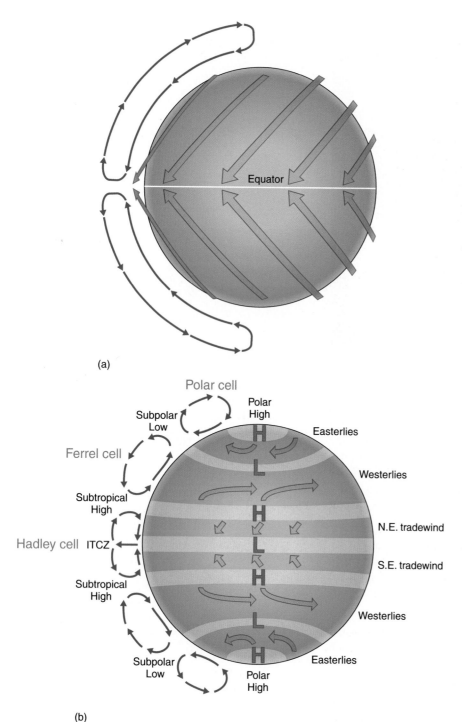

(a)

(b)

FIGURE 8–2

The single-cell (a) and three-cell (b) models of atmospheric pressure and wind. In the single-cell model, air expands upward, diverges toward the poles, descends, and flows back toward the equator near the surface. In the three-cell model, thermally driven Hadley circulation is confined only to the lower latitudes. Two other cells (more theoretical than real) exist in each hemisphere, the Ferrel and polar cells.

atmosphere. We then consider wind and pressure patterns at sequentially smaller spatial and temporal scales. The chapter concludes with a discussion of air–sea interactions.

Single-Cell Model

Scientists have sought to describe general circulation patterns for centuries. As early as 1735 a British physicist, George Hadley (1685–1768), proposed a simple circulation pattern called the **single-cell model** to describe the general movement of the atmosphere. One of his primary goals was to explain why sailors so often found winds blowing east to west in the lower latitudes. (Winds blowing east-to-west or west-to-east are referred to as **zonal winds**; those moving north-to-south or south-to-north are called **meridional**[1]). Hadley's idealized scheme, shown in Figure 8–2a, assumed a planet covered by a single ocean and warmed by a fixed Sun that remained overhead at the equator. Hadley suggested that the strong heating at the equator caused a circulation pattern in which air expanded vertically into the upper atmosphere, diverged toward both poles, sank back to the surface, and returned to the equator. Hadley did not think winds would simply move north and south, however. He believed instead that the rotation of Earth would deflect air to the right in the Northern Hemisphere and to the left in the Southern Hemisphere, leading to the east–west surface winds shown in the figure.[2]

Hadley's main contributions were to show that differences in heating give rise to persistent large-scale motions (called *thermally direct circulations*) and that zonal winds can result from deflection of meridional winds. His idea of a single huge cell in each hemisphere was not so helpful, however.

A somewhat more elaborate model does a better—though still simplified—job of describing the general circulation. This **three-cell model** (Figure 8–2b) was proposed by U.S. meteorologist William Ferrel (1817–1891) in 1865.

The Three-Cell Model

The three-cell model divides the circulation of each hemisphere into three distinct cells: the heat-driven **Hadley cell** that circulates air between the tropics and subtropics, a **Ferrel cell** in the middle latitudes, and a **polar cell**. Each cell consists of one belt of rising air with low surface air pressure, a zone of sinking air with surface high pressure, a surface wind zone with air flowing generally from the high-pressure belt to the low-pressure belt, and an air flow in the upper atmosphere from the belt of rising air to the belt of sinking air. Though more realistic than the single-cell

---

[1]The wind is seldom purely zonal or purely meridional, but instead usually moves in some intermediate direction. In that case we will think of the wind as having both a zonal and a meridional component. It is possible for the two components to be equal (as in a southwesterly wind), but in general one component will be larger than the other.

[2]Hadley lived from 1685 to 1768, before Gaspard de Coriolis (1792–1843) quantitatively described the acceleration due to Earth's rotation. Nonetheless, Hadley had a qualitative knowledge of the Coriolis force and incorporated it into his model.

model, the three-cell model is so general that only fragments of it actually appear in the real world. Nonetheless, the names for many of its wind and pressure belts have become well established in our modern terminology, and it is important that we understand where these hypothesized belts are located.

THE HADLEY CELL

Along the equator, strong solar heating causes air to expand upward and diverge toward the poles. This creates a zone of low pressure at the equator called the **equatorial low**, or the **Intertropical Convergence Zone (ITCZ).** The upward motions that dominate the region favor the formation of heavy rain showers, particularly in the afternoon. Heavy precipitation associated with the ITCZ is observable on weather maps and satellite images (Figure 8–3). Notice in the figure that the equatorial low exists not as a band of uniform cloud cover, but rather as a zone containing many clusters of convectional storms. The ITCZ is the rainiest latitude zone in the entire world, with many locations accumulating more than 200 days of rain each year. Imagine how listless you might feel in such an environment, with hot, humid afternoons giving way to heavy rain showers all year long. It is for this reason that the ITCZ is sometimes called the *doldrums*.

Within the Hadley cell, air in the upper troposphere moves poleward to the subtropics, to about 20° to 30° latitude. As it travels, it acquires increasing west-to-east motion, primarily because of the conservation of angular momentum (see *Box 8–1, Physical Principles: Problems with the Single-Cell Model*). This westerly component is so strong that the air circles Earth a couple of times before reaching its ultimate destination in the subtropics. In other words, the upper-level air follows a great spiraling path out of the tropics, with zonal motion much stronger than the meridional component. Among other things, this explains why material ejected by tropical volcanic eruptions spreads quickly over a wide range of longitudes.

Upon reaching about 20° to 30° latitude, air in the Hadley cell sinks toward the surface to form the **subtropical highs**, large bands of high surface pressure. Because descending air warms adiabatically, cloud formation is greatly suppressed and desert conditions are common in the subtropics. The subtropical highs generally have weak pressure gradients and light winds. Such conditions exert minimal impact on long-distance travel today. But in preindustrial days when oceangoing vessels depended on the wind, its prolonged absence could be catastrophic. Ships crossing the Atlantic from Europe to the New World risked getting stranded in mid-ocean while crossing the subtropics. Often among the cargo were cattle and horses to be brought to the New World, and legend has it that the crews of stalled ships threw their horses overboard. The jettisoned cargo has lent its name to what we colloquially call the **horse latitudes**.

In the Northern Hemisphere, as the pressure gradient force directs surface air from the subtropical highs to the ITCZ, the weak Coriolis force deflects the air slightly to the right to form the **northeast trade winds** (or simply the *northeast trades*). In the Southern Hemisphere, the northward-moving air from the subtropical high is deflected to the left to create the **southeast trade winds**. Notice that the trade winds are fairly shallow. Moving upward through the troposphere, the easterly motions weaken and are eventually replaced by westerly motion aloft. Together, the subtropical highs, the equatorial low, the trade winds, and upper-level westerly motions form the Hadley cells. Because it is produced thermally, Hadley circulation is strongest in the winter season, when temperature gradients are strongest.

FIGURE 8–3

The Intertropical Convergence Zone (ITCZ) is observable as the band of convective clouds and showers extending from northern South America into the Pacific on this satellite image.

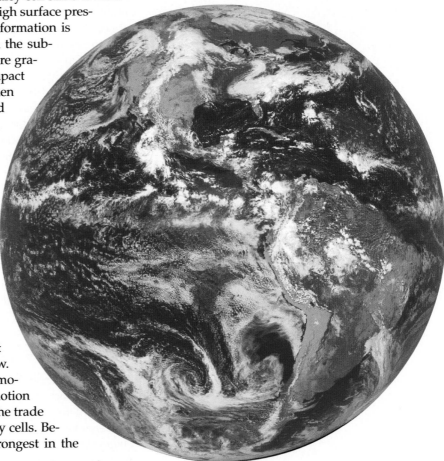

THE FERREL AND POLAR CELLS

According to the three-cell model, the Hadley cell accounts for the movement and distribution of air over about half of Earth's surface. Immediately flanking the Hadley cell in each hemisphere is the Ferrel cell, which circulates air between the subtropical highs and the **subpolar lows**, or areas of low pressure. On the equatorial side of the Ferrel cell, air flowing poleward away from the Northern Hemisphere subtropical high undergoes a substantial deflection to the right, creating a wind belt called the **westerlies**. In the Southern Hemisphere, the pressure gradient force propels the air southward, but the Coriolis force deflects it to the left—thus producing a zone of westerlies in that hemisphere as well. Unlike the Hadley cell, the Ferrel cell is envisioned as an indirect cell, meaning that it does not arise from differences in heating, but is instead caused by turning of the two adjacent cells. Imagine three logs placed side-by-side, touching one another. If the two outer logs are turned in the same direction, friction will set the middle one in motion. Thus, the Ferrel cell shows the same kind of overturning as the other cells, but for different reasons.

In the polar cells of the three-cell model, surface air moves from the **polar highs** to the subpolar lows. Like the Hadley cells, the polar cells are considered thermally direct circulations. Compared to the poles, air at subpolar locations is slightly warmer, resulting in low surface pressure and rising air. Very cold conditions at the poles create high surface pressure and low-level motion toward the equator. In both hemispheres, the Coriolis force turns the air to form a zone of **polar easterlies** in the lower atmosphere.

THE THREE-CELL MODEL VS. REALITY: THE BOTTOM LINE

Do the wind and pressure belts of the three-cell model adequately describe real-world patterns? The answer is: sort of. We have already seen that the ITCZ is real enough to be observed from space and that many deserts exist in their predicted locations. Furthermore, the trade winds are the most persistent on Earth. We would have to say that the Hadley circulation provides a good account of low-latitude motions. On the other hand, the Ferrel and polar cells are not quite as well represented in reality, though they do have some manifestation in the actual climate.

With regard to surface winds, much of the middle latitudes experience the strong westerly winds depicted by the model, especially in the Southern Hemisphere. Of course, local conditions often override this tendency (in fact, much of the central United States is dominated by a southerly flow during the summer). It is even more difficult to observe a persistent pattern of polar easterlies in the overall wind regime. They emerge in long-term averages, but are not a prevailing wind belt as the trades are.

With regard to upper-level motions, the three-cell model is not realistic at all. For example, where the Ferrel cell implies easterly motion in the upper troposphere, there is overwhelming westerly wind. Moreover, large overturning cells do not exist outside of the Hadley zones. Thus, the three-cell model mainly provides a starting point for a more detailed account. But perhaps its failures aren't surprising, given that it doesn't consider land–ocean contrasts or the influence of surface topography, two factors that surely ought to influence planetary winds and pressure.

Semipermanent Pressure Cells

The three-cell model provides a good beginning for describing the general distribution of wind and pressure, but the real world is not covered by a series of belts that completely encircle the globe. Instead, we find a number of alternating **semipermanent cells** of high and low pressure, as shown in Figure 8–4. They are called *semipermanent* because they undergo seasonal changes in position and intensity over the course of the year. Some of these cells result from temperature differences, and others from dynamical processes (meaning that they are related to the motions of the atmosphere). Among the most prominent features in the Northern Hemisphere during winter (a) are the **Aleutian** and **Icelandic lows**—over the Pacific and Atlantic Oceans, respectively—and the **Siberian high** over central Asia. In summer (b), the best-developed cells are the **Hawaiian** and **Bermuda-Azores highs** of the Pacific and Atlantic Oceans and the **Tibetan low** of southern Asia.

FIGURE 8-4
Sea level pressure for January (a) and July (b).

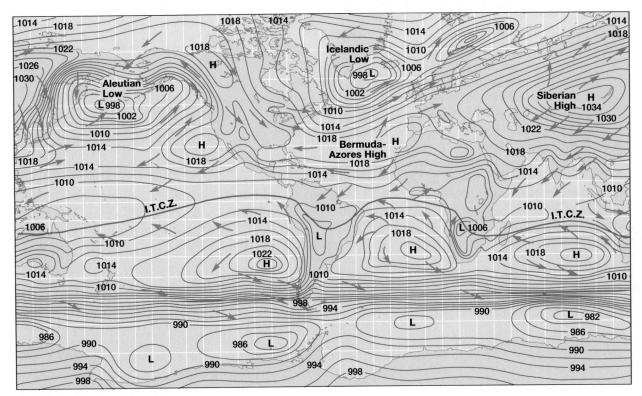

(a) January

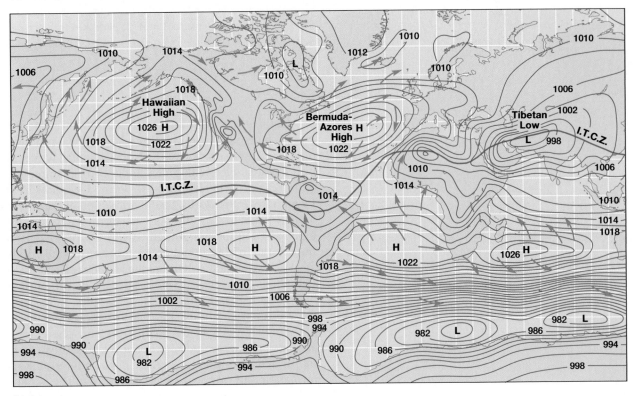

(b) July

Problems with the Single-Cell Model

8-1 PHYSICAL PRINCIPLES

Hadley's single-cell circulation model is a simple one based on the differential heating of Earth's surface. Contrary to Hadley's description, this thermally driven circulation occupies only the range of latitudes between about 30° N and 30° S. Why doesn't the single-cell model cover the entirety of each hemisphere instead of just a zone near the tropics? The answer boils down to the *conservation of angular momentum.*

Just as an object moving in a straight line has "linear" momentum, an object following a curved path has "angular" momentum. If we start a weight twirling at the end of a rope, as in Figure 1, we see the rope sweeping out a growing pie-shaped angle, hence the term *angular momentum*. The amount of angular momentum obviously increases with increasing mass (imagine trying to stop a very heavy weight). Angular momentum also increases with speed, because a larger area is swept out each second. Lastly, angular momentum increases with increasing radius, again because a larger area is swept out per unit time.

Putting this together, we can say that angular momentum is the product of mass, speed, and radius, or

$$A = mvr$$

where A is angular momentum, m is mass, v is speed, and r is the radius of rotation.

If angular momentum in conserved, then A is constant. This simple statement has some serious consequences. Suppose, for example, that we decrease the radius by pulling on the rope. As r decreases, the speed must increase, even though we do not twirl harder. The same phenomenon allows an Olympic diver to twist or flip in the air. After leaving the diving board, the diver's angular momentum is fixed. When she tucks, bringing legs and arms close to her center of mass, she spins rapidly. At the end of the dive, she straightens out, slowing her spin rate to enter the water vertically without much rotation, and is rewarded with a high score (hopefully). It's remarkable to think that divers can't adjust their angular momentum in the air, but rather must leave the board with exactly what's needed for a particular dive.

Now consider the rotating Earth. At the equator, any fixed point on the surface travels a distance equal to the planetary circumference of 40,000 km (24,000 mi) over a 24-hour period. The same applies to a parcel of air that is stationary relative to the surface. At higher latitudes the circumference is smaller, so a parcel at rest travels a shorter distance in a day's time. For example, at 40° the speed is about 31,000 km per day.

Now let's see what happens as a fixed mass of air initially at rest relative to the surface moves northward from the equator. Traveling poleward, r decreases, so v must increase if angular momentum is conserved. At 40° latitude, v increases about 30 percent to 52,000 km/day. In other words, the parcel is moving eastward at 52,000 km/day as it travels around the axis of rotation. At that latitude, the surface is moving at only 31,000 km/day, so the parcel is moving over the surface. Standing on the ground, we would observe a wind speed of 21,000 km/day (52,000 minus 31,000). This is a huge value, equivalent to a wind speed of 243 m/sec, or 547 mph!

According to the single-cell model, upper-level air should flow all the way from the equatorial region to the poles. But if that were the case, the conservation of angular momentum would send the poleward-moving air eastward at unimaginably high speeds—far greater than what we experience in even the most severe tornadoes. Physical considerations prevent the air from maintaining such extreme flows, because even a slight random disruption would cause the wind to break down into numerous eddies with strong north–south motions.

The single-cell model involves another difficulty, also related to conservation of angular momentum. To a large degree, the momentum possessed by the entire Earth–atmosphere system is unchanging—nothing stands outside the planet giving it a shove to increase its spin rate or to slow it down. (We are ignoring gravitational

The size, strength, and locations of the semipermanent cells undergo considerable change from summer to winter. During the winter, a strong Icelandic low occupies a large portion of the North Atlantic, while the Bermuda-Azores high appears as a small, weak anticyclone. During summer, the Icelandic low weakens and diminishes in size, and the Bermuda-Azores high strengthens and expands. Even more dramatically, the Siberian high of the winter in interior Asia gives way to the Tibetan low of summer. As we will see later in this chapter, the seasonal shift in the distribution of semipermanent cells plays a major role in one of Earth's most important circulation patterns—the monsoon of southern and southeastern Asia.

We mentioned earlier that the Hadley cell is fairly easy to see in the real world, with the ITCZ appearing over much of the equatorial regions and the subtropical highs existing over much of the subtropics. But as you can also see in Figure 8–4, the subtropical highs exist primarily over the oceans (as the Hawaiian and Bermuda-Azores highs over the Northern Hemisphere) and not over land. Despite the absence of pronounced high sea level pressure over the subtropical land masses, the air in the middle troposphere does undergo sinking motions that inhibit cloud formation and promote desert conditions. The Sahara Desert of northern Africa, variations caused by roving planets and other small factors.) But angular momentum is transferred between Earth and atmosphere by friction whenever the air moves past the surface. Where there is westerly (from the west) wind, the atmosphere transfers momentum to the surface as it "pushes" in the direction of rotation. Similarly, easterly (from the east) winds imply a transfer of momentum from the surface to the atmosphere.

The single-cell model calls for easterly surface winds everywhere. If this were to occur, the planet would everywhere supply westerly momentum to the atmosphere, which would bring the atmosphere to rest within a week or two. To avoid this problem, easterly winds must be balanced by westerly winds somewhere else. More precisely, averaged over the whole Earth, momentum transferred from the surface to the atmosphere must equal momentum transferred from the atmosphere to the surface. Thus, a single-cell circulation would not be possible on Earth—or on any other rotating planet.

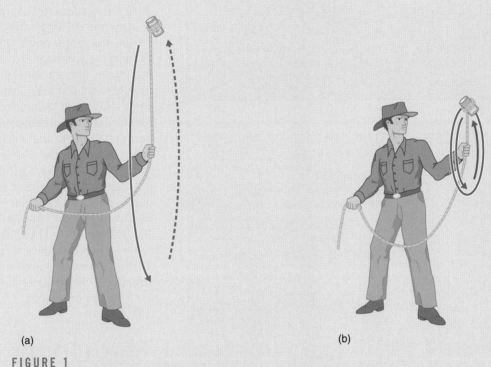

(a) (b)

FIGURE 1
Twirling a weight at the end of a rope (a) illustrates the conservation of angular momentum. As the man pulls the rope closer to his body, shortening the length of the rope (b), the speed of rotation increases in order to maintain a constant angular momentum.

the interior desert of Australia, and the deserts of the southwestern United States and northwest Mexico clearly reflect this sinking process.

As the solar declination changes seasonally, so does the zone of most intense heating. Knowing that the Hadley cells are thermal, we might expect the associated pressure and wind belts to move seasonally, and indeed they do. Although with a lag of several weeks, the ITCZ, subtropical highs, and trade winds all follow the "migrating Sun." This movement has a major impact on many of the world's climates—and on the people who inhabit them.

For example, many areas along the equator are dominated by the ITCZ year round and experience no dry season. Areas located near the poleward margins of the ITCZ, however, are subject to brief dry seasons as the zone shifts equatorward. Compare, for example, the average rainfall patterns for Iquitos, Peru (3° S), and San Jose, Costa Rica (9° N). Iquitos is located close enough to the equator so that it is perennially influenced by the ITCZ, but San Jose has a relatively dry period from January to March, when the low-pressure system is displaced to the south.

Similarly, some areas located on the equatorward edge of the subtropical highs are dry for most of the year, except briefly when the system shifts poleward

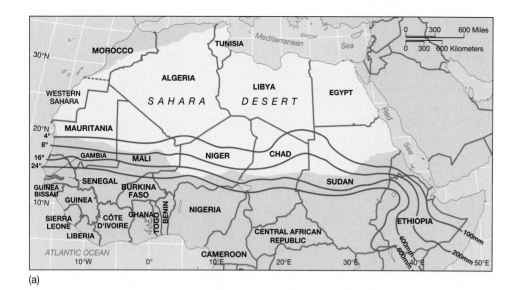

(a)

(b)

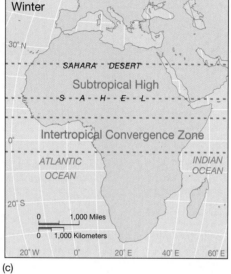

(c)

FIGURE 8–5

The Sahel is a region of Africa bordering the southern Sahara Desert (a). During the summer (b), the ITCZ usually shifts northward and brings rain to the region. For much of the year, the ITCZ is located south of the Sahel, and the region receives little or no precipitation (c).

during the summer. This condition exists in the Sahel of Africa, the region bordering the southern margin of the Sahara Desert (Figure 8–5a). Unlike the Sahara, which is dry all year, the Sahel normally experiences a brief rainy period each summer as the ITCZ enters the region (b). During the rest of the year, the descending air of the Hadley cell leads to dry conditions (c).

The migration of the Hadley cell has long supported a traditional lifestyle in which African herders followed the northward and southward shifting rains. During the 1960s and 1970s, the population of the region increased dramatically, which led to overgrazing and set the stage for catastrophe when a multiyear drought hit the area. Millions of head of livestock died from lack of food and water, which in turn led to the deaths of tens of thousands of people. During the early 1990s, low rainfall again plagued the Sahel—this time along the eastern part of the region in Somalia. Coupled with the existing political and social instability that eventually gave rise to civil war in Somalia, the drought led to the starvation of an estimated 300,000 people. Thus, the existence of these cells—and variations in the way they develop during unusual years—are more than mere abstractions. They have real-world ramifications for millions of people.

The Upper Troposphere

In Chapter 4 we saw that pressure decreases more rapidly with altitude where the air is cold. We also know that temperature in the lower troposphere generally decreases from the subtropics to the polar regions. These two principles are critical in understanding the distribution of wind and pressure in the upper troposphere.

Figure 8–6 maps the global distributions of the mean height of the 500 mb surface (a convenient level representing conditions in the middle troposphere) for January (a) and July (b). In both months, the height of the 500 mb level exhibits a strong tendency to decrease toward the polar regions, due to the lower temperatures found at higher latitudes. In January, the average height of the 500 mb surface over much of the southern United States is about 5670 m (18,600 ft), while over northern Canada it decreases to less than 5300 m (17,400 ft). A similar but less extreme change occurs in July as well.

Three features stand out from the maps in Figure 8–6. First, for both January and July the 500 mb heights are greatest over the tropics and decrease with latitude. Second, the gradient in height is greater in the hemisphere experiencing winter (the Southern Hemisphere in July and Northern Hemisphere in January). Third, at all latitudes the height of the 500 mb level is greater in the summer than during the winter. All three of these features result from the general distribution of temperature in the lower-middle atmosphere; areas of warm air have greater 500 mb heights.

(a) January

(b) July

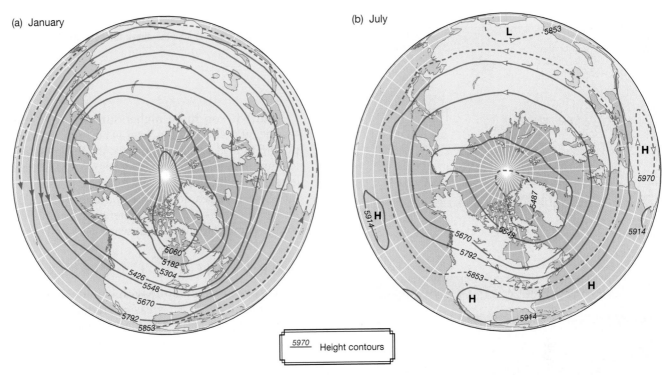

5970   Height contours

FIGURE 8–6

The mean heights of the 500 mb levels for January (a) and July (b). The pattern is mostly zonal with decreasing heights toward the poles.

WESTERLY WINDS IN THE UPPER ATMOSPHERE

Recall from Chapter 4 that height differences correspond to pressure differences, and that when the 500 mb surface slopes steeply, a strong pressure gradient force exists. We can therefore infer from the 500 mb maps that there is always a pressure gradient force across the middle latitudes trying to push the air toward the poles. Of course, in the absence of friction, the winds do not blow poleward, but rather blow parallel to the height contours, from west to east. The pressure gradient force is strongest in winter (the height contours are closely spaced), so the upper-level westerlies are strongest in winter. What does this mean to you? For one thing, it explains why most midlatitude weather systems migrate from west to east. In other words, it tells us why a storm over Chicago might find its way over the East Coast a day or two later, but seldom if ever does such a storm make the reverse trip.

The predominance of westerly winds in the upper troposphere also affects aviation. For example, a commercial aircraft going from Chicago to London has an estimated flight time of about 7.5 hours, while the return trip normally takes an hour longer because it must overcome headwinds. The difference in flight time would be even greater were it not for the fact that airlines route their planes to take advantage of tailwinds and avoid headwinds.

Wind speeds generally increase with height between the surface and the tropopause. Partly this is because of decreasing friction, but more importantly, the pressure gradient force is typically stronger at high altitudes. As illustrated in Figure 8–7, the surfaces representing the 900, 800, and 700 mb levels all slant downward to the north, but not by the same amount. Higher surfaces slope more steeply, which means that the pressure gradient force is greater. You may recall from Chapter 4 that the pressure gradient force is directly proportional to slope, without regard to density.

FIGURE 8–7

Latitudinal temperature gradients cause pressure surfaces to slant poleward. In this example, we assume a constant gradual decrease in temperature with latitude. The intensity of the pressure gradient force remains constant from one latitude to another but increases with altitude.

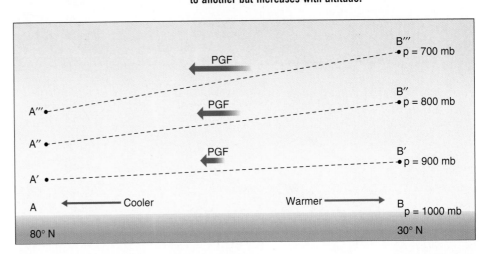

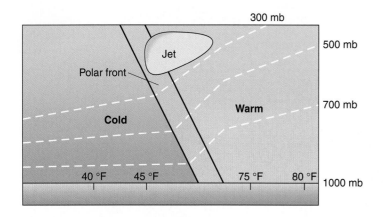

FIGURE 8–8
The polar jet stream is situated above the polar front near the tropopause.

But why do those higher surfaces slope more steeply? The air is warmer at point B, so the layer of air from 900 mb to 1000 mb is thicker at point B than at point A. Similarly, the thickness from 800 mb to 900 mb is greater at point B. In other words, the height change from B' to B'' is greater than the change from A' to A''. It must be, therefore, that the 800 mb surface slopes downward more than the 900 mb surface ($B'' - A''$ is greater than $B' - A'$). The difference in heights between successive surfaces continues to increase upward, leading to stronger winds.

THE POLAR FRONT AND JET STREAMS

The gradual change in temperature with latitude depicted in Figure 8–7 does not always occur in reality. Instead, areas of gradual temperature change often give way to narrow, strongly sloping boundaries between warm and cold air. One such boundary, the **polar front**, is shown in Figure 8–8.

Outside of the frontal zone, the changes in temperature with latitude are gradual (as they were in Figure 8–7), and the slopes of the 900, 800, and 700 mb levels respond accordingly. But within the front, the slope of the pressure surfaces increases greatly because of the abrupt horizontal change in temperature. With steeply sloping pressure surfaces there is a strong pressure gradient force, resulting in the **polar jet stream**. Thus, we see the jet stream as a consequence of the polar front, arising because of the strong temperature gradient. At the same time, the jet stream reinforces the polar front. In Chapter 10, we will see that a jet stream is necessary to maintain the temperature contrast across the front.

Jet streams can be thought of as meandering "rivers" of air, usually 9 to 12 km (30,000 to 40,000 ft) above sea level. Their wind speeds average about 180 km/hr (110 mph) in winter and about half that in summer, though peak winds can exceed twice these values. Like rivers on land, jet streams are highly turbulent, and their speeds vary considerably as they flow. Unlike rivers on land, they have no precisely defined banks. Furthermore, a single jet stream will often diverge at some point and fork off into two distinct jets. Thus, the locations and boundaries of these features on weather maps are often difficult to accurately pinpoint.

The polar jet stream greatly affects daily weather in the middle latitudes. Nearer the equator is another prominent jet, the **subtropical jet stream**, associated with the Hadley cell. As the upper-level air flows away from the ITCZ, the conservation of angular momentum imparts ever-growing westerly motion. When moving toward the northeast, the subtropical jet stream can bring with it warm, humid conditions. Figure 8–9 shows the flow of moisture associated with a subtropical jet stream.

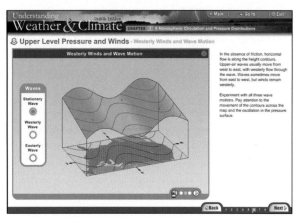

TUTORIAL
Upper-Level Winds and Pressure

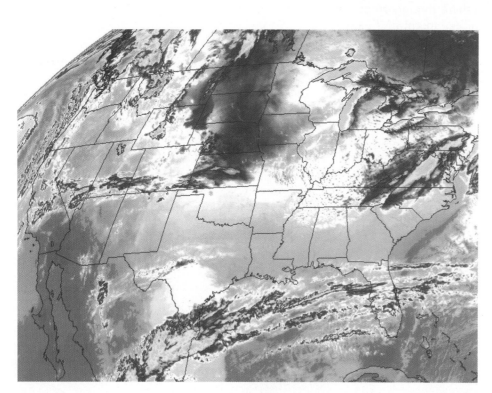

FIGURE 8–9
The subtropical jet stream appears in this infrared satellite image as the band of cloudiness extending from Mexico through Florida.

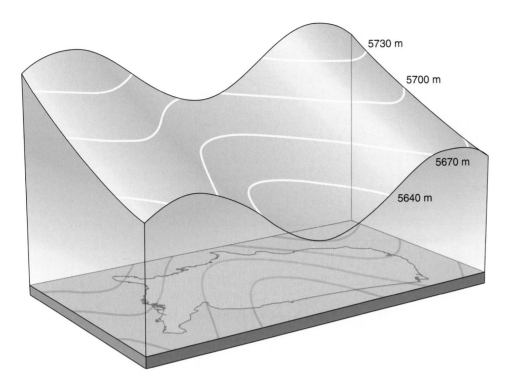

FIGURE 8–10
A hypothetical drawing of the 500 mb surface. Heights decrease from south to north but also rise and fall through the ridges and trough. Vertical changes are highly exaggerated in the figure. Actual height changes are very small compared to the size of the continent.

TROUGHS AND RIDGES

Although on average 500 mb heights decrease toward the poles, at any given time significant departures from the general trend will exist. Typically undulations, or waves, are superimposed on the overall decrease in height toward the poles. Figure 8–10 is a cartoon view of this, showing an axis of low height in the middle of the United States, flanked by "mountains" of high heights on either side. The valley of low heights is called a *trough*, and the upward bulges are called *ridges*. Also shown on the diagram are height contours—notice that they too have a wavelike character. The air flows parallel to the contours, so there is wavelike motion to the air stream as well.

Figure 8–11 shows simplified contour maps depicting the relationship between 500 mb heights and ridges and troughs. No east–west height changes (no trough, no ridge) occur in (a), and the flow is completely zonal. In (b), on the other hand, a trough is in the midsection of the country. Going from point 1 to 2, there is a height decrease from 5610 m to 5580 m. Going from 2 to 3, heights increase again. In other words, going from 1 to 3 requires us to cross a valley (trough) of low heights. We see that height contours are displaced toward the equator in troughs and toward the pole in ridges. We also see that air winding its way poleward around ridges and equatorward around troughs will have a meridional component as well as a zonal component. In fact, when waves are pronounced, we say the flow is "meridional," as opposed to "zonal" when the flow is nearly all westerly.

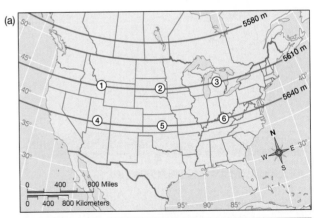

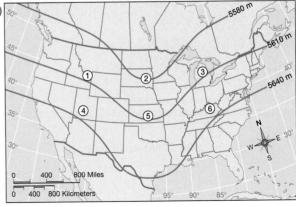

FIGURE 8–11
Troughs occur in the middle troposphere where the 500 mb height contours dip equatorward. In (a) positions 1–3 have the 500 mb level at 5610 m. Farther to the south, at positions 4–6, the 500 mb level is at 5640 m. In (b) the contour lines are in the same position over the East and West Coasts as they were in (a), but they shift equatorward over the central portion of the continent. Thus, positions 2 and 5 have lower pressure than the areas east and west of them. The zone of lower pressure over the central part of the continent is a trough.

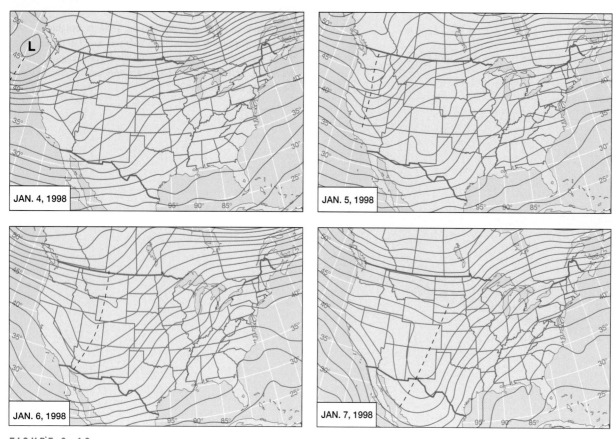

FIGURE 8–12

A sequence of 300 mb maps showing the migration of Rossby waves at 24-hour intervals.

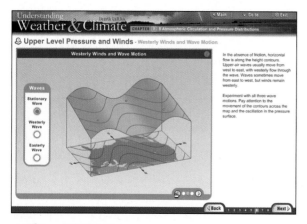

TUTORIAL
Upper-Level Winds and Pressure

ROSSBY WAVES

We've seen that ridges and troughs give rise to wavelike flow in the upper atmosphere of the middle latitudes. The largest of these are called *long waves*, or **Rossby waves**.[3] Usually, there are anywhere from three to seven Rossby waves circling the globe. Like other waves, each has a particular *wavelength* (the distance separating successive ridges or troughs) and *amplitude* (its north–south extent). Though Rossby waves often remain in fixed positions, they also migrate west to east (Figure 8–12), or on rare occasions from east to west.

Rossby waves undergo distinct seasonal changes from summer to winter. They tend to be fewer in number, have longer wavelengths, and contain their strongest winds during winter. The latter two characteristics—wind speed and wavelength—affect the rate at which Rossby waves migrate downwind (see *Box 8–2, Physical Principles: The Movement of Rossby Waves*).

Rossby waves exert a tremendous impact on day-to-day weather, especially when they have large amplitudes. They are capable of transporting warm air from subtropical regions to high latitudes or cold polar air to low latitudes. Because upper-level air tends to change temperature only slowly in the absence of strong vertical motions, Rossby waves can bring anomalous temperatures to just about any place within the middle to high latitudes. This is illustrated by Figure 8–13, which shows a strong Rossby wave over North America on September 22, 1995. Record-breaking low temperatures for the date were observed over much of the central United States as the wave brought cold air from the far north. Farther upwind, a southwesterly flow brought mild air to the extreme northwest of North America, with Fairbanks, Alaska, basking in temperatures in the mid-20s Celsius (mid-70s Fahrenheit).

[3]Named for Carl Gustov Rossby, who contributed much pioneering research on upper-level air flow in the early 1900s.

The southeastern part of North America is geographically similar to Asia. Like Asia, it lies to the north of a warm water body, the Gulf of Mexico, and it undergoes a seasonal reversal of the surface winds due to warming and cooling of the land mass. But despite this similarity, the southern United States does not have a strong monsoon climate. Part of the reason for this is that the smaller size of North America results in weaker oscillations in pressure than those of Asia. More important, however, is the fact that the southern United States does not have a major east–west mountain chain comparable to the Himalayas. In fact, the area is essentially flat, except for the relatively low southern Appalachians that extend southwest-to-northeast. The Appalachians do not create a strong orographic barrier to the southerly wind flow of summer, and therefore they do not promote the extreme rainfall seen in south Asia. They also do not block the passage of winter storms from the north, as do the Himalayas, nor do they promote persistent upper-level convergence. With all of that, it's easy to see why winter is hardly a dry season in eastern North America.

The desert of the southwestern United States experiences what residents often call the *Arizona monsoon*.[4] In spite of its name, the Arizona monsoon bears little resemblance to the real thing. It occurs each summer when warm, moist air from the south combines with strong surface heating to trigger scattered thundershowers. Unlike the intense precipitation of the Asian monsoon, that of the Arizona monsoon does little to alleviate the desert conditions of the American Southwest.

FOEHN, CHINOOK, AND SANTA ANA WINDS

Foehn (pronounced "fern" like the common plant) is the generic name for synoptic-scale winds that flow down mountain slopes, warm by compression, and introduce hot, dry, and clear conditions to the adjacent lowlands. Although the term *foehn* strictly applies to winds coming from the Alps of Europe, we generally use it to describe this type of wind anywhere in the world. In Europe, foehns develop when midlatitude cyclones approach the Alps from the southwest. The air rotates counterclockwise toward the center of low pressure and descends the northern slopes. These winds bring unseasonably warm conditions to much of northern Europe during the winter, when they are most prevalent.

When winds warmed by compression descend the eastern slopes of the Rocky Mountains in North America, they are called **chinooks**. Low-pressure systems east of the mountains cause these strong winds to descend the eastern slopes at speeds that can exceed 150 km/hr (90 mph) when funneled through steep canyons. Like their European counterparts, chinooks are most common during the winter when midlatitude cyclones routinely pass over the region.

Sometimes the presence of a large mass of cold, dense air near the base of a mountain range prevents a chinook from flowing all the way down the slope. The hot air then overrides the cold air, and no warming is observed near the surface. If the chinook strengthens sufficiently, however, it can push the cold air out of its path, and the foothill region undergoes a rapid temperature increase. But if the hot winds weaken even momentarily, the cold air can return to the foothills and bring another sudden change in temperature—this time a nearly instantaneous cooling. Such reversals can take place repeatedly over a short period of time, with each bringing another rapid and sometimes extreme temperature change. Rapid City, South Dakota, for example, once experienced three such cycles over a 3-hour period, with temperature changes as large as 22 °C (40 °F) occurring with each shift.

Chinook winds can be a blessing to ranchers in the western Great Plains who rely on them to melt the snow that covers their rangelands. To others, the rapid temperature oscillations can be a source of misery. Imagine leaving your house in the morning when the temperature is 0 °C (32 °F), getting out of your car when it is 38 °C (100 °F), going for lunch when the temperature is down to −10 °C (14 °F),

---

[4]Also called the *Southwest monsoon* or the *Mexican monsoon*.

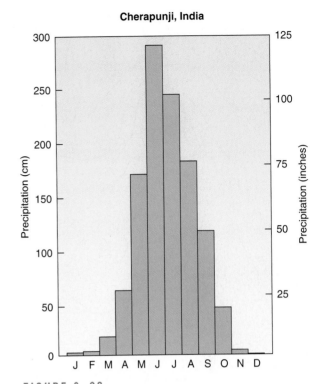

FIGURE 8–20

Monthly mean precipitation at Cherapunji, India, highlights the sudden increase in precipitation that occurs when the south winds of the monsoon begin. Over much of the monsoon region, there is an abrupt increase in precipitation in May or June.

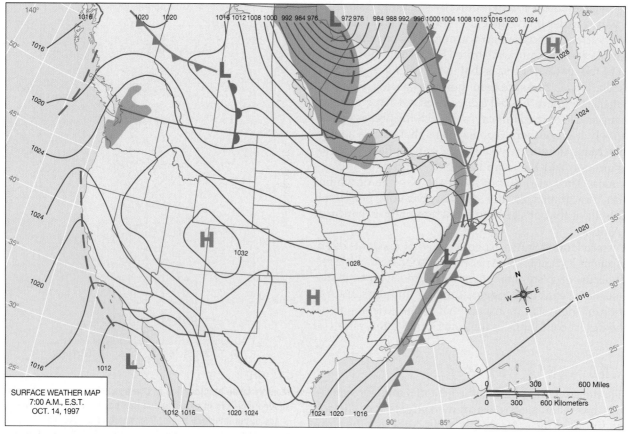

FIGURE 8–21

A weather map showing a high-pressure system over the
Rockies, causing a Santa Ana wind over southern California.

and returning from lunch when it is up to 35 °C (95 °F). Such changes have been
known to occur from chinooks, and there is anecdotal evidence (although nothing
has been proven) suggesting an increase in violent crime, depression, and suicide
during such episodes. Other problems not related to human discomfort can also
arise from chinooks. During the 1988 Winter Olympics at Calgary in Alberta,
Canada, chinook winds melted the snow cover and forced a postponement of the
ski competition for several days.

In parts of the western Great Plains, chinooks are frequent enough and strong
enough to increase the average winter temperatures, as exemplified by Rapid City
and Sioux Falls, South Dakota. Rapid City is situated in the foothills of the Black
Hills and commonly experiences chinooks, whereas Sioux Falls is several hundred
kilometers to the east and well out of their range. But despite the fact that its elevation is 500 m (1650 ft) greater, Rapid City has an average January temperature 4
°C (7 °F) greater than that of Sioux Falls, thanks to frequent chinook winds.

The **Santa Ana winds** of California are similar to foehns and chinooks, but
they arise from a somewhat different synoptic pattern. These winds, common in
the fall and to a lesser extent in the spring, occur when high pressure develops over
the Rocky Mountains (Figure 8–21). Air flowing away from the high pressure descends the western slopes and warms by compression, just as air flowing along the
eastern slopes does for the chinooks. The difference is that the Santa Ana winds
occur in response to a large area of high pressure causing air to flow out of the
Rockies, whereas the chinook forms in response to air flowing across the range.

During Santa Ana conditions, the sinking air can warm by 30 °C (54 °F) or
more and attain temperatures in excess of 40 °C (104 °F) near the coast. Contrary
to what some people believe, Santa Anas are *not* warm because they pass over hot
desert surfaces—it is compression, and compression only, that causes their high
temperatures. In fact, during a well-developed Santa Ana, the coastal areas of California are usually hotter than interior desert locations such as Las Vegas, Nevada.

FIGURE 8–22

Much of coastal California is covered by chaparral, a vegetation type adapted to the summer-drought conditions of the region.

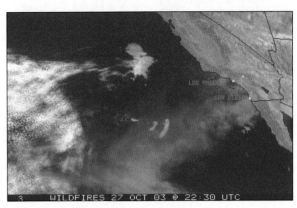

WEATHER IN MOTION
Southern California Fires

Santa Ana winds often contribute to the spread of tremendously destructive fires in California (see *Box 8–4, Focus on the Environment: Wildfires*). The natural vegetation of the region is dominated by an assemblage of species collectively referred to as *chaparral* (Figure 8–22), which is dry and highly flammable. When Santa Anas develop, the combination of hot, dry winds, low humidity, and an abundant source of fuel can set the stage for a major conflagration. Southern California, and particularly San Diego and San Bernardino counties, experienced this type of disaster in October 2003 (Figure 8–23). Two fires in San Diego County destroyed more than 2400 homes and killed 16 people; in San Bernardino County more than 1100 homes were completely burned and six people died.

Such fires are not restricted to southern California. In October 1991, Santa Ana–like winds fanned a wildfire through the Oakland Hills, east of the San Francisco Bay area. It too destroyed more than a thousand homes and killed 24 people.

KATABATIC WINDS

Like foehn and Santa Ana winds, **katabatic winds** warm by compression as they flow down slopes. Unlike foehns and Santa Anas, however, katabatic winds do not result from the migration of surface and upper-level weather systems. Rather, they originate when air is locally chilled over a high-elevation plateau. The air becomes dense because of its low temperature and flows downslope. The two best locations for such winds are along the margins of the Antarctic and Greenland ice sheets.

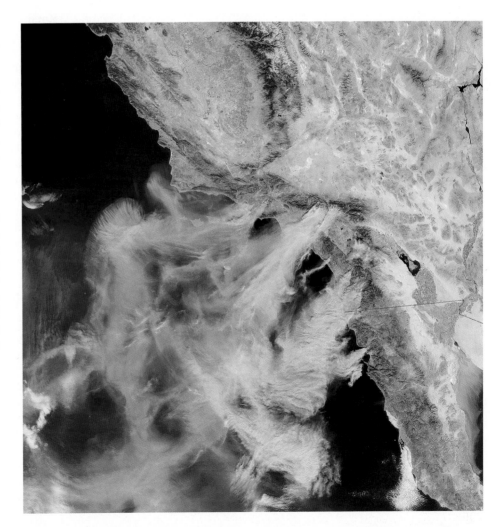

FIGURE 8–23

A satellite image of southern California during the October 2003 fires. Santa Ana winds caused the fires to spread rapidly westward during the onset of the fires and created the large plumes of smoke that were blown offshore. Few clouds appear in this image except over a couple of bays along Baja California and over the extreme bottom portion.

Wildfires

8–4 FOCUS ON THE ENVIRONMENT

In the summer and fall of 2002, the landscape of the western United States was particularly dry due to an extended drought. That situation, coupled with the accumulation of potential fuel that had built up over the years since previous fires, set the potential for devastating wildfires. Many ecologists believe that fire suppression, though it may save homes and large areas of woodland in the short run, creates the conditions for uncontrollable fires. According to this argument, allowing fires to burn simply allows the ecosystem to maintain its normal conditions and precludes the buildup of highly volatile material that ultimately sets off massive conflagrations.

In 2002 the worst fears were indeed realized. By the first day of summer, hundreds of thousands of acres of land had already burned across the West. Unusually high temperatures and strong winds set the stage for the first of the

most destructive fires in June in western Colorado. A couple weeks later, two major fires in Arizona joined together to form a massive wall of flames. And before summer was over, enormous, long-lived fires had also broken out over Oregon. All three fires were the worst ever in the history of the respective states.

Part of what makes massive fires such as these so devastating is that they create their own weather in a way that fosters more rapid spreading. Intense flames create strong thermal updrafts that lower the surface air pressure. This sets up strong pressure gradients, and the resulting strong winds help transport burning embers over great distances. It is little wonder that such fires are so difficult to contain and are often able to jump fire lines.

In one regard, fires sometimes set up weather conditions that actually assist fire crews. The

same convection that can set up strong pressure gradients can also trigger rain showers that help extinguish the flames.

It is ironic, but not surprising, that while the major fires of the spring and summer of 2002 were burning in the West, other regions were coping with severe flooding. Rivers topped their banks in the upper Midwest in the spring. In the summer it was Texas that endured extremely high floodwaters, especially near San Antonio. Why is it not surprising that floods occurred in the central United States while the West was coping with drought? The answer is related to the Rossby waves, discussed earlier in this chapter. When these waves assume favored positions for extended periods of time, some portions of the wave favor high precipitation while others promote dry conditions. This is discussed further in Chapter 10.

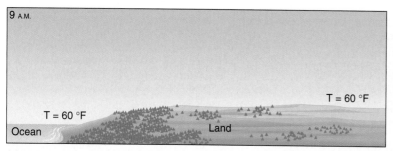

(a)

(b)

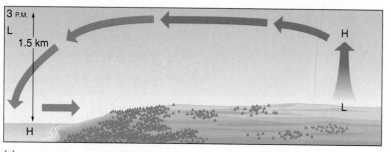

(c)

Katabatic winds usually occur as light breezes, but when funneled through narrow, steep canyons they can attain speeds in excess of 100 km/hr (60 mph). They happen sporadically because it can take some time for the air over the plateau to chill sufficiently. As soon as a mass of cold air forms over a plateau, gravity pulls it downslope and sets the wind in motion. When the cold air has been depleted, the wind ceases.

Much of coastal Antarctica is characterized by these alternating gusts and lulls of wind. And they can be very strong; one site, Cape Denison, occasionally experiences gusts up to 200 km/hr (120 mph) and holds the distinction of having the greatest average recorded wind speed on Earth.

Katabatic winds are not restricted to Greenland and Antarctica. They also flow out of the Balkan Mountains toward the Adriatic coast, where they are called *boras*. In France they are called *mistrals* as they flow out of the Alps and into the Rhone River Valley.

SEA AND LAND BREEZE

Near coastal regions or along the shores of large lakes, the differential heating and cooling rates for land and water form a diurnal (daily) pattern of reversing winds. During the daytime (especially in summer), land surfaces warm more rapidly than the adjacent water, which causes the air column overlying the land to expand and rise upward (Figure 8–24). At a height of about 1 km, the rising air spreads outward, which causes an overall reduction in the surface air pressure. Over the adjacent water less warming takes

FIGURE 8–24
The development of a sea breeze.

FIGURE 8–25
This image of Hawaii shows the effect of a sea breeze.
Heating of the land causes the air to expand upward. Coastal
air flowing toward the interior is lifted as it passes over the
mountains, causing orographic cloud cover.

place, so the air pressure is greater than that over land. The air over the water moves toward the low-pressure area over the land, which sets up the daytime **sea breeze** (remember, winds are always named for the direction *from* which they blow). (See Figure 8–25.)

As a sea breeze encroaches landward, a distinct boundary exists between the cooler maritime air and the continental air it displaces. This boundary, called the **sea breeze front**, usually produces a small but abrupt drop in temperature as it passes. This does not mean that the temperatures will not continue to rise after the sea breeze front moves on—only that there is a temporary lull in the rate of warming. Farther inland more heating can occur before the sea breeze passes, and temperatures become higher than those along the coastal strip.

Note that though the rising air in the heated column creates low pressure at the surface, it also creates higher pressure in the middle atmosphere. But we know that pressure always decreases with height. So how can this be? Remember that the terms *high pressure* and *low pressure* are relative—that is, they mean that the pressure is higher or lower than surrounding air *at the same level*. Thus, in this example the surface pressure over land is less than that over the adjacent ocean because the outflow of air above the 1-km level reduces the total amount of air over the surface. In the middle atmosphere, however, the rising of air from below increases the amount of mass above a particular level, and thereby increases the pressure relative to that of the surrounding air (though the pressure is still less than that below).

At night, when the land surface cools more rapidly than the water, the air over the land becomes dense and generates a **land breeze**. That is, lower land temperatures make for higher surface pressure and offshore flow. Compared to land breezes, sea breezes are usually more intense and last for a longer period of time each day. Sea breezes tend to be strongest in the spring and summer when the greatest daytime temperature contrasts occur between land and sea. A typical

TABLE 8-1	Average Wind Speed and Direction—Los Angeles, CA							
	WINTER		SPRING		SUMMER		FALL	
TIME (PST)	DIR	SPEED (m/s)	DIR	SPEED (m/s)	DIR	SPEED (m/s)	DIR	SPEED (m/s)
4 A.M.	ENE	1.0	E	0.5	WSW	0.4	ENE	0.6
10 A.M.	ENE	1.4	WSW	1.9	WSW	3.2	WSW	1.0
1 P.M.	WSW	2.5	WSW	5.3	WSW	5.5	WSW	4.5
4 P.M.	WSW	3.5	WSW	5.5	WSW	5.8	W	5.1
10 P.M.	NE	0.5	W	1.6	WSW	2.4	W	0.7
1 A.M.	ENE	1.0	—	0	WSW	1.0	NNE	0.2

Source: California Air Resources Board.

FIGURE 8-26

A valley breeze (a) forms when daytime heating causes the mountain surface to become warmer than nearby air at the same altitude. The air expands upward and the air flows from the valley to replace it. Nocturnal cooling makes the air dense over the mountain and initiates a mountain breeze (b).

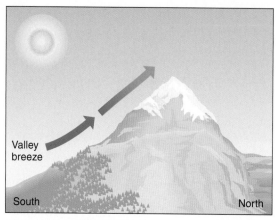

(a)

(b)

sea/land breeze pattern is described in Table 8–1, which shows average wind characteristics for Los Angeles, California.

The sea/land breeze type of circulation is not confined to coastlines but also occurs along the shores of large lakes, in which case it creates daytime **lake breezes**. Such systems occur along the margins of the Great Lakes in the United States and Canada, but there they occupy a narrower zone than do those along the Pacific or Atlantic coasts.

VALLEY AND MOUNTAIN BREEZE

A diurnal pattern of reversing winds similar to the land/breeze system also exists among mountains and valleys. During the day, mountain slopes oriented toward the sun heat most intensely. The air over these sunny slopes warms, expands upward, and diverges outward at higher altitudes in much the same way as it does over inland areas when a sea breeze develops. The **valley breeze** occurs when air flows up from the valleys to replace it (Figure 8–26a).

At night, the mountains cool more rapidly than do low-lying areas, so the air becomes denser and sinks toward the valleys to produce a **mountain breeze** (Figure 8–26b). Mountain breezes are usually just about as intense as valley breezes but tend to be somewhat gustier. Whereas valley breezes blow fairly continuously at speeds below 15 km/hr (10 mph), the nighttime air may be still for several minutes and then suddenly flow downslope.

Air–Sea Interactions

Earlier in this chapter we described how atmospheric circulations propel ocean currents. In a similar manner, the oceans exert an important influence on the input of heat and moisture to the atmosphere. Warm surface waters heat the overlying atmosphere by the transfer of sensible and latent heat. The addition of this heat, in turn, affects atmospheric pressure. Thus, the atmosphere and the ocean are linked as a complex system.

Compared to those in the atmosphere, oceanic motions and temperature changes are exceedingly slow. Temperatures in the lower troposphere can change tens of degrees in a matter of hours, while oceanic temperatures are very stable. Because the ocean surface changes so slowly, information about the current distribution of temperature can be a useful tool for atmospheric scientists in making long-term weather forecasts. In this section we examine some of the important interrelationships between the ocean and the atmosphere.

EL NIÑO, LA NIÑA, AND THE WALKER CIRCULATION

Before its dramatic reappearance in 1983, the phenomenon known as **El Niño** was largely unknown to the public. But that changed in the early part of the year when the unusually warm waters in the eastern Pacific Ocean that mark an El Niño helped spawn a series of powerful storms in southern California. The storms not only caused severe flooding but also generated heavy surf that caused extensive coastal property damage and completely washed the sand away from many beaches. Another recent (1997–98) El Niño was also unusually strong, again leading to episodes of heavy surf, landslides, and flooding in southern California. In addition, precipitation across the southern tier of states was well above normal and severe storms were more frequent; some storms spawned very damaging tornadoes. Residents of the northern United States and eastern Canada also experienced anomalous weather conditions with unusually mild temperatures during the winter. During the fall and winter of 2002–2003, another weaker El Niño developed over the eastern Pacific.

High water temperatures promote two conditions favorable for major storm activity: increased evaporation into the air and reduced air pressure. There is therefore little doubt that the unusual episodes of 1983 and 1997–98, in which the surface waters were as much as 6 °C (11 °F) warmer than normal, played a role in the formation and passage of the storms. So what exactly is El Niño and how does it form?

At 2- to 7-year intervals (every 40 months on average), the surface waters of the eastern Pacific, especially near the coast of Peru, become unusually warm. (The tendency for this warming to occur near the Christmas season led to the name El Niño, a reference to the Christ child.) But El Niño is much more than a simple oceanic warming; it results from a complex interaction with the atmosphere in what is called the **Walker circulation** (Figure 8–27).

Under normal conditions, the trade winds move warm surface waters near the equator westward, causing higher temperatures and even a difference in sea level—about half a meter greater in the western Pacific compared to the eastern Pacific. At the same time, the upwelling of colder water from below replaces the warm water migrating westward in the eastern Pacific. Warmer water in the western Pacific leads to higher air temperatures, lower surface air pressure, and more convective precipitation. An El Niño develops when the trade winds weaken or even reverse and flow eastward. The warm water normally found in the western Pacific gradually "sloshes" eastward as a slowly moving wave, in part due to the higher sea level, and eventually makes its way to the coast of North and South America. The eastern Pacific warms, upwelling weakens, and the warm surface layer deepens both there and throughout the tropical Pacific basin. The change in sea surface conditions linked with the change in the atmospheric pressure distribution is called the **Southern Oscillation**. The El Niño and Southern Oscillation are closely intertwined and atmospheric scientists refer to their combined occurrences as **ENSO events**.

TUTORIAL
ENSO

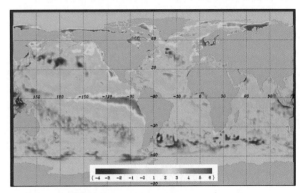

WEATHER IN MOTION
El Niño
La Niña

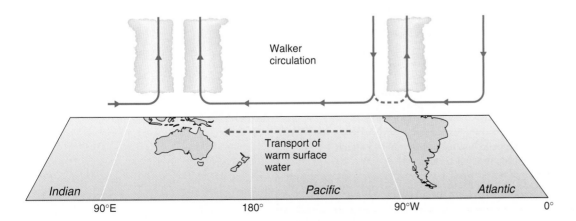

FIGURE 8–27
The Walker circulation involves a westward flow of surface air over the equatorial Pacific. (Although many texts show return flow in closed loops, recent evidence indicates this is wrong. Sinking air is supplied by convergence in the upper atmosphere, not eastward flow in the upper atmosphere across the Pacific basin.) The surface flow drags warm surface waters into the western Pacific. When the westerly flow weakens or reverses, the warm waters to the west migrate eastward and cause an El Niño.

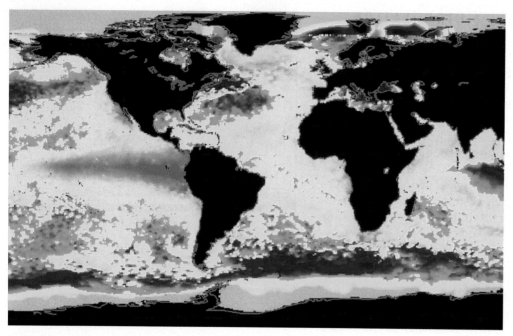

FIGURE 8–28
The 1997–98 El Niño event contained a very large area of much above-normal sea surface temperatures over the tropical east Pacific (shown in bright red). The image is based on satellite data obtained on November 3, 1997.

FIGURE 8–29
The average sea surface temperature differences from normal observed during the November through March period for eight El Niño episodes. The most prominent feature is the +2 °C (4 °F) increase in temperatures in the equatorial tropical east-Pacific.

Figure 8–28 shows the sea surface temperature (SST) anomalies associated with the very strong El Niño of 1997–98. The colors on the figure depict temperature anomalies, the differences in temperature from normal conditions. Of course, no two ENSO events look exactly alike. The shape and exact locations of the pools of warm water vary between events, as do the magnitudes of the SST anomalies. Figure 8–29 shows the overall average distribution of November–March SST anomalies for eight average-to-large El Niño events. Figure 8–30 illustrates the variability in El Niño situations by plotting temperature anomalies for six different events.

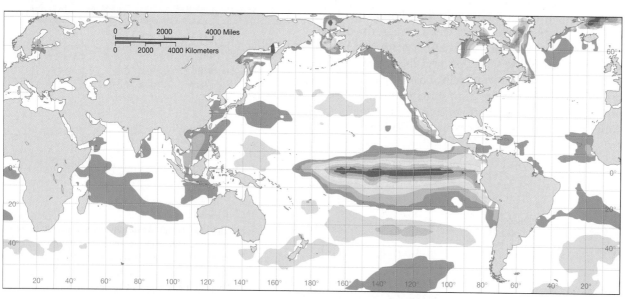

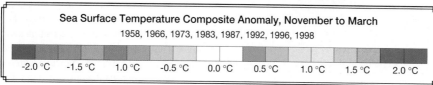

Sea Surface Temperature Composite Anomaly, November to March
1958, 1966, 1973, 1983, 1987, 1992, 1996, 1998

-2.0 °C   -1.5 °C   1.0 °C   -0.5 °C   0.0 °C   0.5 °C   1.0 °C   1.5 °C   2.0 °C

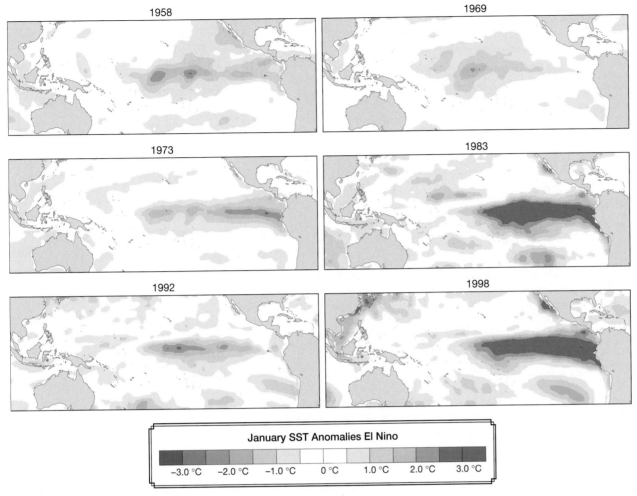

FIGURE 8–30
Every El Niño event looks somewhat different from all others. This figure depicts the observed sea surface temperature anomalies for six El Niño events.

When an El Niño dissipates, it can be followed either by a return to normal sea surface conditions, or by further cooling of the tropical eastern Pacific. The resultant pattern, the reverse of an El Niño, is called a **La Niña** (Figure 8–31). See *Box 8–5, Physical Principles: What Causes El Niños and the Southern Oscillation?*, for a discussion of the factors that lead to changing ENSO conditions. Also, animations depicting the onset and demise of theoretical and observed El Niño and La Niña events are provided in the tutorial on the CD that accompanies this book.

In order to study ENSO events and their effects on conditions elsewhere, a quantitative measure of ENSO activity is needed—an index that captures the occurrence and magnitude of an event. Several such indices have been used over the years. Some are based on sea surface temperatures in defined regions of the eastern equatorial Pacific. Another, the **Southern Oscillation Index (SOI)**, is defined as the monthly sea level pressure departure from normal at Tahiti minus the departure from normal at Darwin, Australia. Thus, if the sea level pressure for a particular month at Tahiti is 1 mb above normal for that month, and the pressure at Darwin is 1 mb below the monthly norm, the SOI for that month is 2. Since Tahiti is east of Darwin, a positive SOI indicates a stronger than average pressure gradient from east to west, and a La Niña situation exists. A negative SOI indicates an El Niño. Note that there is no minimum SOI value, positive or negative, that absolutely delineates the existence of El Niño or La Niña from neutral conditions.

Though the SOI is a simple numeric value, it can be used to provide a historical perspective on ENSO occurrences through time. The fact that it is based on sea level

DID YOU KNOW?

The Atlantic Ocean has its own version of El Niño events. The processes responsible for Pacific El Niños also operates in the Atlantic and similarly gives rise to anomalously warm surface water in the eastern part of the ocean. But Atlantic El Niños are smaller, occur more frequently, and are not correlated with their Pacific cousins.

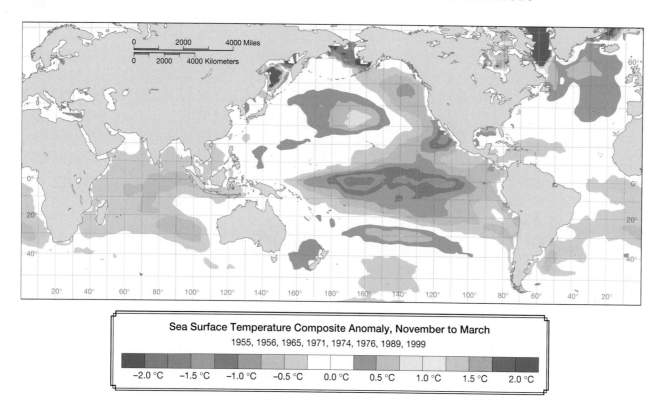

FIGURE 8–31

The average sea surface temperature differences from normal observed during the November through March period for eight La Niña episodes. Note the large area of cooler-than-normal temperatures over much of the eastern Pacific, extending poleward along the North and South American coasts. Cold waters are also found in the Indian Ocean.

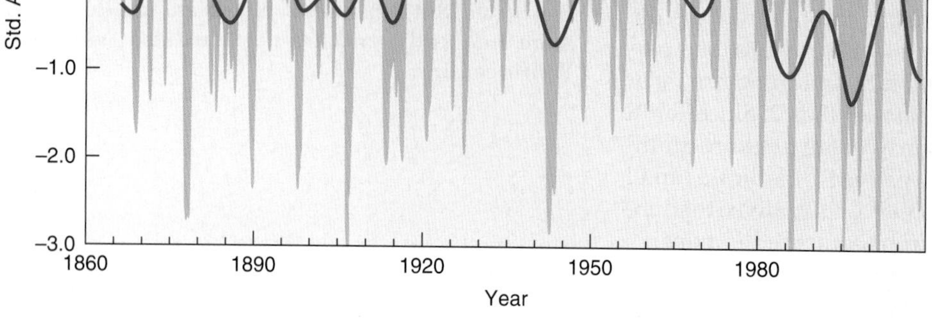

FIGURE 8–32

Time series plot of the Southern Oscillation Index (SOI). Positive values indicate La Niña conditions; negative values are El Niños. Several prominent El Niños have occurred since the 1980s.

pressure data (which have been recorded at Darwin and Tahiti for well over a century), instead of the shorter record of sea surface temperatures, allows for an analysis of ENSO for a longer time period than can be performed using any index based on water temperature. Figure 8–32 plots monthly values of the SOI since the late 1860s, with positive areas (in blue) indicating La Niña conditions and negative areas (in red) representing El Niños. Though there have been some periods in which El Niños and La Niñas were more frequent or less frequent than others, it is clear that both have occurred with substantial regularity over the last century and a half. Moreover, close analysis of the data suggests that ENSO events have been more frequent in recent years than in the mid-1800s.

Some useful observations have been made regarding the impact of El Niño events on seasonal weather. The central coast of California, for example, appears to have a greater likelihood of unusually high amounts of precipitation when El Niño is present, while the northwestern United States and Canadian Pacific coast regions tend to be unusually dry. The effects of El Niño are not restricted to the western coast of North America, however. As you have seen, the large-scale patterns of the atmosphere are strongly influenced by the position of Rossby waves. When high-pressure or low-pressure systems exist in some locations, they affect not only local weather conditions but also the overall size, shape, and position of the entire Rossby wave pattern. The establishment of an upper-level trough over the Pacific Coast, for example, will

What Causes El Niños and the Southern Oscillation?

There are actually two parts to this question: Why do these anomalies arise? That is, what causes an El Niño or La Niña to become established? and What causes the reversals, or oscillations, between the two extremes? An answer to the first question was proposed more than 35 years ago in a pioneering study by Jacob Bjerknes, and has been essentially confirmed by more recent analysis and data. Unlike some other types of climate change, both anomalies arise without any external forcing. That is, they do not result from a non-climatic event such an asteroid impact, a volcanic eruption, or a change in the Sun's radiation. Rather, they result from processes internal to the atmosphere–ocean system. They are, in fact, classic examples of a *positive feedback* mechanism, in which relatively small changes are continually reinforced and amplified.

Consider, for example, the La Niña, where the strengthening trades push more warm water westward. In the western Pacific the growing pool of warm water promotes more uplift, precipitation, and lower pressure. Lower pressure in the west intensifies the pressure gradient and strengthens the trades, which in turn leads to further transport of warm water. At the same time, export of warm water cools the eastern Pacific, enhances upwelling, and raises the thermocline. All of these processes promote further growth of the anomaly. Similar reasoning, but in reverse, explains why El Niños develop. In this case warmer water in the eastern Pacific weakens the trades, which leads to less heat export by westward-moving water. With less heat exported, warmth in the eastern Pacific intensifies, further reducing the pressure gradient and reducing export of heat. Thus we see that once the system begins to drift toward either condition, there will be a tendency for the anomaly to intensify and become firmly established. This line of reasoning favors two very different states. In the El Niño state there is a thick pool of anomalously warm water in the eastern Pacific lying above a relatively deep thermocline. At the other extreme the eastern Pacific is anomalously cold, the thermocline is shallow, and the trades are strong.

If positive feedback explains the appearance of El Niños and La Niñas, what leads to their breakdown? That is, once established, why don't they persist indefinitely? Why is there oscillation from one state to the other? Much less is known with certainty about this, and the topic remains a matter of intense interest within the scientific community. However, there is little doubt that for oscillations to arise naturally in the system, there must be delayed negative feedback processes that are out of phase with the positive feedbacks described above. That is, there must be one or more restorative processes that lag behind processes driving the system toward the extreme states. As an El Niño or La Niña builds, the delayed restoring processes also grow and eventually become large enough to overwhelm the amplifying processes. When that happens, the system moves away from the extreme state (El Niño or La Niña) toward more normal conditions. But the positive feedbacks are still in play, and if they are large enough, the system moves to the other extreme. So, in general terms, an oscillation between the extremes arises from the joint effects of positive and lagged negative feedbacks.

A number of candidate processes have been proposed as delayed negative feedbacks. One possibility is that during an El Niño, Rossby waves generated in the east-central Pacific travel westward and are reflected backward from the western ocean boundary as small waves on the ocean surface. These so-called Kelvin waves bring a deeper thermocline to the eastern Pacific, counteracting the El Niño. Similar waves form with La Niña, but they bring a shallower thermocline to the eastern Pacific, and thus similarly move the system toward more normal conditions. Another possibility, termed the *recharge oscillator*, contends that the buildup and discharge of warm water from the tropical Pacific gives rise to the oscillation. The idea is that while an El Niño builds, the entire tropical Pacific experiences a gradual "recharge" of heat as the thermocline deepens. Some time during the El Niño proper the excess warmth is discharged to higher latitudes, the thermocline becomes shallower, and the system moves toward the other extreme. This institutes another cycle of recharge, followed by another flushing of heat to the extratropics (the middle and high latitudes) with the ensuing El Niño.

Another possibility is that during an El Niño the central Pacific atmosphere warms because of enhanced convection and condensation, and this in turn generates low-pressure cells on either side of the equator. Initially these cyclones amplify the El Niño, but eventually they pump cold water eastward and thereby destroy the event they helped construct. Still other mechanisms have been suggested as important in the oscillations, including forcing by disturbances unrelated to the ENSO. No single process has been shown as responsible for ENSO events, and many experts believe that multiple processes are involved to varying degrees from one episode to the next.

Support for all these ideas comes from computer models that are able to reproduce ENSO-like cycles. But there is no agreement about what controls the period of oscillation, why it averages about 4 years instead of some other period, or even why the period of oscillation is variable, with ENSOs appearing every 2 to 7 years. Answers to these and other questions about ENSO are important both for understanding our present climate and for anticipating the nature and impact of future climate changes.

promote the development of a ridge farther to the east. For this reason, certain weather conditions in the eastern United States are set up through **teleconnections**, the relationships between weather or climate patterns at two widely separated locations. El Niños favor the formation of ridge-trough patterns that often bring rainy conditions to the southeastern United States and mild, dry conditions to the northeastern United States and eastern Canada.

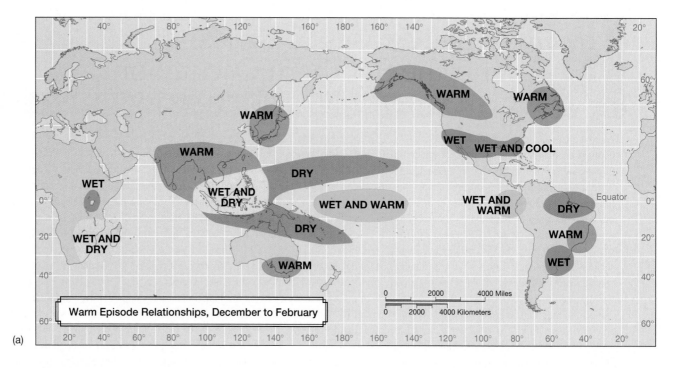

(a)

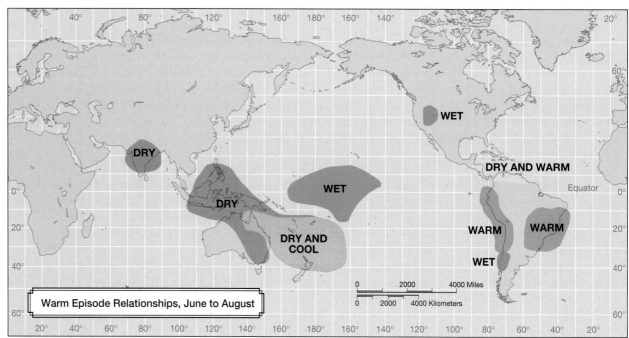

(b)

FIGURE 8–33

Worldwide climatic conditions associated with the occurrence of El Niño—(a) and (b)—and La Niña—(c) and (d)—
events, for the periods of December–February and June–August. While the indicated conditions often appear during ENSO
events, they do not occur with all El Niños or La Niñas.

These teleconnections also occur outside North America; El Niños tend to promote enhanced precipitation over coastal Ecuador and Peru, the central Pacific, parts of the Indian Ocean, and eastern equatorial Africa. The western tropical Pacific, Australia, India, southeastern Africa, and northeastern South America usually undergo decreased precipitation with the occurrence of El Niño events. Figure 8–33 depicts some of the worldwide climatic patterns generally associated with El Niños and La

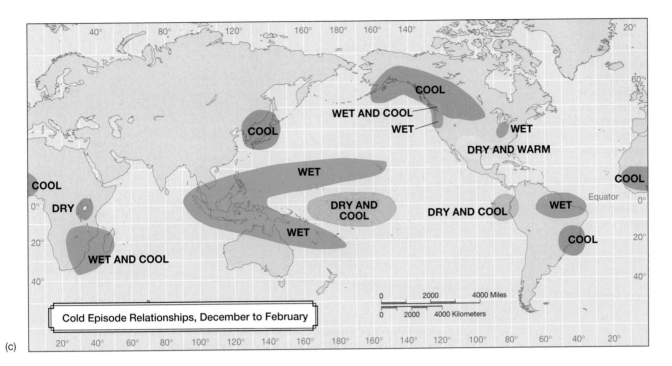

(c)

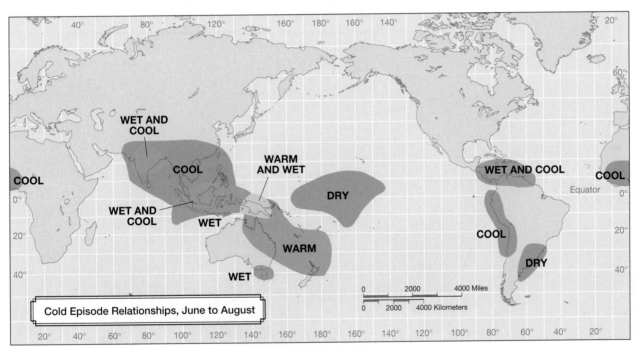

(d)

Niñas, and Figure 8–34 shows the average temperature and precipitation conditions that have occurred over the United States during recent ENSO episodes.

Like its warm-water counterpart, a La Niña tends to favor distinct (but different) regional and teleconnection patterns, including dry conditions along the California coast, the southern United States, and Peru. Enhanced precipitation tends to occur in the Pacific Northwest and western Canada, Indonesia, the Caribbean, and south Asia.

DID YOU KNOW?

Coral reefs provide information on ocean water temperatures that can be used to analyze the frequency and intensity of El Niño events of the distant past. One such study has suggested that the last century or so has witnessed the most intense El Niños of the last 130,000 years, with those of 1982–83 and 1997–98 possibly being the greatest to have occurred over that time span. It is possible, though not conclusively proven, that this change could be a response to global warming over the last century.

FIGURE 8-34

El Niño and La Niña events tend to promote temperature and precipitation responses across the conterminous United States. During eight recent El Niño events, November–March temperatures have tended to be higher than normal in the north-central United States and lower than normal across the southern tier of states (a). At the same time, coastal California and the southeast generally have wet conditions, while relative drought tends to occur in the Pacific Northwest and over some of the Ohio River Valley (b). La Niñas tend to promote warm winter conditions over much of the southeast (especially coastal Texas) and parts of the northwest and the north-central United States (c), along with wetness in the northwest and relative drought in the southeast (d).

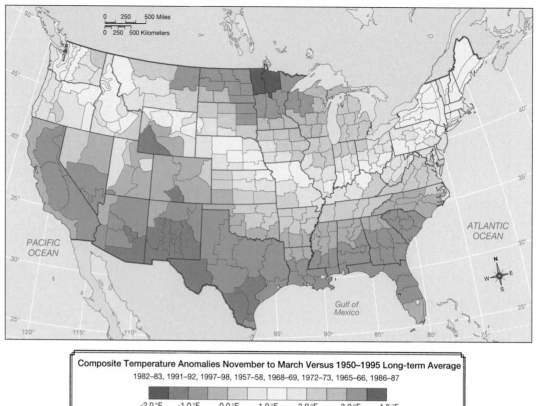

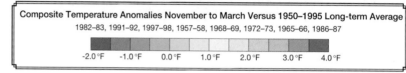

Composite Temperature Anomalies November to March Versus 1950–1995 Long-term Average
1982–83, 1991–92, 1997–98, 1957–58, 1968–69, 1972–73, 1965–66, 1986–87

-2.0 °F -1.0 °F 0.0 °F 1.0 °F 2.0 °F 3.0 °F 4.0 °F

(a) El Niño

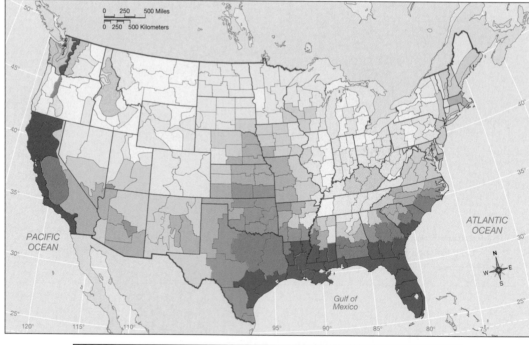

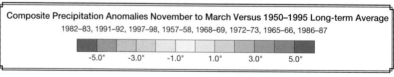

Composite Precipitation Anomalies November to March Versus 1950–1995 Long-term Average
1982–83, 1991–92, 1997–98, 1957–58, 1968–69, 1972–73, 1965–66, 1986–87

-5.0" -3.0" -1.0" 1.0" 3.0" 5.0"

(b) El Niño

TUTORIAL
ENSO

The fact that certain types of weather tend to occur in the presence of El Niños or La Niñas should not be taken to mean that some particular type of weather will always accompany the event. If the relationship between sea surface temperatures and atmospheric patterns were that well defined, long-range weather forecasting would be very much easier than it is. One of the reasons individual El Niños and

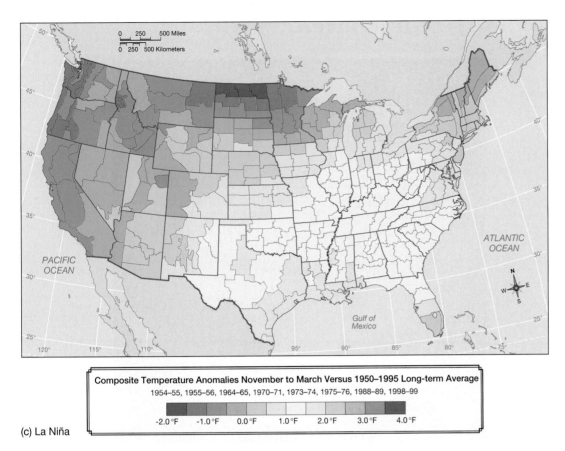

Composite Temperature Anomalies November to March Versus 1950–1995 Long-term Average

1954–55, 1955–56, 1964–65, 1970–71, 1973–74, 1975–76, 1988–89, 1998–99

-2.0 °F   -1.0 °F   0.0 °F   1.0 °F   2.0 °F   3.0 °F   4.0 °F

(c) La Niña

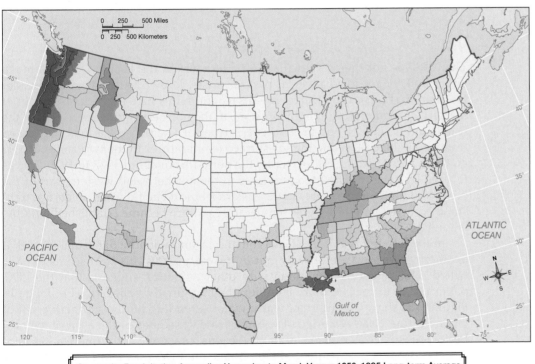

Composite Precipitation Anomalies November to March Versus 1950–1995 Long-term Average

1954–55, 1955–56, 1964–65, 1970–71, 1973–74, 1975–76, 1988–89, 1998–99

-5.0"   -3.0"   -1.0"   1.0"   3.0"   5.0"

(d) La Niña

La Niñas do not always produce the same kind of atmospheric response is that no two are exactly the same. Each has its own size, shape, and location, and therefore influences atmospheric patterns differently. Not only that, but the atmosphere is far too complex to be governed entirely by any single factor, such as oceanic water temperatures.

FIGURE 8–35

FIGURE 8–35

Pacific sea surface temperatures obtained in June 2002 suggest that a reversal in the Pacific Decadal Oscillation has begun. If so, this represents a major shift from the pattern in existence between 1976 and the late 1990s.

PACIFIC DECADAL OSCILLATION

The ENSO is not the only oscillation pattern across the Pacific Ocean. A much larger and longer-lived reversal pattern exists, known as the **Pacific Decadal Oscillation** (PDO). The PDO represents a pattern in which two primary nodes of sea surface temperature exist, a large one in the northern and western part of the basin, and a smaller one in the eastern tropical Pacific. At periods that range from about 20 to 30 years, the sea surface temperatures in the two zones undergo fairly abrupt shifts. For example, the period from about 1947 to 1976–77 was marked by generally low temperatures in the northern and western part of the ocean and high temperatures in the eastern tropical Pacific. Then an abrupt transition occurred and the reverse temperature pattern became established until the late 1990s.

In the late 1990s TOPEX/Poseidon and Jason 1 satellite data suggested that another reversal was underway, but in the early 2000s this shift seems to have halted. This phenomenon, along with other evidence, suggests that the PDO pattern may in fact be more complex than originally believed. The image in Figure 8–35 shows a pattern of warm water extending off the southern coast of Japan in June 2002. A wedge of warmer waters also lies in the eastern tropical Pacific. Mostly cooler waters exist off the west coast of North America. Because of the enormous heat content of the oceans, the PDO may exert a direct, major impact on the pressure distribution of the atmosphere. A history of the PDO is shown in Figure 8–36, plotting a useful indicator of the phenomenon, the *PDO index*, over the last century. Positive values indicate an episode with warm waters in the east tropical Pacific, and negative values representing cold waters in that area.

Recent research has also suggested that the phase of the PDO influences the impact that El Niño events have on climate. When the PDO is in a "warm phase" (high temperatures in the eastern tropical Pacific), El Niño's impacts on weather are more pronounced than when the PDO is in a cold phase (low temperatures in the eastern tropical Pacific).

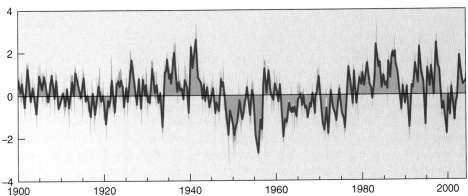

FIGURE 8–36
The PDO Index is a commonly used indicator of the strength and mode of the Pacific Decadal Oscillation. Positive values of the index, plotted for the years 1900–2004, indicate warm waters in the eastern tropical Pacific. In the late 1990s it appeared that a shift from positive to negative was occurring, but this shift seems to have abruptly ended in the early 2000s.

ARCTIC OSCILLATION

Multiyear oscillations are not restricted to the Pacific Ocean. An important one exists in the Atlantic. It is called the **Arctic Oscillation** (AO), and is closely related to another observed pattern called the **North Atlantic Oscillation** (NAO). Like the PDO, the AO alternates between a warm phase and a cold phase in sea surface temperatures. During the warm phase, there is relatively low pressure over polar locations and higher-than-normal pressure at lower latitudes. A weaker polar high means cold winter air masses do not penetrate as far south in the eastern United States, making for warm winters there. At the same time, cold air masses are more frequent in Newfoundland and Greenland, so winters are colder. The higher-than-normal pressure over the central Atlantic during this phase results in stronger westerly winds, steering warm (and moist) air to northern Europe. During the cold phase, higher-than-normal pressure dominates over the pole, with lower pressure in the central Atlantic. The eastern United States experiences cold winters, as do northern Europe and Asia, while Newfoundland and Greenland are relatively warm. The AO includes both a north–south oscillation in pressure and an east–west oscillation in temperature, with Greenland and Europe experiencing opposite conditions.

The Greenland–Europe temperature seesaw has been known for centuries, and climatologists have studied it for decades as part of the NAO. What's new is a realization that the temperature oscillation is part of a hemisphere-wide phenomenon and that it is closely linked to changes in the Arctic Ocean. In a warm phase, Atlantic water pushes farther into the Arctic, and as a result sea ice thins (by more than 1 m during the most recent warm phase). During a cold phase, low-level winds are strong, which effectively isolates the Arctic Ocean from warmer, saltier water to the south and promotes a thicker layer of sea ice.

The AO switches phase irregularly, roughly on a time scale of decades. This tends to produce a run of similar winter conditions. For example, Hitler's invasion of Russia during World War II bogged down in the extreme cold-phase winter of 1941–42, which itself followed two severely cold winters. (Ironically, in planning the campaign, he relied on a forecast of normal or mild conditions, which was largely based on the mistaken idea that three cold winters in succession would never occur.) It's interesting that the last 20 years or so have seen an unusually warm phase, exceeding anything observed in the last century. This might be further evidence of human-induced warming, or it might simply be the latest excursion in a long-observed natural cycle.

Summary

Although air pressure is not the first thing that comes to mind when we think of weather, it is the driving force for atmospheric motions. Horizontal changes in air pressure distribution not only initiate the wind but also influence the vertical motions that promote or inhibit cloud formation.

While the three-cell model of wind and pressure belts, or bands, provides a good foundation for understanding the general circulation, in reality the atmosphere is dominated by a number of semipermanent cells of pressure.

The circulation of the upper troposphere is largely the result of latitudinal differences in temperature, with a strong tendency for westerly winds outside of equatorial latitudes. The westerlies dominate, but other atmospheric movements occur in the upper troposphere. The most prominent are the Rossby waves, the large, looping motions of air associated with the jet streams. Adding further complexity to the upper-level winds are short waves and other eddying motions.

Although this chapter (and indeed this book) deals primarily with the atmosphere, weather and climate phenomena interact with the oceans. The large-scale winds of the atmosphere initiate the large, slow-moving currents of the ocean that likewise influence the input of energy and water vapor into the air. The most famous interaction between the oceans and the atmosphere is the ENSO event (El Niño–Southern Oscillation).

One of the most significant patterns of wind flow is the monsoon of Asia, the seasonal reversal of winds that are dry in the winter and extremely wet in the summer. Smaller-scale winds can also affect weather and climate. Foehn winds occur near many mountain ranges and bring hot, dry conditions. In North America, foehn winds called *chinooks* and *Santa Anas* warm the Great Plains and the West Coast, respectively. Daily sea/land breeze patterns are prominent features of coastal zones, while mountain and valley breezes produce similar patterns inland.

The circulation of the atmosphere is strongly intertwined with sea surface conditions across the oceans. We discussed three important patterns of atmospheric/oceanic behavior: the ENSO phenomenon, the Pacific Decadal Oscillation, and the Arctic Oscillation.

Key Terms

general circulation   page 213

single-cell model   page 214

zonal wind   page 214

meridional winds   page 214

three-cell model   page 214

Hadley cell   page 214

Ferrel cell   page 214

polar cell   page 214

Intertropical Convergence Zone (ITCZ; equatorial low)   page 215

subtropical highs   page 215

horse latitudes   page 215

northeast trade winds   page 215

southeast trade winds   page 215

subpolar lows   page 216

westerlies   page 216

polar highs   page 216

polar easterlies   page 216

semipermanent cells   page 216

Aleutian low   page 217

Icelandic low   page 217

Siberian high   page 217

Hawaiian high   page 217

Bermuda-Azores high   page 217

Tibetan low   page 217

polar front   page 222

polar jet stream   page 222

subtropical jet stream   page 222

Rossby wave   page 224

ocean currents   page 226

Ekman spiral   page 227

North Equatorial Current   page 228

South Equatorial Current   page 229

Equatorial Countercurrent   page 229

Gulf Stream   page 229

North Atlantic Drift   page 229

Canary Current   page 229

Labrador Current   page 229

East Greenland Drift   page 229

West Greenland Drift   page 229

upwelling   page 230

global scale   page 230

synoptic scale   page 230

mesoscale   page 230

microscale   page 230

monsoon   page 232

monsoon depressions   page 232

foehn winds   page 233

chinook winds   page 233

Santa Ana winds   page 234

katabatic winds   page 235

sea breeze   page 237

sea breeze front   page 237

land breeze   page 237

lake breeze   page 238

valley breeze   page 238

mountain breeze   page 238

El Niño   page 239

Walker circulation   page 239

Southern Oscillation   page 239

ENSO events   page 239

La Niña   page 241

Southern Oscillation Index (SOI)   page 241

teleconnections   page 243

Pacific Decadal Oscillation   page 248

Arctic Oscillation   page 249

North Atlantic Oscillation   page 249

Review Questions

1. Describe the single-cell and three-cell models of the general circulation.

2. What is the Hadley cell and where is it found?

3. Of the pressure and wind belts described in the three-cell model, which have the strongest basis in reality?

4. Why do the trade winds flow from the northeast and southeast instead of directly from the east?

5. What are the Ferrel and polar cells?

6. Describe the distribution of semipermanent cells and their seasonal changes in location and size.

7. What is the Sahel? Describe its seasonal cycle of rainfall and explain its origin.

8. Describe the average patterns of the 500 mb level for January and July. What causes the patterns?

9. Explain why upper-atmospheric winds outside the tropics have a strong westerly component on average.

10. Why does the equatorward bending of height contours for the 500 mb level imply the presence of a trough?

11. Explain how temperature patterns lead to the development of the polar jet stream.

12. Describe the distribution of Rossby waves and their impact on daily weather.

13. What is the Ekman spiral?

14. What is upwelling? How is it caused?

15. Describe the scope of global, synoptic, mesoscale, and microscale wind systems.

16. Describe the wind patterns associated with the monsoon of southern Asia.

17. Describe foehn winds.

18. How do katabatic winds differ in origin from foehn winds?

19. What causes sea/land and mountain/valley breezes to develop?

20. What is an El Niño and how is it related to the Walker circulation?

21. Describe the Pacific Decadal Oscillation and the Arctic Oscillation.

Critical Thinking

1. Figure 8–2a depicts the classic description of the single-cell model. Can you think of any reason the arrows should not be directed in a straight line toward the equator?

2. The three-cell model of circulation places the center of the equatorial low right along the equator. Do you think that the varying solar declination through the course of the year would be able to shift the center of the low all the way to the tropics of Cancer and Capricorn?

3. Which of the belts depicted in the three-cell model is likely to exhibit the greatest temperature gradients?

4. Examine Figure 8–4 and determine which of the semipermanent cells have the highest and lowest surface air pressures. How do the strengths of the winter and summer cells in the Northern and Southern Hemispheres compare to each other?

5. Figure 8–4 shows that pressure gradients, and therefore wind speeds, are strong in the middle-high latitudes of the Southern Hemisphere (such as at the southern tip of South America). Why doesn't a similar feature exist in the Northern Hemisphere?

6. If the western Great Plains were to have very low temperatures and the East Coast relatively warm conditions, what inferences would you make about the position and amplitude of the Rossby wave pattern?

7. Weather forecasters often use a type of weather map depicting the thickness of the 1000–500 mb layer of the atmosphere. What information would this map convey to the meteorologist?

8. Why isn't upwelling an important process of the Gulf Stream?

9. Why do ocean surface temperature patterns change so slowly, especially when compared to atmospheric patterns?

10. El Niño and La Niña conditions help meteorologists make seasonal forecasts for the southeastern United States—thousands of kilometers away. What atmospheric mechanism permits such extrapolations?

Problems and Exercises

1. Go to **http://weather.noaa.gov/fax/nwsfax.html**. In the first group of maps (labeled "standard barotropic levels"), select 500 mb. Then click on the link under *4a*, called *Height/temp*. The map that will appear is a map of the 500 mb pattern, with solid lines depicting the height of the 300 mb level (in tens of meters).

    a. Locate the position of the major troughs and ridges.

    b. Is the current pattern a zonal or meridional one?

    c. Where is the jet stream most prominent?

    d. Does today's jet stream appear across the entire region mapped?

    e. How does today's compare to the mean distributions shown in Figure 8–6? Why is the map for today (in all likelihood) less zonal than those in Figure 8–6?

2. If you have a high-speed connection to the Web, go to **http://grads.iges.org/pix/movie.html** to see an animation depicting the change in the 500 mb pattern as predicted by a numerical model.

    a. Describe the predicted movement of the Rossby waves.

    b. How does the movement of the Rossby waves respond to their wavelengths and internal wind speeds (as implied by the spacing of the contours)?

Quantitative Problems

Rossby wave patterns and other phenomena discussed in this chapter are best understood when actual numbers are applied to them. We suggest you log on to this book's Web site (**http://www.prenhall.com/aguado/**) and work out the quantitative problems for Chapter 8.

Useful Web Sites

http://weather.noaa.gov/fax/nwsfax.html
The National Weather Service site offers many current weather maps. The first group includes standard maps for various pressure levels, including maps for North America and the entire Northern Hemisphere.

http://www.ssec.wisc.edu/data/sst/latest_sst.gif
Satellite image depicting current ocean temperatures.

http://www.cpc.ncep.noaa.gov/products/analysis_monitoring/ enso_update/index.html
Weekly update on ENSO with several animations showing sea surface temperature distributions.

http://topex-www.jpl.nasa.gov/science/el-nino.html
Information on El Niño and La Niña, along with numerous links (including one on the PDO).

http://virga.sfsu.edu/gif/jetsat_00.gif
Satellite image of North America with superimposed arrows depicting strength and position of current jet stream.

http://iri.columbia.edu/climate/ENSO/
An excellent overview of ENSO.

Media Enrichment

Tutorial
Upper-level Winds and Pressure

This tutorial develops the relationship between upper-level pressure and height fields and shows how ridges and troughs are reflected on contour maps. Rossby waves are presented as common features of the midlatitudes, with animations showing simultaneous movement of the wave and prevailing westerly flow through the wave.

Tutorial
ENSO

This tutorial uses animations to show interactions between the atmosphere and ocean of the tropical Pacific. El Niños and La Niñas are contrasted with normal conditions, with both composite and individual episodes depicted. The tutorial also covers connections between these events and North American weather.

Weather Image
Algerian Dust Storm

Viewed from the space shuttle, this North African dust storm appears as a brownish blob. The margin of the dust storm is capped in places by cumulus clouds, formed by the uplift of air ahead of the wind. These storms can continue into the Atlantic Ocean, spreading dust as far away as North and South America.

Weather Image
Southern California Sea Breeze

This satellite view of the Los Angeles area shows the development of orographic clouds along Catalina Island and the Palos Verdes Peninsula. These clouds formed when the sea breeze was lifted as it approached land.

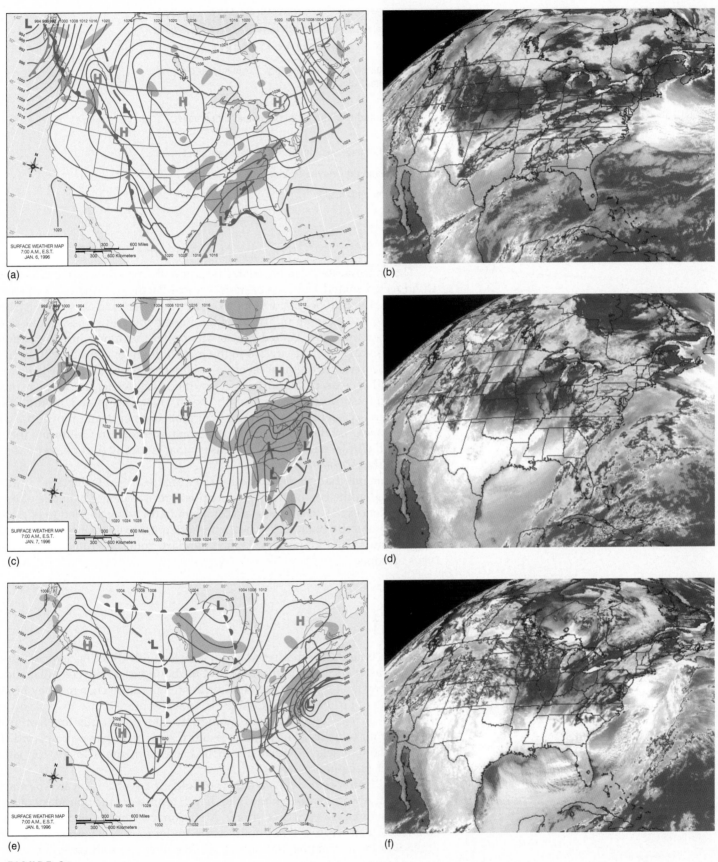

(a)

(b)

(c)

(d)

(e)

(f)

FIGURE 3

The northeaster of January 1996.

TABLE 9–1 Air Masses		
TYPE	**SOURCE REGIONS**	**PROPERTIES AT SOURCE**
Continental Arctic (cA)	Highest latitudes of Asia, North America, Greenland, and Antarctica	Extremely cold and very dry. Extremely stable. Minimal cloud cover.
Continental Polar (cP)	High-latitude continental interiors	Cold and dry. Very stable. Minimal cloud cover.
Maritime Polar (mP)	High-latitude oceans	Cold, damp, and cloudy. Somewhat unstable.
Continental Tropical (cT)	Low-latitude deserts	Hot and dry. Very unstable.
Maritime Tropical (mT)	Subtropical oceans	Warm and humid.

The southwestern United States occasionally experiences the effects of mT air advected inland from the eastern tropical Pacific. Off the coast of southern California, the cold ocean current moderates the overlying air so that it does not have either the extreme moisture or the high temperature associated with air over the southeastern United States. In the late summer, however, moisture sometimes moves northeastward in the form of high-level outflow from tropical storms or hurricanes off the west coast of Mexico. This can lead to a layer of high clouds and increased humidity over southern California. It can also produce local thundershowers. These usually occur over the inland mountains and deserts, but occasionally interrupt the normal summer drought over the coastal region as well. Farther to the east in Arizona and eastern California, the introduction of mT air over the deserts causes what is locally (but not very accurately) called the *Arizona monsoon*.

The general characteristics of air masses are described in Table 9–1, but as we emphasized at the beginning of this chapter, precise categorization is often difficult. The concept of an air mass is most meaningful when applied to large bodies of air separated by boundary zones with horizontal extents of tens (or perhaps a couple hundred) of kilometers. These boundary zones are categorized into four types of fronts.

Fronts

Fronts are boundaries that separate air masses with differing temperature and other characteristics. Often they represent the boundaries between polar and tropical air. They are important not only for the temperature changes they bring but also for the uplift they cause. Cold air is typically more dense than warm air, so when one air mass encroaches on another, the two do not mix together. Instead, the denser air remains near the surface and forces the warmer air upward. These upward motions lead to adiabatic cooling and sometimes to the formation of clouds and precipitation.

A **cold front** occurs when a wedge of cold air advances toward the warm air ahead of it. A **warm front**, on the other hand, represents the boundary of a warm air mass moving toward a cold one. A **stationary front** is usually similar to a cold front in structure but has not recently undergone substantial movement (in other words, it remains stationary). Unlike the other three types of fronts, **occluded fronts** do not separate tropical from polar air masses. Instead, they appear at the surface as the boundary between two polar air masses, with a colder polar air mass usually advancing on a slightly warmer air mass ahead of it. Figure 9–4 includes the symbols used to show the location of these fronts on surface weather maps.

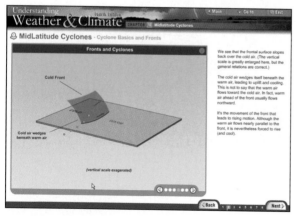

TUTORIAL
Midlatitude Cyclones

Though some fronts are named for the temperature contrasts associated with them, distinct changes in other weather elements also occur along these boundaries. People experiencing the passage of a front not only notice a rapid change in temperature; they may also note a shift in wind speed and direction, a change in moisture content, and an increase in cloud cover.

Figure 9–5 illustrates the overall structure of typical Northern Hemisphere midlatitude cyclones, of which cold, warm, and (sometimes) occluded fronts are an important part (we discuss midlatitude cyclones in detail in Chapter 10). The pressure distribution is somewhat circular in the larger cold sector on the poleward side of the cyclone, leading to a counterclockwise rotation of the air (in the Northern Hemisphere). The cold front separates the cold air that typically comes out of the northwest from the warmer air ahead. Between the cold front and the warm front ahead of it, pressure decreases toward the apex where the two fronts join together, and the wind is typically southwesterly. Let's now look at the structure of fronts in more detail.

COLD FRONTS

We are all familiar with the dramatic shifts in weather that can occur quickly with the approach and passage of a cold front. During the winter, they can disrupt life for millions of people as they cause intense flooding from heavy rain or white-out conditions from blowing snow.

Figure 9–6 shows the structure of a typical cold front, with its upper boundary sloping back in the direction of the cold air. The vertical scale in this figure is highly exaggerated and makes the slope of the front appear far steeper than it really is. In reality, the typical cold front has a surface slope of about 1:100, meaning that its surface rises only 1 m for every 100 m of horizontal extent.

Observe also that the forward edge of the front is not flat but curves backward, due to the decrease in friction with increased height. As the front advances, friction from the surface slows the lowest portion of the cold air. But the effect diminishes with distance from the surface, so the air at higher levels advances more rapidly. This gives the cold front its characteristic profile, with a steeper slope near the leading edge at the surface. Cold fronts move at widely varying speeds, ranging from a virtual standstill to about 50 km/hr (30 mph).

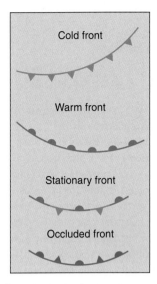

FIGURE 9–4
The symbols used to represent the four types of fronts on weather maps.

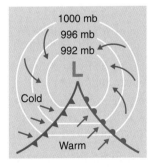

FIGURE 9–5
A typical midlatitude cyclone. Cold and warm fronts separated by a wedge of warm air meet at the center of low pressure. Cold air dominates the region outside the wedge separating the cold and warm fronts.

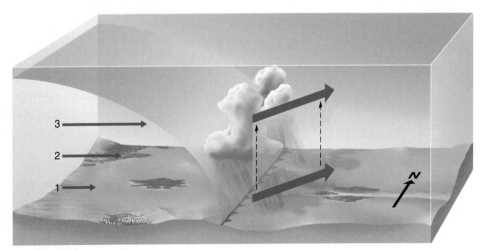

FIGURE 9–6
Cold fronts typically move more rapidly and in a slightly different direction from the warm air ahead of them. This causes convergence ahead of the front and uplift of the warm air that can lead to cumuliform cloud development and precipitation. In this example, the cold air (in blue) advances from west to east (notice that the wind speed depicted by the thin arrows increases with height). The warm air (in red) is blowing toward the northeast. The cold air wedges beneath the warm air and lifts it up.

The Pineapple Express

Most precipitation along the west coast of North America results from storms that cross the Pacific Ocean between November and April, and most of these storms pass through the Gulf of Alaska. Unlike the winter storms that affect the central part of Canada and the United States, the Pacific storms are dominated by maritime polar air, which seldom brings extremely low temperatures to the region. Thus, the precipitation associated with them almost always occurs as rain along the immediate coastal region and as rain or snow in inland (especially high-elevation) locations.

The snow line (the elevation marking the change from rain to snow) for these storms depends on the temperature. It usually varies between several hundred and a few thousand meters above sea level. The colder the storm, the lower the snow line. Mountain snowpack is critical to every eco-

nomic aspect of California. Water is stored in the watersheds as snow melts in the springtime and flows into the large reservoir system that provides drinking water and the irrigation supply. Precipitation that falls as rain evaporates more quickly than snow and is more likely to be lost before it can replenish the reservoirs.

When air temperatures remain above 0 °C long enough, the snow begins to melt. In southern California, this can be a matter of hours or days after snowfall. But in the upper reaches of the Sierra Nevada, snow can remain on the ground for months before it begins to melt in the spring.

In some years, the majority of Pacific storms do not approach the West Coast from the northwest but instead travel eastward from the vicinity of the Hawaiian Islands. This storm track is referred to locally as the *Pineapple Express*. When storms

take this path repeatedly, as they did during the El Niño winters of 1982–83 and 1992–93, the greatest amounts of precipitation are concentrated farther to the south than during more normal years. Instead of a general pattern of increasing precipitation with latitude, the southern part of the coast might receive greater amounts than the northern coast. Furthermore, because the Pineapple Express passes over warmer waters than do storms from the Gulf of Alaska, the air tends to be warmer and more humid, and the height of the snow line increases considerably. As a result, even though this type of storm track can lead to heavy precipitation in the mountains, a smaller percentage of it accumulates as snow—much to the chagrin of local residents who depend on the snow for skiing and other recreational activities.

The cloud cover we usually observe with the passage of a cold front results from the convergence of the two opposing air masses. Differences in wind speed and direction allow cold northwesterly wind to converge on the warm air ahead of it and displace it upward. Furthermore, the air ahead of a cold front tends to be unstable and therefore easily lifted. This promotes development of cumuliform clouds along these boundaries. With their large vertical extent, cumuliform clouds can often produce intense precipitation. However, because of the limited horizontal extent and rapid movement of the frontal wedge, such precipitation is often of short duration.

The production of surface weather maps is now almost entirely automated and performed by computer. But meteorologists still plot the location of fronts manually, because identifying them is often a subjective process. Meteorologists will look for the following five features to determine cold front positions:

1. Significant temperature differences between adjacent regions, with lower temperatures behind the cold front
2. Dew point differences, with drier air in the cold sector
3. Bands of cloud cover and precipitation, which nearly coincide with the position of the front
4. Narrow zones where wind direction changes, typically northwesterly in the cold sector to southwesterly in the adjacent warm region
5. Boundaries separating regions where the atmospheric pressure has decreased over the 3-hour period (typically ahead of the cold front) and those where the pressure has increased over the previous 3 hours (behind the cold front)

While one might logically expect that the best indicator of the position of a cold front is the shift in temperatures, that is not always true. When a cold front advances rapidly, the temperature does not decrease as rapidly behind the front as does the humidity. Therefore, the distribution of dew points is sometimes the better indicator of frontal position.

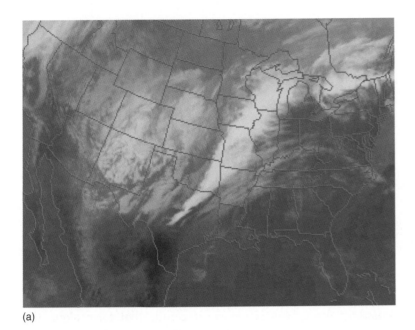

(a)

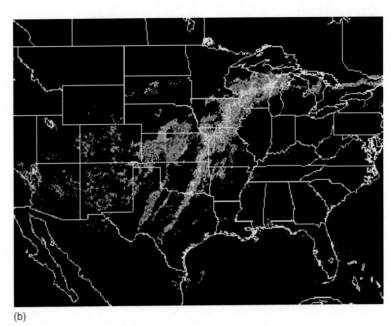

(b)

FIGURE 9–7

A midlatitude cyclone. The cloud cover in the satellite image (a) appears to be a continuous, uniform band. The radar composite map (b) reveals that the cloud cover is, in fact, marked by areas of varying precipitation intensities.

Though the position of cold fronts are plotted as a line on surface maps, they are three-dimensional and usually extend upward past the 500 mb level. Nonetheless, they are not plotted on upper-level weather maps, in part because there are not enough data available aloft to adequately determine their location and in part because their surface position is more significant than their position aloft.

When seen from space, the cloud bands ahead of cold fronts often appear to contain a fairly uniform distribution of cloud cover. Closer examination, however, reveals that the cloud bands often contain pockets of thicker cloud cover and more intense precipitation. For example, Figure 9–7 shows a frontal zone extending across the central United States on March 31, 1998. The portion of the front that extends from northeastern Missouri to central Texas is the cold front. The satellite view (Figure 9–7a) provides a compelling example of how the clouds are aligned parallel to the front, with a sharp transition between the warm, moist air ahead of the front and much drier, colder air behind. But as is often the case, this image gives a false impression of uniform precipitation.

The radar composite map (Figure 9–7b) gives us a much better description of the intensity of precipitation along the front. Weather radar imagery is obtained by emitting energy with wavelengths of about 10 cm (4 in.). Clouds composed of large droplets scatter some of the electromagnetic energy back toward the radar unit, which displays the position and other characteristics of the cloud. The color on the radar display indicates the intensity of the returned energy, with bright red portions indicating large droplets and heavy precipitation.

In this instance, four pockets of very heavy precipitation are located in central Missouri, eastern Kansas, eastern Oklahoma, and central Texas. Notice that at the time of the map, no precipitation is indicated along the front in northern Texas and central Oklahoma. However, those areas were hardly dry for the duration of the frontal passage. In fact, central Oklahoma received more rain than any other location, with recorded amounts above 4 cm (1.6 in.). This tendency for spotty, short-lived precipitation is typical of strong cold fronts, a feature that places them in distinct contrast to warm fronts, which we discuss next.

WARM FRONTS

Warm fronts separate advancing masses of warm air from the colder air ahead. As with cold fronts, the differing densities of the two air masses discourage mixing, so the warm air flows upward along the boundary. This process is called **overrunning**.

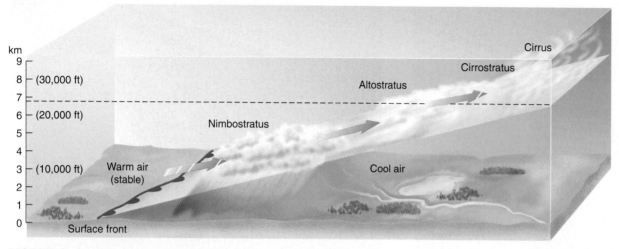

FIGURE 9–8

A warm front. Overrunning leads to extensive cloud cover along the gently sloping surface of cool air.

FIGURE 9–9

Warm fronts have gentler sloping surfaces and do not have the convex-upward profile of cold fronts. Surface friction decreases with distance from the ground, as indicated by the longer wind vectors away from the surface (a). This causes the surface of the front to become less steep through time (b).

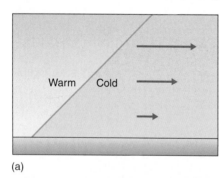

(a)

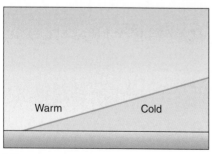

(b)

Figure 9–8 illustrates the typical sequence of clouds that form along the frontal surface. The warm air flows up along the frontal boundary in much the same way that air rises above a mountain slope. But the gradual slope of the frontal surface leads to a very gradual progression of cloud types. As the air rises along the boundary, adiabatic cooling first leads to formation of low-level stratus clouds. As the air continues to rise, a sequence of higher clouds develops, with nimbostratus, altostratus, cirrostratus, and, finally, cirrus occurring in that order. As the front moves eastward or northeastward, the leading segment of the cloud sequence (the cirrus) is seen first, followed by the continually thickening and lowering cloud cover. Thus, even an amateur forecaster can predict an episode of continuous light rain from a warm front a day or two in advance, simply by observing the sequence of clouds as they pass overhead.

Warm fronts are about half as steep as cold fronts (their slopes are about 1:200), which causes the lifting of the warm air to extend for greater horizontal distances than for cold fronts. Although the clouds above the warm front cover a greater horizontal extent than the clouds above a cold front, they usually consist of smaller droplets and have a lower liquid water content. This results in less intense precipitation than is usually associated with cold fronts. For this reason, warm fronts in the summer are considered a friend to farmers; they provide a light, steady rain that nourishes crops but does not produce flash floods or severe storms.

The overrunning air above a warm front is usually stable, which normally leads to the formation of wide bands of stratiform cloud cover. Though the rising air may sometimes be unstable and lead to formation of cumuliform clouds, this is the exception rather than the rule. Precipitation along the front tends to be light, but the wide horizontal extent and typically slow movement of warm fronts (usually about 20 km/hr—12 mph) allows rain to persist over an area for up to several days.

The clouds along a warm front exist in the warm air above the wedge of dense, cold air. Thus, when rain falls from the base of these clouds, the falling droplets pass through the cold air below on their way to the surface. If the falling droplets are substantially warmer than the air they fall through, they can evaporate rapidly and form a frontal fog.

If the air is cold enough during the passage of winter warm fronts, the drops freeze on their way down to form sleet, or they solidify on contact with the surface to form freezing rain. So summer warm fronts are generally beneficial, while in winter they can cause widespread problems.

Figure 9–9 shows the typical change in the slope of a warm front over time. Because wind speeds are greater aloft than near the surface ahead of the front, the upper portion of the wedge advances more rapidly than does the part near the surface, and the slope of the frontal boundary becomes less steep.

Meteorologists use much the same rules for identifying warm front positions that they do for cold fronts, but with some differences. One must first look for a zone where warmer air advances toward cooler air (the opposite of cold fronts). Dew points typically increase behind the position of the warm front; winds commonly shift from southwesterly ahead of the front to southeasterly behind it; and cloud cover and precipitation bands are common. Also, the zone ahead of the warm front generally undergoes decreasing air pressures while the area immediately behind the front typically has stable air pressure.

STATIONARY FRONTS

More often than not, the boundaries separating air masses move gradually across the landscape; at other times, they stall and remain in place for extended time periods. Nonmoving boundaries are called *stationary fronts*. They are identical to cold or warm fronts in terms of the relationship between their air masses. As always, the frontal surface is inclined, sloping over the cold air.

Determining whether a front is stationary is somewhat subjective. For example, if the front advances at a rate of 1 km/hr (0.6 mph), does that constitute enough movement for it to be nonstationary? If not, what about 2 km/hr? In practice, a meteorologist will make the designation by looking at the previous one or two surface weather maps (compiled at 3-hour intervals). If there has been no movement during that period, the front is considered stationary.

If this sounds simple, keep in mind that the exact position of a front at any point in time is difficult to pinpoint. This is in part because fronts are zones of transition rather than abrupt boundaries; they are located at no precise line. Furthermore, maps are compiled from a finite number of weather stations. A front may undergo some movement that is not detected because it does not move across an observing station.

OCCLUDED FRONTS

The most complex type of front is an occluded front (sometimes called an *occlusion*). The term *occlusion* refers to closure—in this case, cutting off a warm air mass from the surface by the meeting of two cold fronts. According to the original model of cyclone evolution, occluded fronts form when a faster-moving cold front "catches up" to the warm front ahead. The warm air rises. At the surface, one mass of cold air merges with another, so there is a smaller difference in temperature (and likewise dew point) from one side of an occluded front to the other. A warm air mass is present, but it is aloft, pinched off from the surface, and therefore not reflected in surface temperature.

Figure 9–10a shows a typical midlatitude cyclone prior to occlusion, with the cold and warm fronts intersecting at the center of low pressure. When the cold front meets the warm front ahead of it, the segments of the two fronts closest to each other become occluded, as shown in (b). The warm air does not disappear but instead gets lifted upward, away from the surface. The occluded front becomes longer as more and more of the cold front converges with the warm front. Eventually, the cold front completely overtakes the warm front (c), and the entire system is occluded.

In this occlusion, the air behind the original cold front was colder than that ahead of the warm front. This is an example of a *cold-type occlusion*, originally considered to be the more common type because air in the western part of the storm is imported from colder, higher latitudes. In British Columbia and the Pacific Northwest of the United States, *warm-type occlusions* are more common. In these, relatively mild mP air behind a Pacific cold front moves onshore and meets up with more frigid cP air.

This description of the occlusion process is essentially what was proposed by the leading meteorologists of the early twentieth century—and it probably does apply to many midlatitude cyclones. However, in many cases, perhaps most,

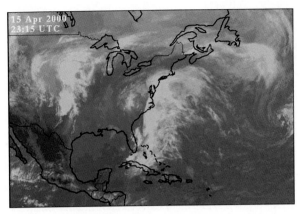

WEATHER IMAGE
Eastern U.S. Water Vapor

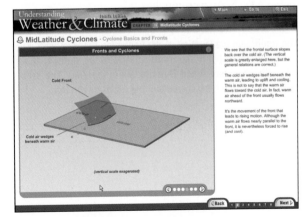

TUTORIAL
Midlatitude Cyclones

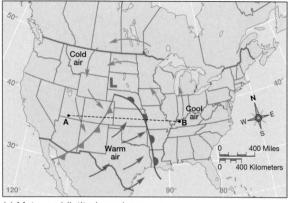

(a) Mature midlatitude cyclone

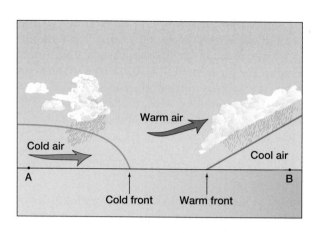

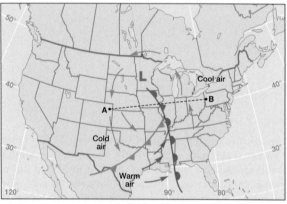

(b) Partially occluded midlatitude cyclone

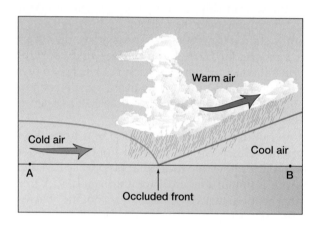

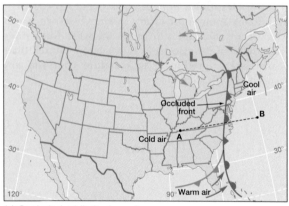

(c) Occluded midlatitude cyclone

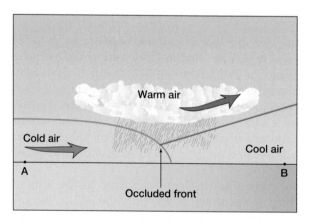

FIGURE 9–10

The traditional explanation of the occlusion process, with a cold front overtaking a warm front. The panels on the left show the placement of the fronts relative to the surface, along with the location of a sample cross-section (the dashed line with the letters *A* and *B* at the end of the line). The panels on the right show the profiles of the fronts across the transect shown on the corresponding panel on the left.

structures resembling an occluded front form in ways unrelated to the traditional explanation. For example, occlusions sometimes occur when the circular core of low pressure near the junction of the cold and warm fronts changes shape and stretches backward, away from its original position (Figure 9–11). In (a), the cold and warm fronts are joined at the dashed line. At some later time (b), the cold and warm fronts have the same orientation with respect to each other as they did in

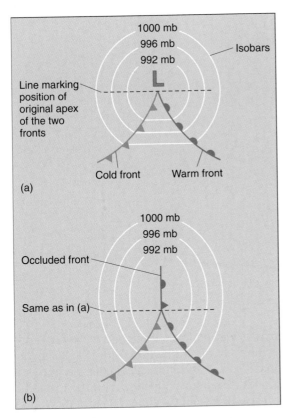

FIGURE 9–11
Some occluded fronts form when the surface low elongates and moves away from the junction of the cold and warm fronts.

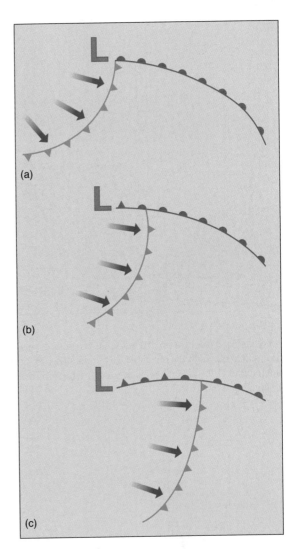

FIGURE 9–12
Some occlusions occur when the intersection of the cold and warm fronts slides along the warm front.

(a), but both have been pulled back beyond the dashed line. The circular isobar pattern of (a) becomes elongated to form a trough over the occluded region. Thus, the structure of the occluded front is the same as that proposed in the traditional model, but the process responsible for its formation is somewhat different. In other cases, the cold front moves eastward relative to the warm front, so that the point where the fronts intersect moves progressively farther east (Figure 9–12). The occluded front appears as a relic portion of the warm front, situated to the west of the new intersection of the cold and warm fronts.

Yet another way occluded fronts can form is by a combination of distinct surface and upper-level processes. At low levels, a cold front overtakes and merges with a warm front, while aloft a rapidly moving upper-level frontal zone overtakes and joins with the surface front. Again the result is an occlusion, but not for reasons outlined in the classical description. Finally, we should note that cold-type occlusions are now thought to be rare events, with warm-type occlusions far more common for all midlatitude cyclones, whether originating over ocean or land.

DRYLINES

Though fronts are the most common type of boundary separating air masses, we find another type of boundary in the spring and summer in Texas and the southern Great Plains. **Drylines** are areas where mT and cT air masses reside next to each other. Though the temperatures on either side of the boundary may not differ enough to create a front, the changes in water vapor content across the dryline can be substantial.

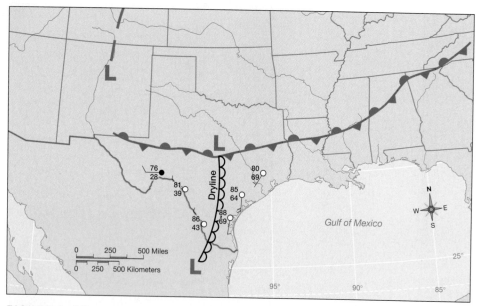

FIGURE 9–13

A dryline over Texas. West of the line the humidity is low, with dew points ranging from the upper 20s to the 40s (°F). East of the line the air humidity is greater, owing to the flow of air from the Gulf of Mexico.

The dryline forms as warm moist air advects westward and meets up with cT air moving in from the west or southwest. Figure 9–13 shows a dryline moving across Texas at noon CST on March 30, 2002. The temperatures and dew points (in °F)[3] at the selected weather stations are shown on the top left and bottom left of the station models, respectively. The dryline clearly demarcates the moist air from the Gulf and the drier air to the west.

Drylines tend to have greater temperature differences across their boundaries during the day than at night. The difference in temperature usually gets smaller overnight as the lower water vapor content of the cT air promotes more rapid cooling than occurs for the mT air to the east. By the early morning the greater cooling rate allows the cT air temperatures to approach those of the mT air, reducing the temperature contrast. During the day, more rapid warming occurs west of the line.

Table 9–2 shows the change in temperature and dew point observed at an automated weather-observing system near Austin, Texas. A substantial drop in the humidity occurred as the dryline passed between 2 and 3 P.M., though no increase in temperature occurred. It is noteworthy that thunderstorms had been going on in the area during the morning hours because drylines often promote thunderstorms with the potential for tornadoes or other severe weather. The mechanism by which this happens will be discussed further in Chapter 11.

---

[3]We are using °F rather than °C in this instance because U.S. surface weather maps still rely on the Fahrenheit scale.

TABLE 9–2   **Temperatures and Dew Points Prior to and Following the Passage of a Dryline**

TIME (CST)	TEMPERATURE °F (°C)	DEW POINT °F (°C)
10 A.M.	66 (19)	64 (18)
11 A.M.	66 (19)	62 (17)
12 P.M.	71 (22)	64 (18)
1 P.M.	77 (25)	66 (19)
2 P.M.	78 (26)	62 (17)
3 P.M.	78 (26)	46 (8)
4 P.M.	80 (27)	46 (8)
5 P.M.	75 (24)	39 (4)

Summary

We have seen that the atmosphere tends to arrange itself into large masses with relatively little horizontal change in temperature and humidity. Air masses form over particular source regions—large areas of high or low latitudes of either continental or maritime locations. The five major types of air masses are continental polar, continental arctic, maritime polar, continental tropical, and maritime tropical.

The boundaries of these masses, called *fronts*, are particularly important to meteorology—not only because their passage causes abrupt changes in the temperature and humidity, but because they promote uplift and cloud formation. The four types of fronts are cold, warm, stationary, and occluded. They are components of weather systems called *midlatitude cyclones*, which we discuss in the next chapter. Drylines are similar to fronts in that they separate air masses, though the dryline marks a humidity boundary rather than a temperature boundary. Drylines are preferred regions for severe storm activity, a topic discussed in Chapter 11.

Key Terms

air masses   page 255

fronts   page 255

source regions   page 256

continental polar air masses
page 256

continental arctic air masses
page 258

maritime polar air masses
page 258

northeasters   page 259

continental tropical air
masses   page 259

maritime tropical air masses
page 259

cold front   page 264

warm front   page 264

stationary front   page 264

occluded front   page 264

overrunning   page 267

drylines   page 271

Review Questions

1. What are the requirements for an area to serve as a source region?

2. Where are the primary air mass source regions in North America?

3. Describe the characteristics of cA, cP, cT, mT, and mP air masses.

4. Of the five types of air masses, which are the hottest, driest, coldest, and most damp?

5. What is the primary difference between arctic and polar air masses?

6. Which of the air mass types are likely to be stable or unstable?

7. Describe the changes that occur when a continental air mass migrates out of its source region.

8. Describe the structure of cold, warm, stationary, and occluded fronts.

9. What is overrunning?

10. Why do cold fronts have steeper slopes than warm fronts?

11. How do the alternative models of the occlusion process differ from the traditional model?

12. What is the difference between warm-type and cold-type occlusions?

13. What are drylines and why are they important?

Critical Thinking

1. Contrast the formation of air masses in the Northern and Southern Hemispheres.

2. Explain the limitations and benefits of classifying air into distinct masses.

3. Continental polar air masses can migrate into Florida during the winter but not into northern India. Why not?

4. What parts of North America are likely to experience the most frequent changes in air mass during the summer and winter?

5. Distinct temperature changes can be detected across narrow regions that are not associated with fronts. What can cause such conditions to exist?

6. The southwestern United States experiences what is locally referred to as a monsoon. Can we say the same thing about Florida or Texas?

7. Warm fronts are extremely rare over southern California. Why?

8. What does the presence of a continental polar air mass tell you about the height of the 500 or 300 mb levels?

9. Where in North America might you expect to see the collision of mT and cT air masses? What time of year would this be most likely?

10. Which types of fronts are most and least likely to have inversions?

Problems and Exercises

1. The following temperatures and dew points are observed. What are the likely types of air masses present for each?

   a. T = 29 °C (85 °F)   Dew Pt = 19 °C (66 °F)

   b. T = −18 °C (0 °F)   Dew Pt = −21 °C (−5 °F)

   c. T = 3 °C (38 °F)   Dew Pt = 0 °C (32 °F)

   d. T = 38 °C (100 °F)   Dew Pt = −4 °C (−25 °F)

2. Assume that a cold front has a slope of 1:100 and that the height of the 700 mb level is 1500 m. How far behind the surface position of the front will the 700 mb position be?

Quantitative Problems

The Web site for this book (**http://www.prenhall.com/aguado/**) offers some quantitative problems that can help reinforce the concept of air masses and fronts. We encourage you to go to the Chapter 9 page of the Web site and answer the problems.

Useful Web Sites

http://www.udel.edu/dgs/Publications/pubsonline/OFR40/1998damagingstorms.htm
Some background on the effects of northeasters.

http://weather.unisys.com/surface/sfc_front.html
Map of current position of surface fronts across North America.

Media Enrichment

Weather in Motion
Satellite Movie of the January 6–8, 1996, Blizzard

This movie shows the blizzard of 1996 as seen by color infrared satellite imagery. The bright red portions show the regions of most intense precipitation. As you can see, the zone of heaviest storm activity moves northward along the Atlantic coast. As that portion of the storm approaches New England, a distinct flow of moisture rotates westward toward the Great Lakes, feeding moisture into the system and promoting exceptionally strong blizzard conditions. By the end of the movie, a large, sweeping arc of heavy moisture associated with maritime polar air extends well into the continental interior. As we will see in Chapter 10, this sweeping zone clearly illustrates one of the three distinct flows associated with the conveyor belt model of midlatitude cyclones.

Weather Image
The Perfect Storm

This GOES 7 color-enhanced infrared image (taken 1200 UTC October 30, 1991) shows an enormous midlatitude cyclone that wreaked havoc along the entire Atlantic coast. This storm was called "the Perfect Storm" by the National Weather Service and was the subject of a best-selling book and popular movie. An excellent example of a "nor'easter," the storm was an extremely strong extratropical cyclone that developed during the autumn along the East Coast of the United States due, in part, to the contrast between continental polar (cP) air masses from Canada and milder, maritime air masses along the Atlantic Coast. This storm developed along a cold front located off the east coast of the United States. A strong upper air low and remnant moisture from Hurricane Grace caused the storm to intensify explosively. During the storm, hurricane-force winds were reported on Cape Cod, Massachusetts, and flooding and record high tides occurred all along the mid-Atlantic and New England coastlines.

Weather Image
Pacific Storm Water Vapor

This GOES satellite water vapor image shows a midlatitude cyclone located near the coast of the western United States. The image was taken on December 31, 1996. This storm brought maritime polar (mP) air into the Pacific Northwest.

Weather Image
Continental Tropical Air Mass

This image shows atmospheric water vapor content over northern Africa at 23:15Z on April 15, 2000. Notice that much of the atmosphere over the Sahara Desert is very dry in comparison to other regions in the image. Like the southwestern United States and northern Mexico, northern Africa is a source region for continental tropical (cT) air masses. This region has very little available surface water and minimal vegetation. In addition, surface temperatures are very high due to northern Africa's subtropical location. Consequently, air masses over northern Africa—and other continental tropical source regions—are extremely hot, dry, and often cloud-free. In this image, there are some clouds (most likely due to thunderstorms) along the Ivory Coast, where moisture from the Atlantic Ocean has penetrated northward.

Weather Image
Eastern U.S. Water Vapor

This image shows atmospheric water vapor content over the eastern United States at 23:15Z on April 15, 2000. This image comprises data from the water vapor channels of four weather satellites in geosynchronous orbit (GOES satellites). Light areas correspond to relatively wet regions (often clouds), and dark areas correspond to relatively dry regions. Notice the region of high water vapor content along the eastern coast of the United States. This cloudiness is due to the presence of a stationary front located along the East Coast.

PART FOUR
Disturbances

Overhead view of buried cars on an interstate highway after a severe five-day snowstorm, Buffalo.

CHAPTER

10 | MIDLATITUDE CYCLONES

CHAPTER OUTLINE

L ess than a week before the end of 1999, as people everywhere fretted about the Y2K computer bug, much of Europe was shaken by a natural event unrelated to the turning of the calendar. The most powerful storm in 50 years hit Europe with winds that topped 190 km/hr (110 mph). It brought power outages to nearly two million households, shut down three nuclear power plants, and crippled much of France's air and land transportation systems. Worse still, 97 people from across western Europe died because of the storm, mostly from being hit by falling or flying debris. When asked about the situation, a worker cleaning up the debris, David Chézeaud, said, "Since I was born, I've never seen anything like that storm. All over Paris there are problems like this."

Among the European countries, France was the worst hit—both in terms of loss of life and material damage. Many of Paris's most famous landmarks were badly damaged, and 10,000 trees were toppled in Versailles. Germany, Switzerland, and Great Britain also sustained major damage and numerous fatalities.

Although European forecasters saw the storm approaching from the Atlantic and predicted that it would reach the continent, they did not expect its winds to be as severe as they were by the time it made landfall. As is the case with the west coast of North America, weather forecasting in Europe is complicated by the paucity of weather stations over the expanse of ocean to the west of the continent—a fact sometimes overlooked by the public. No sooner had the criticism started to mount over the understated forecast—two days after the storm—than another, even stronger, storm arrived, killing at least 22 more people. Thus, one very rare event was followed very shortly by a second. This rare combination of events should remind us that regardless of the technological sophistication of the twenty-first century, we are subject to the random whims of nature.

The two storm systems brought hurricane-force winds but were distinctly different from hurricanes. Unlike hurricanes, which originate over tropical waters, these *midlatitude cyclones* originate in the middle or high latitudes and are marked by well-defined fronts separating two dissimilar air masses. This chapter describes the midlatitude cyclones that are a common feature of the weather outside the tropics.

Polar Front Theory

From 1914 to 1918, the world experienced one of history's great catastrophes, World War I. The advent of new weapons, including the machine gun and mustard gas, made it nearly impossible for opposing armies to gain large tracts of ground from their opponents. Until then, an army attacking with rifles and bayonets could charge its opponents with some reasonable chance of success. However, against a foe dug into trenches and armed with the latest weapons, such maneuvers were almost doomed to failure. Thus, the war zone remained stagnant for long periods, since neither army could advance across the *front*, or battle line.

While the war was going on in western Europe, Vilhelm Bjerknes (pronounced *bee-YURK-ness*) established the Norwegian Geophysical Institute in the city of Bergen. With several colleagues,[1] including his son Jacob, Bjerknes developed a modern theory of the formation, growth, and dissipation of midlatitude cyclones, storms that form along a front in middle and high latitudes. Bjerknes observed the systems forming along a boundary separating polar air from warmer air to the south. Comparing that boundary to the one separating the opposing armies in western Europe, he called his model the **polar front theory** (also called

TUTORIAL
Midlatitude Cyclones

---

[1]Tor Bergeron, who discovered the ice crystal process for the formation of precipitation (Chapter 7), was among the scientists in this group.

◀ **Severe flooding in the Midwest.**

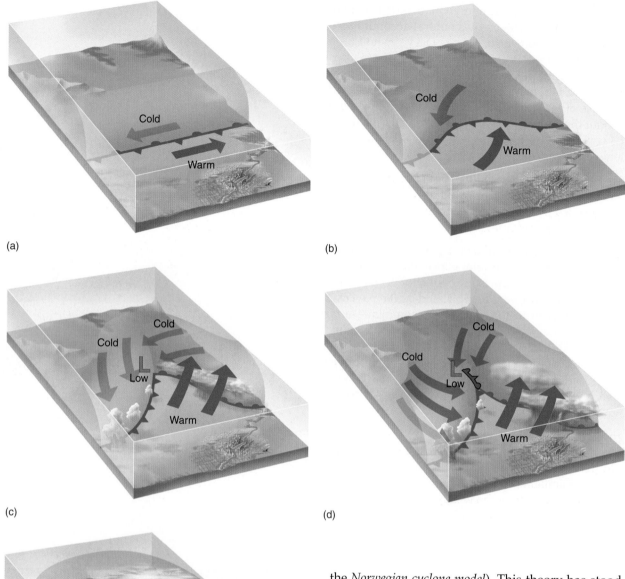

(a)

(b)

(c)

(d)

(e)

FIGURE 10–1

The life cycle of a midlatitude cyclone. The stationary polar front (a) separates opposing masses of cold and warm air. Cyclogenesis first appears as a disruption of the linear frontal boundary (b); as the cyclone becomes mature, distinct warm and cold fronts extend from a low-pressure center (c) and (d). Occlusion occurs as the center of the low pulls back from the warm and cold fronts (e).

the *Norwegian cyclone model*). This theory has stood the test of time remarkably well, and though we now have far more observational information available than did Bjerknes (especially for the middle and upper troposphere), we still describe the life cycle of a midlatitude cyclone in much the same way that the scientists of the Bergen school did decades ago.

Midlatitude cyclones are large systems that travel great distances and often bring precipitation—and sometimes severe weather—to wide areas. Lasting a week or more and covering large portions of a continent, they are familiar as the systems that bring abrupt changes in wind, temperature, and sky conditions. Indeed, all of us who live outside the tropics are well acquainted with the effects of these common events.

The Life Cycle of a Midlatitude Cyclone

The Bergen meteorologists were perfectly placed to witness the atmosphere along the polar front, and they used their observations to describe the formation of midlatitude cyclones, a process called **cyclogenesis**, along this boundary. Although many cyclones do originate along the polar front, they also form in other areas, especially downwind of major mountain barriers. We

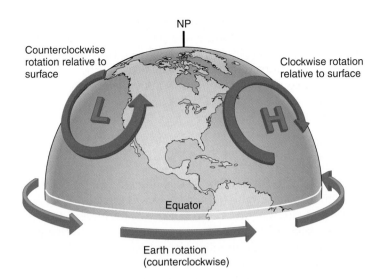

FIGURE 10–5

Earth vorticity and relative vorticity of air. As the Northern Hemisphere rotates counterclockwise, it generates Earth vorticity. Relative vorticity is the rotation of air relative to the surface, without regard to the planet's rotation. Absolute vorticity is the sum of the two.

the bottom left of the figure). Between points 3 and 5, it rotates clockwise. The rotation of a fluid (such as air) is referred to as its **vorticity**.[3] The figure shows vorticity changes in the moving air, relative to the surface. Viewed from space, there is an additional component of vorticity arising from Earth's rotation on its axis. The overall rotation of air, or its **absolute vorticity**, thus has two components: **relative vorticity**, or the vorticity relative to Earth's surface, and **Earth vorticity**, which is due to Earth's daily rotation about its axis.

Relative vorticity depends on air motions with respect to Earth's surface, while Earth vorticity is a function solely of latitude—the higher the latitude, the greater the vorticity—with zero vorticity at the equator.[4] If the flow of air relative to the surface is in the same direction as the rotation of Earth itself (counterclockwise in the Northern Hemisphere), relative and Earth vorticity complement each other and increase the total or absolute vorticity (Figure 10–5). For this reason, counterclockwise rotation in the Northern Hemisphere is said to have *positive vorticity*, to be consistent with the convention used for the Coriolis force (see Chapter 4). Air rotating clockwise possesses *negative vorticity*.

Figure 10–6 shows the trough from Figure 10–4 in greater detail so we can examine the air's vorticity. In segment 1, the air flows southeastward with no change in direction or speed. Because it undergoes no rotation, the air has zero relative vorticity. In segment 2, the air continuously turns to its left to yield positive relative vorticity. In segment 3, the air flows continuously toward the northeast and has no relative vorticity. Thus, the trough has three distinct regions: two with zero relative vorticity and one with positive relative vorticity. Two transition zones separate the regions of maximum and minimum (zero) relative vorticity. Across transition zone *A*, vorticity increases as the air flows, whereas across transition zone *B* the relative vorticity decreases. (The Rossby wave shown here covers a limited range of latitude. As a result, Earth vorticity changes only slightly in Figure 10–6, and changes in absolute vorticity correspond closely to changes in relative vorticity.)

At this point you might logically ask, "So what?" The answer is that vorticity changes in the upper troposphere lead to pressure changes near the surface. Let us see how. As you learned in Chapter 8, angular momentum is conserved in the absence of any outside forces. As a cowboy's twirling rope is pulled in, the reduction

FIGURE 10–6

The change in vorticity along a Rossby wave trough. As the air flows from positions 1 to 3, it undergoes little change in direction and thus has no relative vorticity. From positions 4 to 6, it turns counterclockwise and thus has positive relative vorticity. The air flows in a constant direction from positions 7 to 9. Thus, the trough has three segments based on vorticity, separated by two transition zones.

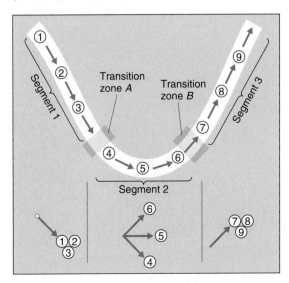

---

[3]For this discussion, we are only concerned with rotation with respect to the local vertical (such as on a merry-go-round). *Vorticity* can also refer to rotation around horizontal axis (such as with a horizontal roll of paper towels).

[4]Earth vorticity is proportional to the sine of the latitude. Thus, Earth vorticity is only 16 percent greater at latitude $55°$ (sin $55° = 0.819$), than at $45°$ (sin $45° = 0.707$), This means that over the midlatitudes where Rossby waves are most likely to occur, the Earth vorticity changes with latitude are fairly small.

in the area swept by the rope causes it to twirl faster. The same thing happens when the vorticity or rotation of a parcel of air changes. That is, as the horizontal area occupied by an air parcel decreases by convergence, its vorticity or spin must increase, as in transition zone *A*. Decreasing vorticity, as in transition zone *B*, leads to divergence. This very important relationship can be summarized in the simple equation

$$-\frac{1}{\zeta}\frac{\Delta\zeta}{\Delta t} = div$$

where $-\dfrac{1}{\zeta}\dfrac{\Delta\zeta}{\Delta t}$ is the standardized change (here, the decrease) in absolute vorticity (ζ) with respect to time (*t*), and *div* = divergence. Conversely, an increase in absolute vorticity with respect to time leads to convergence. (For simplicity, we limit our discussion of divergence and convergence to horizontal changes in area.)

Divergence in the upper atmosphere, caused by decreasing vorticity, draws air upward from the surface and provides a lifting mechanism for the intervening column of air. This, in turn, can initiate and maintain low-pressure systems at the surface (Figure 10–7). Conversely, increasing upper-level vorticity leads to convergence and the sinking of air, which creates high pressure at the surface. Surface low-pressure systems resulting from upper-tropospheric motions are referred to as **dynamic lows** (also called *cold core lows*)—distinct from the **thermal** (*warm core*) **lows** caused by localized heating of the air from below. Cold core lows at the surface typically exist beneath regions of decreasing vorticity in the upper atmosphere, just downwind of trough axes. They are therefore associated with low pressure aloft, though the center of the low is not located directly above the surface low. In contrast, warm core lows have high pressure aloft. This arises because they have a relatively weak decline in pressure with altitude due to their higher temperatures. Thus, at some height in the middle troposphere their pressure becomes greater than that of the surrounding air, and a zone of high pressure exists. (Hurricanes, discussed in Chapter 12, are classic examples of warm core lows.)

Figure 10–8 shows a typical relationship between the distribution of 500 mb heights (representative of the pressure pattern in the middle troposphere) and absolute vorticity. The areas of greatest vorticity (shaded purple) occur along the two trough axes (in this case, over northern California and the lower Mississippi Valley). Downwind of these zones, vorticity decreases very rapidly. Thus, as air flows away from the vorticity maxima, upper-level divergence occurs, which in turn

FIGURE 10–7
Upper-level convergence and divergence along favored positions on a Rossby wave create high and low pressure at the surface.

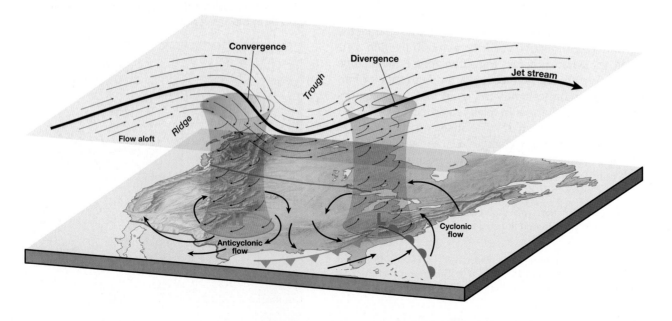

ones, and the same number of people now occupy a greater area.

Because wind speed on an upper-level weather map is directly proportional to the spacing of height contours, we can use these maps to identify regions of speed divergence. Specifically, speed divergence occurs where contour lines come closer together in the downwind direction. In Figure 1c, the wind speed, indicated by the length of the arrows, increases in the direction of flow and causes speed divergence.

Speed convergence occurs when faster-moving air approaches the slower-moving air ahead. In the example of runners in the race, convergence might occur behind a muddy part of the track that slows the runners. The fastest runners, who have pulled ahead of the others, are the first to encounter the muddy spot. As they slog through the muck, the trailing runners have the opportunity to catch up. The entire pack bunches up in a smaller area and convergence occurs. A similar phenomenon occurs in Figure 1d.

DIFFLUENCE AND CONFLUENCE

A second type of divergence and convergence, **diffluence** and **confluence**, occurs when air stretches out or converges horizontally due to variations in wind direction. In Figure 2a, a certain amount of air is contained in the shaded area between points 1 and 3. As it passes to the region between points 2 and 4, the same amount of air occupies a greater horizontal area. This is diffluence, a pattern that commonly appears wherever vorticity changes cause divergence. Confluence is shown in Figure 2b.

Close inspection of Figure 2 reveals an interesting relationship between the different types

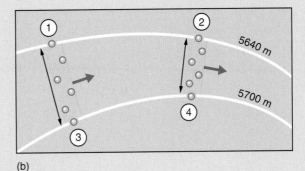

(a)

(b)

FIGURE 2
Diffluence and confluence.

of convergence and divergence. Diffluence occurs where height contours on an upper-level map spread apart in the upwind direction (confluence occurs where they converge). But as the height contours spread apart, the horizontal pressure gradient decreases downwind and the air slows down. This, of course, creates *speed convergence*. Stated another way, one type of divergence (diffluence) is accompanied by a type of convergence (speed convergence).

How does divergence actually occur downwind of a trough axis when diffluence and speed convergence occur simultaneously? The answer is that in most instances the diffluence downwind of trough axes is slightly greater in magnitude than the speed convergence, so the sum of the two yields a net divergence.

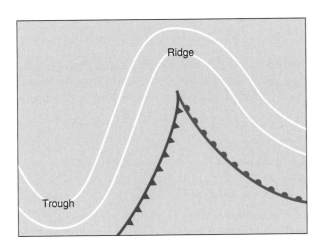

FIGURE 10–10
The effect of differing vertical pressure gradients on either side of warm and cold fronts leads to upper-tropospheric troughs and ridges.

Short Waves in the Upper Atmosphere and Their Effect on Surface Conditions

While Rossby waves play a role in establishing regions of upper-level divergence and convergence, they are not the only waves in the atmosphere that do so. The atmosphere also contains smaller eddies. Some of these, called **short waves**, are smaller ripples superimposed on the larger Rossby waves. These eddies migrate downwind within the Rossby waves and exert their own impact on the life cycle of midlatitude cyclones. Depending on where they are located within the Rossby waves, they can either enhance or reduce the local divergence or convergence.

The formation of short waves depends on **temperature advection**, the horizontal transport of warm or cold air by the wind. Because air in the upper troposphere is well removed from the direct source of atmospheric heating (the surface), the temperature changes we experience from day

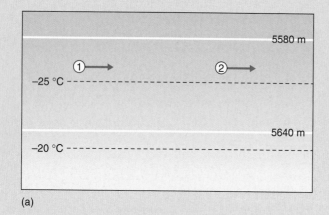

(a)

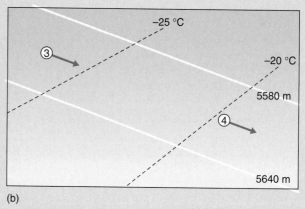

(b)

FIGURE 1

A barotropic atmosphere (a) exists where the isotherms (the dashed lines showing the temperature distribution) and height contours (solid lines) are aligned in the same direction. No temperature advection occurs when the atmosphere is barotropic. A baroclinic atmosphere occurs where the isotherms intersect the height contours. Cold air advection is occurring in (b), warm air advection in (c).

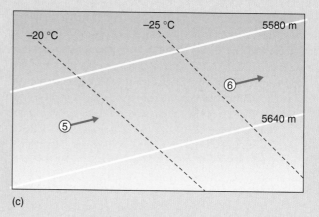

(c)

patterns are influenced in turn by temperature conditions near the surface. (See *Box 10–3, Forecasting: Short Waves in the Upper Atmosphere and Their Effect on Surface Conditions*, for a more detailed analysis of upper-level and surface phenomena.)

An Example of a Midlatitude Cyclone

The descriptions and examples of troughs, ridges, cyclones, anticyclones (each described in Chapter 4) and fronts (Chapter 9) represent idealized patterns. But the real atmosphere seems not to have learned the lessons presented in meteorology to night are barely perceptible in the upper atmosphere. Upper-level air changes temperature very slowly as it moves from one region to another, and air flowing horizontally from a warm region can retain its high temperature as it moves into a region otherwise occupied by cold air. We refer to the horizontal movement of relatively warm air as **warm air advection**. The opposite, of course, is called **cold air advection**. Both types of temperature advection appear on Rossby waves and, as we will see, affect the development of surface cyclones.

A useful method for detecting warm and cold air advection on a map of the upper atmosphere is to compare the orientation of height contours and isotherms. Figure 1 illustrates three possible patterns of 500 mb height levels and temperature advection. In Figure 1a, parallel height contours (solid lines) are aligned in a west-to-east direction so that a geostrophic wind flows from west to east. The isotherms (the dashed lines showing the temperature distribution at the 500 mb level) run parallel to the height contours and indicate a northward decrease in temperature. Because the air flow is parallel to the height contours (and in this case the isotherms), the temperature is the same (just less than −25 °C) at positions 1 and 2, and there is neither cold nor warm air advection. When the height contours

and isotherms are in alignment, the atmosphere is said to be **barotropic**.

In Figure 1b, the height contours are parallel to each other, as are the isotherms. But in this instance the height contours and isotherms intersect each other, and the temperature increases from position 3 (below −25 °C) to position 4 (above −20 °C). This is an example of cold air advection, wherein a parcel of colder air is transported from 3 to 4. The opposite situation, warm air advection, occurs in Figure 1c, where the temperature decreases in the direction of airflow. When the height contours and the isotherms intersect, as in both (b) and (c), the atmosphere is said to be **baroclinic**.

Refer to Figure 2 and observe the two baroclinic zones at positions 1 (cold advection) and 2 (warm advection). Where cold advection exists, the entering air is denser than the air ahead of it because of its lower temperature. This gives it a negative buoyancy that causes it to sink downward, bringing cold air toward the surface. Cold air advection typically occurs behind a cold front, thereby enhancing the temperature contrast found on either side of the front. Where warm advection occurs, entering air is warmer and more buoyant than the air ahead of it and therefore rises. The warm and cold air advection thus cause vertical motions similar to those associat-

ed with statically unstable air (Chapter 6). This situation is called **baroclinic instability**.

In addition to undergoing rising or sinking motions, the air in areas of warm or cold air advection undergo a slight turning—to the right in areas of cold air advection and to the left in regions of warm air advection. These motions are what cause the ripples (short waves) to form on the Rossby waves. When a short wave is located downwind of a Rossby wave trough axis, the divergence is enhanced and surface cyclones intensify.

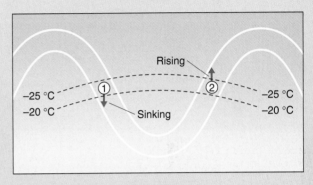

FIGURE 2

Warm and cold air advection around a Rossby wave. The air at position 1 flows from colder to warmer air, resulting in cold air advection. Along a zone of cold air advection, sinking motions and a turning of the air to the right tend to take place. Warm air advection occurs at position 2 along with a rising of the air.

textbooks, and real-life conditions often depart markedly from these idealized examples. The following example from 1994 is a fairly typical midlatitude cyclone making its trek across North America.

APRIL 15

Figure 10–11 shows the surface and 500 mb weather maps and a satellite image of North America on April 15. Clear skies, low humidities, and light winds dominate the western United States and southwest Canada. In contrast, the midlatitude cyclone over the north-central part of the United States has brought strong winds, overcast skies, and heavy rain showers. The surface wind in the warm sector flows northward out of the southern states and turns somewhat to the northwest as it approaches the center of the low pressure. North and west of the system, the air rotates counterclockwise around the low. Temperatures in the warm sector are typically in the 60s to low 70s °F, considerably greater than those to the west of the cold front.[5]

---

[5]Observed temperatures, dew points, wind velocities, and sea level pressures are depicted on the surface map for a few stations. Note that the temperatures and dew points (°F) are shown on the upper and bottom left, respectively, of the so-called *station models*. A line coming out of the circle shows the direction from which the wind is blowing, and the number of short and long tick-marks attached near the end of the line represents the wind speed. *Appendix C: Weather Map Symbols* provides more detailed information.

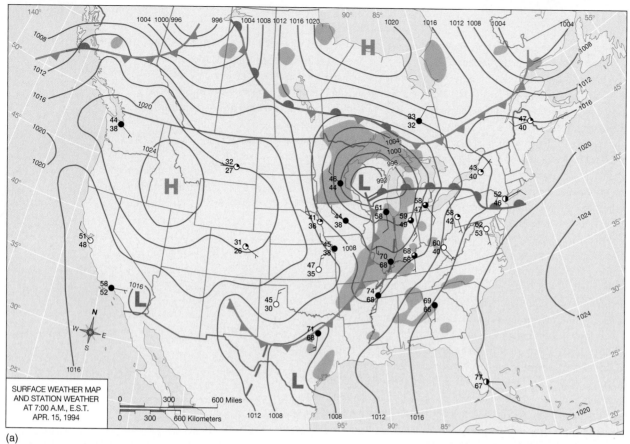

(a)

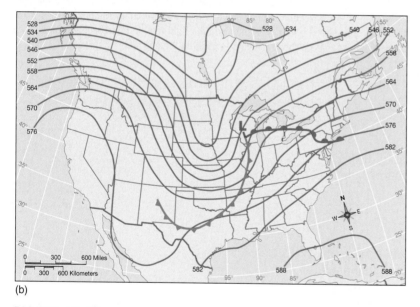

(b)

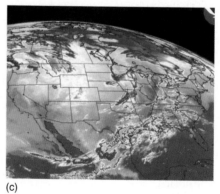

(c)

FIGURE 10-11

Weather maps of the surface (a) and 500 mb level (b), and a satellite image (c) for April 15, 1994, 7 A.M. EST. Note that the positions of the surface fronts have been superimposed on the 500 mb map.

At the 500 mb level (b), a well-defined trough exists to the west of where the surface cold front is located. Strong westerly winds in excess of 100 km/hr (60 mph) occur over the western portion of the Canadian–United States border and then become northwesterly as they enter the trough. Downwind of the trough axis, the air again flows eastward toward the North Atlantic.

Just downwind of the trough axis, the vorticity decreases over northern Missouri, Iowa, and Minnesota. As we would expect, diffluence in this region leads to net divergence aloft and low pressure at the surface. Out west, downwind from the ridge axis, upper-level convergence leads to high pressure at the surface centered over central Idaho.

FIGURE 10–16
Typical winter storm paths.

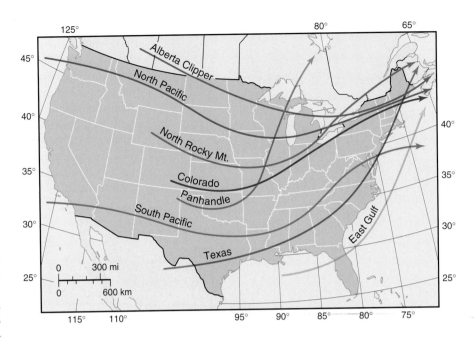

direction as the 500 mb flow, at about one half the speed. Keep in mind, however, that the 500 mb level wind pattern changes through time, so predicting the track of a cyclone involves more than just examining the current upper-level flow and assuming a constant movement parallel to the current height contour pattern; one must also predict the change in the 500 mb pattern.

Many midlatitude cyclones have their origin over the north Pacific off the coast of Japan. Upon reaching the Aleutian Islands of Alaska, the systems can die out, migrate toward the southeast, or continue on an eastward path across British Columbia. The most likely path the cyclones take upon reaching North America varies with the season, with northern treks favored in the summer, and movement toward the southeast more likely in the winter.

Upper-level winds are about twice as vigorous on average in the winter than in the summer. During the winter, net radiation decreases rapidly with increasing latitude, giving rise to a stronger latitudinal temperature gradient than in summer (Chapter 3). This results in greater pressure gradients (and winds) in the upper atmosphere. It is no surprise that midlatitude cyclones generally move faster in the winter.

Though a winter midlatitude cyclone can take many different paths across North America, two are particularly common: the Alberta Clipper and the Colorado low (Figure 10–16). The Alberta Clipper is associated with zonal flow and a polar jet stream that sweeps across southern Canada and the northern United States. Though it can bring frigid conditions, snowfall is usually light. In contrast, some midlatitude cyclones passing farther to the south over western North America spawn new centers of low pressure as they pass over the central Rocky Mountains. They then follow a path from the southern Plains toward the northeastern United States and eastern Canada. These storms, usually warmer and containing greater amounts of water vapor in the air, often produce extremely heavy snowfall.

Storms occasionally have their genesis well to the south of the polar front (in contrast to the original model of the early Norwegian meteorologists) and can track northward along the eastern United States. Such storms often have strong uplift and high water vapor contents—conditions favorable for the development of extremely heavy snowfalls. Such conditions led to the "storm of the century" in March of 1993, which produced strong winds and record-breaking snow accumulations over the eastern third of the United States.

MIGRATION OF SURFACE CYCLONES RELATIVE TO ROSSBY WAVES

For a midlatitude cyclone to form, there must be upper-level divergence. If there is more divergence aloft than convergence near the surface, the surface low deepens and a cyclone forms. If the convergence at the surface exceeds the divergence aloft, the low gradually fills until it ceases to exist.

Although the optimal place for midlatitude cyclones to develop is just beneath the zone of decreasing vorticity aloft, they don't usually remain in a fixed position relative to the upper-level trough. Instead, they are usually pushed along so that they migrate in the same direction (and at about half the speed) as the

DID YOU KNOW?

At one point the famous "storm of the century" that hit the eastern United States and Canada recorded a central barometric pressure of 960 mb—lower than that associated with 80 percent of the hurricanes and tropical storms that make landfall in the United States. It was also an unusually large midlatitude cyclone that affected 26 of the 50 United States, spawning blizzards that shut down virtually every airport on the eastern seaboard.

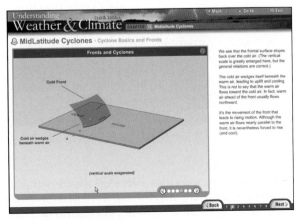

TUTORIAL
Midlatitude Cyclones

winds at the 700 mb level—typically about 3 km (2 mi) above the surface. Figure 10–17 shows the movement of a surface cyclone relative to an upper-level trough and ridge pattern (for simplicity, we assume that the position of the Rossby wave containing the trough and ridge remains fixed in time, though this is not usually the case for extended periods). In (a) the surface low associated with a cyclone exists under the zone of upper-level divergence. The divergence aloft provides the uplift that maintains the surface low. At the same time, however, the upper-level winds are moving the surface system northeastward relative to the Rossby wave. Thus the cyclone moves away from the region of maximum divergence aloft and eventually locates beneath the zone of upper-level convergence (b). At this point air descends toward the center of the low, the requisite uplift that maintained the

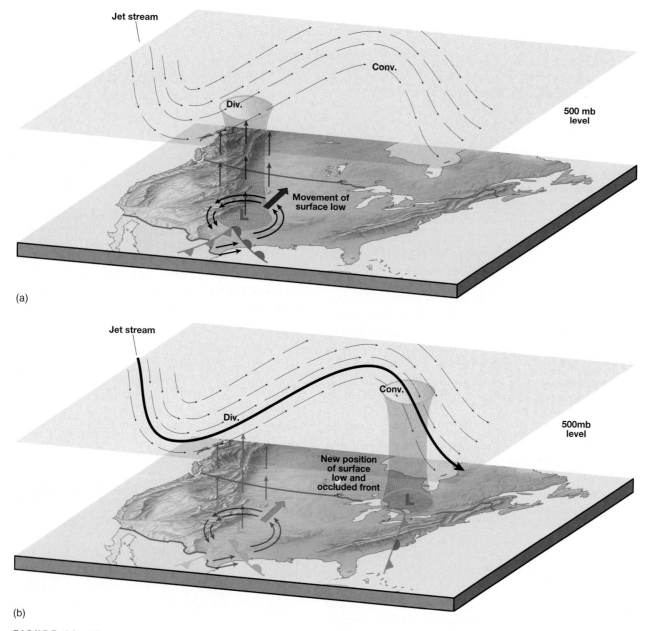

(a)

(b)

FIGURE 10–17

Surface high- and low-pressure centers can migrate relative to the Rossby wave aloft. In (a), the surface low exists below the area of upper-level divergence, and the resultant uplift maintains or strengthens the cyclone. Surface systems are generally guided by upper-level winds, so eventually the center of the low might be displaced to the zone where upper-level convergence occurs (b). The sinking air fills into the low, causing its demise. This process occurs as the cyclone undergoes the transition from its mature to occluding stages.

surface low ceases, and the subsiding air causes the surface pressure to increase. The cyclone eventually dies out as its air pressure becomes nearly the same as that surrounding it. Note that this process occurs simultaneously with the occlusion process.

The Modern View—Midlatitude Cyclones and Conveyor Belts

We have examined the structure of midlatitude cyclones at the surface and in the middle troposphere, but we should not overlook the fact that cyclones are three-dimensional entities. In other words, the 500 mb level (or any other level) is not separate from the surface portion—it is simply a different part of the same system. The **conveyor belt model** (Figure 10–18), which we look at next, provides a better depiction of the three-dimensional nature of midlatitude cyclones.

This model describes the midlatitude cyclone in terms of three major flows. The first flow, the *warm conveyor belt*, originates near the surface in the warm sector and flows toward and over the wedge of the warm front. As the warm belt flows

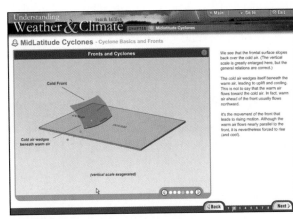

TUTORIAL
Midlatitude Cyclones

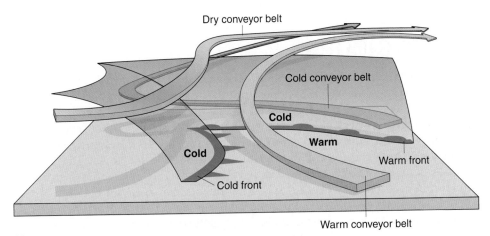

(a)

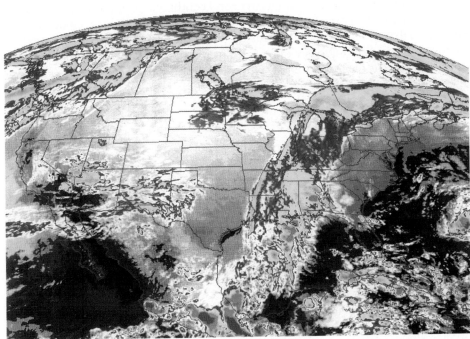

(b)

FIGURE 10–18
The conveyor belt model of midlatitude cyclones.

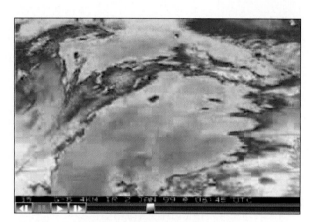

up the frontal surface, adiabatic cooling leads to condensation and precipitation. Moreover, as the air rises into the middle troposphere, it begins to turn to its right and become incorporated into the general westerly flow downwind of the upper-level trough. The cloud cover associated with the rising warm conveyor belt appears prominently as the bright, wide band extending from south to north in (b).

The *cold conveyor belt* lies ahead (north) of the warm front. It enters the storm at low levels as an easterly belt flowing westward toward the surface cyclone. But like the warm conveyor belt, it too ascends as it flows, turns anticyclonically (clockwise in the Northern Hemisphere), and becomes incorporated into the general westerly flow aloft. Although it originates as cold (and therefore relatively dry) air, the cold conveyor belt gains moisture from the evaporation of raindrops falling from the warm conveyor belt above. The cold conveyor belt extends in (b) from northern Michigan to eastern South Dakota.

The final component of the three-dimensional circulation is the *dry conveyor belt* that originates in the upper troposphere as part of the generally westerly flow. This broad current brings the coldest air into the cyclone, and it is important in maintaining the strong temperature contrast across the cold front. The upper-level air sinks slightly as it approaches the cold front from the west, but then it rises and merges with the general upper-level flow. The dry conveyor belt separates the cloud bands from the warm and cold conveyor belts, which helps give the cloud distribution its distinctive comma-shaped appearance.

Anticyclones

So far we have said little about anticyclones, but they are as much influenced by upper-level conditions as are cyclones, and they also have an impact on weather. (Recall from Chapter 4 that these phenomena are areas of high pressure around which the wind blows clockwise in the Northern Hemisphere.) While cyclones can bring heavy precipitation and strong winds, anticyclones foster clear skies and calm conditions because the cool air within them slowly sinks toward the surface.

That should not be taken to mean that anticyclones are always associated with wonderful weather. Indeed, outbreaks of continental polar (cP) air over the eastern United States are associated with anticyclones behind southward- or southeastward-moving cold fronts. Furthermore, anticyclones often tend to remain over a region for a long time, which can lead to droughts. Finally, anticyclones over the Rocky Mountains can lead to Santa Ana wind conditions over the West Coast.

This chapter has presented the characteristics of midlatitude cyclones and anticyclones that influence weather outside the tropics. We know, however, that the atmosphere often undergoes violent types of weather, usually on smaller time and space scales than those associated with midlatitude cyclones. In Chapter 11 we'll meet thunderstorms and tornadoes, phenomena that can cause considerable damage and loss of life.

Summary

Although much of our knowledge of midlatitude cyclones comes directly from the work of Norwegian meteorologists in the early twentieth century, recent insights into upper-level winds have greatly increased our understanding. Midlatitude cyclones and anticyclones both depend on a close interaction between processes occurring in the upper and lower troposphere. Counterclockwise rotation in the Northern Hemisphere has positive vorticity, while clockwise rotation has negative vorticity. The greatest positive vorticity occurs around the trough axis of a Rossby wave. The decreasing vorticity immediately downwind of the axis causes divergence, which leads to the formation of cyclones at the surface. At the same time, the fronts associated with midlatitude cyclones help form the Rossby waves that create upper-level convergence and divergence. Thus, a constant interaction takes place between the upper and lower atmosphere.

Traditionally, the distribution of clouds along frontal boundaries has been linked to convergence and overrunning on either side of the fronts. The modern approach explains the cloud distributions as the result of three separate air flows, or conveyor belts. A cold conveyor belt flows toward the center of low pressure ahead of the warm front. As it approaches the low pressure, it rises, and adiabatic cooling produces the wide band of cloud cover. Similarly, a rising conveyor belt of warm air flows ahead of the cold front to provide another band of cloud cover. Both the warm and cold conveyor belts turn anticyclonically near the center of low pressure and join the upper-level westerly flow. A dry conveyor belt in the middle atmosphere flows above cold fronts.

Surface cyclones and anticyclones migrate in the direction of the mid-tropospheric (700 mb) winds of Rossby waves. Cyclones thus move from regions of upper-level divergence (which help maintain or intensify the surface low pressure) to regions of upper-level convergence (which weaken the cyclones).

Key Terms

polar front theory   page 279
cyclogenesis   page 280
mature cyclone   page 281
occlusion   page 282
Rossby wave   page 284
vorticity   page 285
absolute vorticity   page 285

relative vorticity   page 285
Earth vorticity   page 285
dynamic low   page 286
thermal low   page 286
speed divergence   page 290
speed convergence   page 290
diffluence   page 291

confluence   page 291
short waves   page 292
temperature advection
  page 292
warm air advection   page 293
cold air advection   page 293
barotropic   page 293

baroclinic   page 293
baroclinic instability
  page 293
cutoff low   page 296
conveyor belt model
  page 301

Review Questions

1. Define *cyclogenesis*. Where does it most commonly occur according to the original polar front theory?

2. Describe the isobar and wind patterns associated with mature midlatitude cyclones.

3. Where is precipitation most likely to be found within midlatitude cyclones?

4. Where within a midlatitude cyclone are overrunning, convergence, and instability likely to serve as precipitation-inducing processes?

5. What are Earth, relative, and absolute vorticity?

6. Why is counterclockwise rotation in the Northern Hemisphere said to have positive vorticity?

7. Where are the zones of positive, negative, and zero vorticity in a typical ridge and trough pattern?

8. In what part of a ridge and trough system do you find the areas of decreasing and increasing vorticity? Why is their existence important?

9. How do dynamic and thermal lows differ from each other?

10. Where are upper-level divergence and convergence most likely to occur?

11. What type of upper-level condition typically lies above and behind a cold front?

12. Where are upper-level ridges generally located relative to midlatitude cyclones?

13. What are diffluence and speed divergence? How do they differ from confluence and speed convergence?

14. If the upper-level air flow over North America is zonal, what can you infer with regard to widespread precipitation conditions?

15. Which type of general upper-level air flow is more likely to occur if minimal temperature variations exist across North America?

16. Describe the three conveyor belts of the conveyor belt model of midlatitude cyclones. How does the conveyor belt model differ from the description of airflow presented in the original polar front model?

17. Explain how the movement of midlatitude cyclones relative to the upper-level airflow contributes to the demise of cyclones.

Critical Thinking

1. Why is the term *polar front theory* probably a misnomer in view of current knowledge about cyclogenesis?

2. A commercial aircraft is flying at the 300 mb level and goes across a midlatitude cyclone over both the cold and warm fronts. What kind of weather changes might the aircraft encounter?

3. After a front is fully occluded, its demise is imminent. Why can't the occluded front persist for several more weeks?

4. Why can't systems similar to midlatitude cyclones develop over the tropics?

5. Vorticity is usually discussed with reference to a vertical axis. What types of Earth, relative, and absolute vorticity conditions would you expect to exist with regard to a horizontal axis?

6. Why don't thermal lows migrate like dynamic lows do?

7. Why do forecasters take particular interest in the distribution of vorticity on 500 mb weather maps?

8. Clear skies often portend warm conditions during the summer but are often associated with very cold conditions in the winter. Explain why this is true.

9. Are Rossby waves likely to have greater representation in the Southern Hemisphere or the Northern Hemisphere? Why?

Problems and Exercises

1. Go to **http://weather.uwyo.edu/upperair/uamap.html** and view the current 500 mb map for North America. Without referring to a surface map, make an educated guess about the position of surface midlatitude cyclones, cold fronts, and warm fronts. Then go to **http://weather.uwyo.edu/surface/front.html** to see how well your educated guess worked out (be sure to highlight analysis for image type).

2. Using the same Web site as for problem 1, print the maps of the surface, 300 mb, 500 mb, 700 mb, and 850 mb levels. What patterns emerge as you move upward?

3. Visit **http://weather.uwyo.edu/surface/front.html** on a daily basis and keep track of existing midlatitude cyclones and their associated fronts. Do the systems correspond well to the life cycle described in this chapter?

Quantitative Problems

This chapter has discussed the importance of temperature differences in the establishment of upper-level troughs and the critical roles of vorticity and divergence in the life cycle of midlatitude cyclones. The Chapter 10 page of this book's Web site (**http://www.** prenhall.com/aguado/**) gives you the opportunity to perform some calculations to help you better understand the material. We recommend you access the site and perform some of the simple but revealing problems.

Useful Web Sites

http://www.giss.nasa.gov/data/stormtracks/
An online atlas of storm tracks from 1961 through 1998. Maps are available for each year's seasonal or monthly storm track frequency, intensity, and individual paths.

http://www.hpc.ncep.noaa.gov/sfc/namfntsfcwbg.gif
Simplified, current surface weather map highlighting frontal systems.

http://weather.unisys.com/archive/index.html
An excellent source of archived surface and upper-air weather maps and satellite images.

Media Enrichment

Tutorial
Midlatitude Cyclones

Beginning with the formation of cyclones, this tutorial demonstrates how surface pressure is related to upper-level divergence and convergence. It puts particular emphasis on the structure of midlatitude cyclones, with numerous three-dimensional diagrams and animations showing frontal surfaces, conveyor belts, and other features of these powerful storms.

Weather in Motion
A Midlatitude Cyclone Passes over the Southeast

This movie shows the movement of a storm over slightly more than two days, beginning at about 1000 UTC (5 A.M. EST) on January 26, 1998. Initially, several bands of clouds approach Mississippi, Tennessee, and Kentucky from the west with a minimal amount of counterclockwise rotation. By 1500 UTC the individual bands merge into a large area of continuous cloud cover that begins to show more discernible rotation. By 0000 UTC on the twenty-seventh, clouds with lower tops can be seen over the Ohio River Valley, moving northeast away from the center of rotation. By about noon UTC, the region of deepest cloud cover extends from South Carolina to Virginia within a broad area of precipitation covering the East Coast. As the movie comes to an end, the rotation into the low pressure center over Kentucky and West Virginia remains strong. This movie reinforces the idea that although the original polar front theory nicely approximates the structure and evolution of midlatitude cyclones, the actual movement of air and moisture within the system is more complex.

Weather in Motion
Tornado-Generating Storm

This midlatitude cyclone on March 20, 1998, produced tornadoes over northern Georgia and western North Carolina. The most violent weather took place along the leading edge of the southern portion of the cold front. This highlights the fact that storm activity is not usually uniform along the length of a frontal boundary.

Weather in Motion
January 1999 Blizzard

This midlatitude cyclone brought blizzard conditions to the north-central United States and southern California. A strong counterclockwise rotation appears over the western part of the system, associated with a strong low-pressure system. The warm conveyor belt appears prominently during the last half of the movie. It is the band of intense cloud cover that flows south to north until it approaches the core of low pressure. The airflow then bends toward the east, where it is incorporated into the westerly airflow of the middle and upper troposphere.

Weather Image
Cyclones Near the Martian Poles

More than 20 years ago NASA's *Viking* orbiter allowed astronomers to detect spiral storms on the surface of Mars. These storms occur near the poles during the Martian summer. In 1999 the Hubble Space Telescope led scientists to the discovery of very large cyclones near the Martian North Pole, three times larger than those previously seen on the planet. Unlike the spiral storms previously observed on Mars, or the midlatitude cyclones of Earth, these storms feature three or more cloud bands surrounding a cloud-free eye, similar to tropical storms on our planet. Such storms, though rare, have also been observed at high latitudes on Earth, sometimes generating extreme winds.

CHAPTER **11** | # LIGHTNING, THUNDER, AND TORNADOES

It is hard to imagine anybody failing to be impressed by the beauty—as well as the danger—brought about by a thunderstorm. Spectacular though they may be, such storms occur about 40,000 times each day. Their frequency varies greatly from place to place, yet virtually every location on Earth is vulnerable to thunder and lightning from time to time.

Lightning can create inconveniences—such as blowing out all the electrical appliances in a house. It can also do considerable damage, such as starting forest fires. And, of course, it can kill; during an average year, about 200 people are killed by lightning in the United States and Canada. But considering that the population of these two countries approaches 300 million people, it is easy to see that your chances of being struck are extremely remote.

However, consider the experience of the McQuilken family on their trip to Sequoia National Park, California, in August 1975. As the sky began to darken, Sean, Michael, and their sister Mary noticed their hair standing on end. Recognizing the apparent comedy of the situation, the boys posed for the photograph shown in Figure 11–1. Hail followed almost immediately. Then lightning struck—literally—and Sean was knocked unconscious. Michael quickly administered artificial respiration, which probably saved Sean's life. Another victim was less fortunate, however. The lightning had apparently forked off, with another branch hitting two nearby people, one of whom was killed.

The effects of lightning and thunder are eclipsed by an even greater menace—tornadoes. We will now examine how, where, and why violent weather occurs, and we'll look at the situations that cause some storms to be weak and others to become destructive and deadly.

◀ Lightning hitting Half Dome, Yosemite National Park in California.

Processes of Lightning Formation

About 80 percent of all **lightning** results from the discharge of electricity *within* clouds, as opposed to discharge from cloud to surface. This **cloud-to-cloud lightning** occurs when the voltage gradient within a cloud, or between clouds, overcomes the electrical resistance of the air. The result is a very large and powerful spark that partially equalizes the charge separation. Cloud-to-cloud lightning causes the sky to light up more or less uniformly. Because the flash is obscured by the cloud itself, it is commonly called **sheet lightning.**

The remaining 20 percent of lightning strokes are the more dramatic events in which the electrical discharge travels between the base of the cloud and the surface. Most of this **cloud-to-ground lightning** occurs when the negative charges accumulate in the lower portions of the cloud. Positive charges are attracted to a relatively small area in the ground directly beneath the cloud. This establishes a large voltage difference between the ground and the cloud base. The positive charge at the surface is a local phenomenon; it arises because the negative charge at the base of the cloud repels electrons on the ground below. Although the term *cloud-to-ground* is used, the same effect occurs in water—and lightning often strikes lakes, rivers, and oceans.

Although a stroke of lightning may come and go in just a few moments, a regular sequence of events must occur for the event to take place. Electrification of a cloud is the initial stage in all lightning. After that, a path must develop through which electrons can flow. Only then is electricity actually discharged to produce a lightning stroke.

CHARGE SEPARATION

All lightning requires the initial separation of positive and negative charges into different regions of a cloud. Most often the positive charges accumulate in the upper reaches of the cloud, negative charges in lower portions. Small pockets of positive charges may also gather near the cloud base (Figure 11–2a). Now the question is: How does this **charge separation** occur in the first place? Nobody knows for sure, because clouds that produce lightning and thunder happen to be particularly inhospitable laboratories. But we do know several facts from which we can get some idea of how charges separate. Lightning occurs only in clouds that extend above the freezing level, and it is also restricted to precipitating clouds. Thus, the ice crystal processes responsible for precipitation must also influence charge separation. Laboratory experiments with artificial clouds and numerical calculations suggest that electrification results from collisions between ice crystals and graupel surrounded by cloud droplets. It seems likely that charges are transferred across thin films of water present on ice crystals and soft hail.

Though we normally don't notice it, solids are often coated with a liquid surface layer just a few molecules thick. This layer consists of molecules only weakly bound to the solid below. It is present even at temperatures well below the freezing point. (Among other things, the presence of this layer explains why ice is so slippery at temperatures well below zero degrees Celsius.[1]) In a cloud, when an ice crystal and soft hail collide, some of the liquid-water molecules on the hailstone's surface migrate to the ice. In fact, evidence exists suggesting that the collision actually increases the tendency for liquid to exist, thereby increasing the likelihood of liquid-water migration to the ice crystal. Along with the water molecules, there is usually a transfer of positive charge from the hailstone to the ice crystal or, equivalently, a transfer of negative charge from the crystal to the hailstone. In this way ice crystals surrender negative ions to the much larger hailstones, which then fall downward toward the cloud base.

DID YOU KNOW?

Worldwide, there are about 4 million lightning discharges per day, resulting in about 2000 injuries and 600 deaths per year.

---

[1]You may have heard that pressure from an ice skate's blade melts enough ice to create a slippery film of water, but that's not correct. The ice is slippery whether you press hard or not.

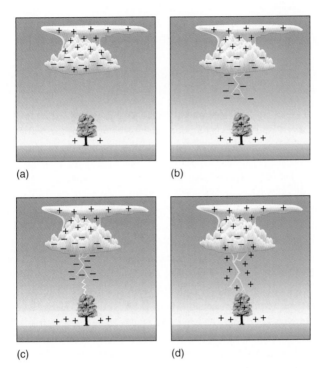

(a)　　　　　　　(b)

(c)　　　　　　　(d)

FIGURE 11-2
The first step in the formation of lightning is the development
of a stepped leader (a). The leader approaches the surface as
a very rapid sequence of steps (b) and (c), until contact is
made with an object at the ground. The flow of electricity
produces the lightning stroke (d).

Depending on the liquid-water content and temperature of the cloud, a *positive* charge is occasionally transferred to the hailstone, leading to an accumulation of negative charges aloft. Very recent research suggests that surface defects also regulate the charge transfer between particles. (See *Box 11–1, Physical Principles: Electricity in the Atmosphere,* for more information on electrical charges in the air.)

RUNAWAY DISCHARGES

For many years scientists thought of lightning as a huge spark of static electricity similar to what one sometimes sees just before touching a doorknob after walking across a carpet. In that case, the voltage gradient becomes so large that the insulating properties of the air are overcome, and a conducting channel forms. It is now believed that lightning does not operate that way. For one thing, the voltage gradients necessary to produce such a discharge are not observed in clouds. Secondly, recent measurements show that X-rays are routinely emitted as part of lightning discharges. There is no known source for these X-rays within the traditional "big spark" view. Many physicists have therefore returned to an idea first proposed in 1925. Lightning is a consequence of electrons that have been accelerated to very high speeds (near the speed of light). Electrons moving faster than a few percent of the speed of light do not experience the resistance felt by ordinary slow-moving electrons. In fact, resistance actually decreases as their speed increases. So, if an electric field accelerates a fast electron, the resistance falls and the electron moves even faster, the resistance decreases further, and so on. Eventually the electron approaches the speed of light and gains a huge amount of energy. Fast-moving electrons create others by colliding with atoms, and this can result in an avalanche of runaway electrons. When a large number of runaway electrons accumulate in a small volume, the energy is released in a so-called **runaway breakdown**. There are still many uncertainties about all this that need to be explained, particularly how initial fast electrons are accelerated (cosmic rays have been suggested). But the fast electron model does explain the existence of X-rays just before lightning flashes, which is something the traditional model cannot do.

Electricity in the Atmosphere

11-1 PHYSICAL PRINCIPLES

Lightning is, of course, an electrical disturbance, much of which can be explained by the basic principles of atmospheric electricity. You know from Chapter 1 that ions (charged particles) are most abundant high in the atmosphere (in the ionosphere, from about 80 to 500 km, or 50 to 300 mi). The upper atmosphere has a positive charge, just as we find near the positive pole of a battery. In the same way that a battery stores energy, electrical charges in the atmosphere represent stored energy and have the potential to do work. For both batteries and the atmosphere, this electrical potential is expressed by voltage, which is simply the energy per unit charge. For example, if a battery is rated at 1.5 volts (V), it means that 1.5 joules are available per coulomb of charge (1.5 J/C). A coulomb (C) is equivalent to the charge carried by about 6×10^{19} electrons. The higher the voltage, the greater the energy release for each coulomb transferred.

In the case of Earth, a huge voltage difference exists between the surface and the ionosphere—about 400,000 volts! This voltage gradient sets up what we call the **fair-weather electric field.** The fair-weather field is always present, even in bad weather, so a better name might be the **mean electric field.** The fair-weather field can be thought of as the background situation, on which extreme events such as lightning are superimposed.

Does electricity flow in response to the voltage gradient of the fair-weather field? Yes, but because air is a good insulator, the current is weak, about 2000 coulombs per second (2000 A) for the entire planet. In North America, individual houses are typically wired for 200 A service, so we see that the atmospheric current is truly very small. Nevertheless, it does represent a continuous leakage, whereby electrons are transferred from the surface, or (equivalently) positive charges are transferred from the atmosphere.

This implies that for the mean electric field to be maintained, it must be continuously replenished. As a matter of fact, lightning discharges in thunderstorms are thought to be the primary recharge mechanism. In other words, cloud-to-ground lightning discharges transfer electrons to the surface, maintaining the voltage difference and the resulting electric field.

In the lower atmosphere, the fair-weather electric field gradient is on the order of 100 V per meter. (Although this might sound impressive, remember that few ions are present, so the total available energy is very low.) Of course for lightning to occur, the field strength must be greatly intensified above the background value. How this happens can be only partly explained today, nearly 250 years after Ben Franklin performed his famous kite experiment.

WEATHER IN MOTION
Lightning

LEADERS, STROKES, AND FLASHES

In cloud-to-ground lightning, the actual lightning event is preceded by the rapid and staggered advance of a shaft of negatively charged air called a **stepped leader** (Figure 11–2b) from the base of the cloud. The leader is not a single column of ionized air; it branches off from a main trunk in several places. Only about 10 cm (4 in.) in diameter, each section of the column first surges downward about 50 m (165 ft) from the base of the cloud in about a millionth of a second (or microsecond). This invisible leader pauses for about 50 microseconds, as electrons pile up at the tip and generate a strong electric field in the surrounding area. The field generates more runaway electrons that surge downward another 50 or so meters in the next step. The downward movement in a rapid sequence of individual steps gives the stepped leader its name. In each step, the newly created runaway electrons collide with air molecules and trigger a flare of X-rays.

When the leader approaches the ground (Figure 11–2c), a spark surges upward from the ground toward the leader. When the leader and the spark connect, they create a pathway for the flow of electrons that initiates the first in a sequence of brightly illuminated **strokes,** or **return strokes** (Figure 11–3). The electrical current, flowing at about 20,000 amperes (A), appears to work its way downward from the base of the cloud, but the stroke actually propagates upward (Figure 11–2d). The conducting path is completed at the surface, from which there is a surge of positive charge upward toward the cloud. The current heats the air in the conducting channel to temperatures up to 30,000 K (54,000 °F), or five times that of the surface of the Sun!

The electrical discharge of the first stroke neutralizes some, but not all, of the negatively charged ions near the base of the cloud. As a result, another leader (called a **dart leader**) forms within about a tenth of a second, and a subsequent stroke emerges from it. We call the combination of strokes a lightning **flash**, the net effect of which is to transfer electrons from the cloud to the ground. While

FIGURE 11–3
Lightning.

most flashes consist of only 2 or 3 strokes, some consist of as many as 20 individual strokes. Because they occur in such rapid succession, they appear to be a single stroke that flickers and dances about.

The total transfer of electrons is not large, only about as many electrons as we use in burning a 100-watt light bulb for half a minute or so. So how is lightning able to split trees and perform other dramatic work? For one thing, in lightning the charge transfer is rapid, and so the electric current is discharged many thousands of times faster than in a household current. (Think of a light bulb as having low current flowing for a relatively long time. The total charge transferred is the same as in lightning, where a huge current flows for just an instant.) Another factor is that the voltage is much larger than in a household circuit, so the energy release is much larger for each electron transferred. Taken together, these facts mean that a huge amount of energy gets released over a very brief period of time, making each stroke extremely powerful. A light bulb would need a month or two to release the same amount of energy.

Positive charges found at the top of thunderstorm clouds can also lead to lightning. When high-level winds are strong, thunderstorm clouds become tilted, with positive charges carried ahead of the storm (Figure 11–4). These positive charges induce negative charges at the surface, resulting in a lightning strike that shoots positive charges to the surface. As a result, it often happens that the first of a storm's lightning strikes are positive, and measurements reveal that they can be many times stronger than the negative strokes that follow. This positive form of lightning is therefore particularly dangerous. It can occur several miles away from the storm, where people do not feel threatened; it tends to have larger peak electrical currents; and it typically lasts longer, making fires more likely. Though an estimated 9 percent of all cloud-to-ground flashes in the United States and Canada originate this way, there is considerable variability across North America. Strokes originating from positive polarity are more common in the upper Midwest of the United States and throughout much of Canada, where they represent perhaps 20 percent of all occurrences.

TYPES OF LIGHTNING

Far less common than strokes and leaders is the bizarre type of electrification called **ball lightning.** Ball lightning appears as a round, glowing mass of electrified air, up to the size of a basketball, that seems to roll through the

FIGURE 11–4
Positive lightning stroke.

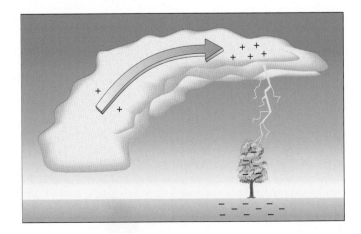

A Personal Account of Ball Lightning

I saw ball lightning during a thunderstorm in the summer of 1960. I was 16 years old. It was about 9:00 P.M., very dark, and I was sitting with my girlfriend at a picnic table in a pavilion at a public park in upstate New York. The structure was open on three sides and we were sitting with our backs to the closed side. It was raining quite hard. A whitish-yellowish ball, about the size of a tennis ball, appeared on our left, 30 yards away, and its appearance was not directly associated with a lightning strike. The wind was light. The ball was 8 feet off the ground and drifting slowly toward the pavilion. As it entered, it dropped abruptly to the wet wood plank floor, passing within 3 feet of our heads on the way down. It skittered along the floor with a jerky motion (stick-slip), passed out of the structure on the right, rose to a height of 6 feet, drifted 10 yards further, dropped to the ground and extinguished non-explosively. As it passed my head, I felt no heat. Its acoustic mission I liken to that of a freshly struck match. As it skittered on the floor it displayed elastic properties (a physicist would call them resonant vibrating modes). Its luminosity was such that it was not blinding. I estimate it was like staring at a less than 10-watt light bulb. The whole encounter lasted for about 15 seconds. I remember it vividly even today, as all eyewitnesses do, because it was so extraordinary. Not until 10 years later, at a seminar on ball lightning, did I realize what I had witnessed.

Source: Graham K. Hubler. Reprinted by permission from Nature. *Copyright 2000, Macmillan Magazines Ltd.*

DID YOU KNOW?

Lightning strokes can extend more than 16 km (10 mi) from the side of clouds. Thus, lightning can hit in an area where clear skies prevail.

air or along a surface for 15 seconds or so before either dissipating or exploding. One form is a free-floating, reddish mass that tends to avoid good electrical conductors and flows into closed spaces or through doorways and windows. Another form is considerably brighter and is attracted to electrical conductors (including people). Various explanations for ball lightning have been offered for at least 150 years, but until recently none could account for all aspects of the phenomenon, and most had glaring weaknesses. The situation improved significantly early in 2000, with the report of experiments involving artificial lightning strikes on soil. It was seen that lightning reduces silicon compounds in the soil to tiny nanoparticles of silicon carbide (SiC), silicon monoxide (SiO), and metallic silicon (Si). Unlike the original silicon compounds, these contain significant chemical energy and are unstable in an oxygen environment. They are ejected into the air, where they cool rapidly and condense into filmy chains and networks. The networks are light, so they float easily in the atmosphere. Most important, they burn brightly as they oxidize, releasing the stored energy in the form of visible light. (See *Box 11–2, Special Interest: A Personal Account of Ball Lightning,* for a firsthand description of this phenomenon.)

St. Elmo's fire is another rare and peculiar type of electrical event. Ionization in the air—often just before the formation of cloud-to-ground lightning—can cause tall objects such as church steeples or ships' masts to glow as they emit a continuous barrage of sparks. This often produces a blue-green tint to the air, accompanied by a hissing sound.

Recent observations and photographs from space shuttle missions have revealed the existence of previously unknown electrical phenomena at the tops of thunderstorms. **Sprites** (Figure 11–5) are very large but short-lived electrical bursts that rise from cloud tops as lightning occurs below. A sprite looks somewhat like a giant red jellyfish, extending up to 95 km (57 mi) above the clouds, with blue or green tentacles dangling from the reddish blob. Sprites accompany only about 1 percent of all lightning events. (Interestingly, military and commercial pilots now admit to having seen sprites before they were observed from shuttle missions, but they did not often report them lest they be accused of having hallucinations.)

Blue jets (Figure 11–6) are upward-moving electrical ejections from the tops of the most active regions of thunderstorms. They shoot upward at about 100 km/sec (60 mph) and attain heights of up to 50 km (30 mi) above the surface. In all likelihood, other types of electrical activity above thunderstorms remain to be discovered.

FIGURE 11–5
A sprite.

THUNDER

The tremendous increase in temperature during a lightning stroke causes the air to expand explosively and produce the familiar sound of **thunder.** Although sound travels rapidly—about 0.3 km (0.2 mi) per second—it is much slower than the speed of light (300,000 km, or 186,000 mi, per second). This difference creates a lag between the flash of light and the sound of thunder; the farther away the lightning, the longer the time lag. You probably know the familiar rule of thumb for estimating the distance of a lightning stroke: Simply count the number of seconds between the stroke and the thunder and divide by three to determine the distance in kilometers (divide by five for the distance in miles).

This method does not work for very distant strokes, those more than about 20 km (12 mi) away. The decrease in the density of air with height causes sound waves to bend upward. At relatively short distances, the amount of bending is negligible. But beyond about 20 km, it is sufficient to displace the sound waves so that they cannot be heard at ground level. Lightning that seems to occur without thunder is sometimes called **heat lightning,** though this term is misleading in that it implies there is something unusual about it. The only oddity is that the sound of distant thunder does not reach the listener.

You have probably noticed that nearby thunder sounds like a loud, brief clap, while more distant thunder often occurs as a continuous rumble. A lightning stroke producing thunder may be several kilometers in length, so one part of it may be significantly farther from a listener than other parts. Thus, thunder makes a continuous sound as it takes longer for the sounds of more distant parts of the stroke to reach the listener. At greater distances, the echoing of sound waves off of buildings and hills can cause the thunder to make a rumbling sound.

Lightning Safety

Despite its splendor, we must not forget that lightning can be lethal, killing an average of 69 people each year in the United States and 7 in Canada. Fortunately, our current understanding of lightning suggests some safety rules.

First and foremost, in the presence of lightning, always take cover in a building, being careful not to make contact with any electrical appliances or telephones. Do not stand under a tree or other tall object that is likely to serve as a natural

FIGURE 11–6
A blue jet.

lightning rod. Lightning hitting a tree can easily flow through it and electrocute or burn those near this particularly vulnerable location. Avoid standing on rooftops, hill crests, or other high areas where lightning requires a shorter path from the cloud base. And of course, don't watch the lightning storm from a pool or hot tub!

Automobiles (other than convertibles) are relatively safe, but not because the rubber tires provide insulation against grounding (as believed by many people). The real reason is that if a car is hit, the electricity will flow around the car body rather than through the interior (or its occupants). The same fact explains why lightning seldom brings down airplanes, even though any particular commercial aircraft is hit on an average of once a year.

We often associate deadly lightning strikes with golfing during a thunderstorm and other foolish behaviors, but the danger is not always easy to avoid. On January 1, 2000, for instance, a single lightning bolt killed a family of six near Mount Darwin, Zimbabwe, in a tragedy eerily similar to one that had occurred a few months earlier in Zimbabwe, when a single strike killed six persons near the city of Gokwe. Lightning deaths are becoming ever more frequent throughout that region as forests are cleared, leaving villages more exposed in open areas. The problem is greatly exacerbated by the use of dry thatch as a roofing material. Soot from cooking fires impregnates the thatch with carbon, making the roof highly conductive and attractive for lightning.

In the United States a substantial decline (greater than 50 percent) has occurred in the number of lightning fatalities since the 1920s. This fact is even more impressive when one considers that since that time the country's population has increased from about 150 million to 270 million. The decline in fatalities is believed to be due in part to a population decline in rural areas, where the risk of being struck is greater. Better public education, improved warnings, advances in medical aid, and modernized electrical systems in homes and other buildings are also believed to have contributed to the decline.

Thunderstorms: Air Mass and Severe

Most lightning events are associated with localized, short-lived storms that dissipate within tens of minutes after forming. These storms, called **air mass thunderstorms,** actually extinguish themselves by creating downdrafts that cut off the supply of moisture into the precipitating clouds. For that reason, they do not normally produce severe weather. On other occasions, however, downdrafts from heavy precipitation actually intensify the storms that generate them. These storms are classified as **severe thunderstorms** and produce the greatest damage and loss of life.

AIR MASS THUNDERSTORMS

Air mass thunderstorms are the most common and least destructive of thunderstorms. They also have very limited lifespans, usually lasting for less than an hour. Despite the name, which implies that these thunderstorms might occupy entire air masses (which are very large), air mass thunderstorms are localized. But the term does make sense when you consider that air mass thunderstorms occur within individual air masses and are well removed from frontal boundaries. Think of it this way: Air mass thunderstorms are contained *within* uniform air masses, but they do not occupy the *entire* air mass.

Our current understanding of air mass thunderstorms is based on the Thunderstorm Project, which examined such events in Ohio and Florida during the late 1940s. An air mass thunderstorm normally consists of a number of individual updrafts (*cells*), each undergoing a sequence of three distinct stages—cumulus, mature, and dissipative (Figure 11–7).

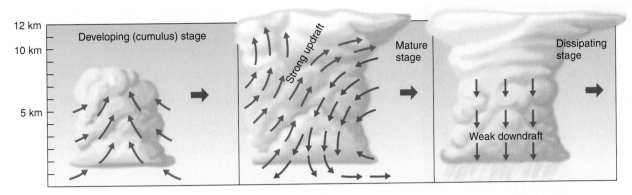

FIGURE 11-7

The cumulus (a), mature (b), and dissipative (c) stages of an air mass thunderstorm.

CUMULUS STAGE

The first stage of an air mass thunderstorm begins when unstable air begins to rise, often by the localized convection that occurs as some surfaces undergo more rapid heating than others. Because air mass thunderstorms frequently occur at night when the air is cooling, we know that other lifting processes can also trigger uplift. Regardless of which process causes uplift, the rising air cools adiabatically to form fair-weather cumulus clouds. These initial clouds may exist for just a matter of minutes before evaporating. Although they do not directly lead to any precipitation, the initial clouds play an important role in thunderstorm development by moving water vapor from the surface to the middle troposphere. Ultimately, the atmosphere becomes humid enough that newly formed clouds do not evaporate but instead undergo considerable vertical growth. This growth represents the **cumulus stage** in the air mass thunderstorm.

Clouds in the cumulus stage grow upward at 5 to 20 m/sec (10 to 45 mph). Within the growing clouds, the temperature decreases with height at roughly the saturated adiabatic lapse rate, and a portion of the cloud extends above the freezing level. Ice crystals begin to form and grow by the Bergeron process. The sky rapidly darkens under the thickening cloud; when precipitation begins to fall, the storm enters its next stage of development.

MATURE STAGE

The **mature stage** of the air mass thunderstorm begins when precipitation—as heavy rain or possibly graupel—starts to fall. As the falling rain or graupel drags air toward the surface, downdrafts form in the areas of most intense precipitation. You can observe this process in your own yard. Simply turn on a garden hose full blast and put your hand just outside the stream of water; you will notice a breeze in the direction in which the hose is pointed. The downdrafts are strengthened by the cooling of the air—by as much as 10 °C (18 °F)—that occurs as the precipitation evaporates.

The mature stage marks the most vigorous episode of the thunderstorm, when precipitation, lightning, and thunder are most intense. Air mass thunderstorms usually consist of multiple cells, located in different parts of the cloud and formed at different times. The top of the cloud extends to an altitude where stable conditions suppress further uplift. Strong winds at the top of the cloud push ice crystals forward and create the familiar anvil shape extending outward from the main part of the cloud.

During the cumulus and mature phases of the storm, an abrupt transition exists between the edge of the cloud and the surrounding unsaturated air. Updrafts dominate the interior of the cloud, while downdrafts occur just outside it. This sets up a highly turbulent situation that encourages entrainment (Chapter 6). The entrainment of unsaturated air causes the droplets along the cloud margin to

FIGURE 11–8

An air mass thunderstorm. The part of the cloud that has a washed-out appearance has become glaciated.

shrink and cool the cloud by evaporation. The outer part of the cloud becomes more dense and less buoyant, thus suppressing further uplift.

DISSIPATIVE STAGE

As more and more of the cloud yields heavy precipitation, downdrafts occupy an increasing portion of the cloud base. When they occupy the entire base, the supply of additional water vapor is cut off and the storm enters its **dissipative stage**. Precipitation diminishes and the sky begins to clear as the remaining droplets evaporate. Only a small portion—perhaps 20 percent—of the moisture that condenses within an air mass thunderstorm actually falls as precipitation. The greatest amount simply evaporates from the cloud.

Figure 11–8 shows an air mass thunderstorm. As is typical for thunderstorms in the mature phase, each tower consists of an individual cell and is in a different part of its life cycle. Notice in particular that some of the storm cloud appears washed-out and less well-defined than the rest. Such areas consist entirely of ice crystals, with no liquid droplets, and are said to be *glaciated*. They are not necessarily colder than other parts of the cloud; they are merely old enough so that all the supercooled droplets have had a chance to freeze.

SEVERE THUNDERSTORMS

By definition, severe thunderstorms have wind speeds that exceed 93 km/hr (58 mph)[2], hailstones larger than 1.9 cm (0.75 in.) in diameter, or spawn tornadoes. The downdrafts and updrafts in severe storms reinforce one another and thereby intensify the storm. This reinforcement usually requires suitable conditions over an area from 10 to 1000 km (6 to 600 mi) across, a size referred to as *mesoscale*. In other words, most severe thunderstorms get a boost from a mesoscale atmospheric pattern that allows the wind, temperature, and moisture fields to "cooperate" and thereby create very strong storms.

Because the conditions favoring severe thunderstorms exist over a rather large area, they typically appear in groups, with several individual storms clustered together. Such clusters of thunderstorms are generally referred to as **mesoscale convective systems (MCSs)**. In some cases, MCSs occur as linear bands called **squall lines**. At other times, they appear as oval or roughly circular clusters called **mesoscale convective complexes (MCCs)**. Regardless of how they are arranged, the individual storm cells of an MCS form as part of a single system: They are not just a grouping of individual storms that happen to be near each other. The storm cells develop from a common origin or exist in a situation in which some cells directly lead to the formation of others. When one or more of the defining characteristics is present, a storm within an MCS is classed as severe.

MCSs can bring intense weather conditions to areas covering several counties. They often have lifespans of up to 12 hours, but in some cases they can exist for as long as several days. They are fairly common in North America, and in some parts of the central United States and Canada, they account for as much as 60 percent of the annual rainfall. Because the surrounding circulation supports an MCS, they lead to much stronger winds and heavier precipitation than is normally found in an air mass thunderstorm.

Severe thunderstorms can also arise from **supercells**, intensely powerful storms that contain a single updraft zone. Though supercells often appear in isolation, they can also occur as a part of an MCS.

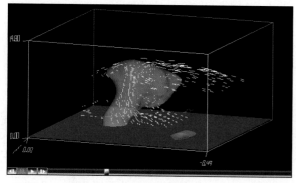

WEATHER IN MOTION

Cloud and 3-D Flow Tracers
Cloud and Vertical Velocity
Vertical Motion Cross Section
Cloud and Horizontal Flow

---

[2]This seemingly odd value was originally designated as 50 nautical miles per hour, or knots.

Certain conditions are necessary for the development of all severe thunderstorms. Among these are wind shear, high water vapor content in the lower troposphere, some mechanism to trigger uplift, and a situation called *potential instability*, described in *Box 11–3, Forecasting: Potential Instability*.

We will now briefly describe MCCs, squall lines, and supercells. Keep in mind that the descriptions of these storm types are quite general, and individual systems might not be easy to categorize. It is also common for storm systems to evolve from one type of system to another.

MESOSCALE CONVECTIVE COMPLEXES

In the United States and Canada, severe weather arises most often from mesoscale convective complexes (MCCs) (Figure 11–9). In the most general sense, MCCs are defined as oval or roughly circular organized systems containing several thunderstorms.[3]

Although not all MCCs create severe weather, they are self-propagating in that their individual cells often create downdrafts, leading to the formation of new, powerful cells nearby. To see how this occurs, imagine a large cluster of thunderstorms. At the surface a flow of warm, humid air comes from the south, and in the middle troposphere the wind flows from the southwest. This setting provides the wind shear necessary for a severe thunderstorm. As we have already seen, the precipitation from each thunderstorm cell creates its own downdraft. The downdraft is enhanced by the cooling of the air as the rain evaporates and consumes latent heat. Upon hitting the ground, the downdraft spreads outward and converges with the warmer surrounding air to form an **outflow boundary** (Figure 11–10).

---

[3]Meteorologists have some precise criteria for classifying a system as an MCC, based on its signature on satellite imagery. We will simply apply the term to organized systems of thunderstorms clustered in a pattern that is closer to circular than linear.

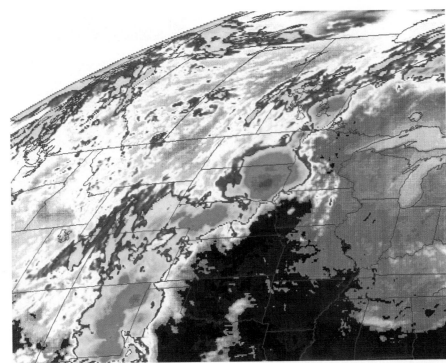

FIGURE 11–9
Satellite image showing a mesoscale convective complex over eastern South Dakota.

FIGURE 11–10
A radar image highlighting two outflow boundaries.

Potential Instability

In Chapter 6, we discussed the concept of static stability. Air with strong temperature lapse rates were said to be statically unstable. Another type of instability that influences vertical air motions, called **potential instability**, arises when a layer of warm, dry air rests above one that is warm and humid. Lifting the two layers can cause the temperature lapse rate to increase, thus making the air statically unstable.

Consider the inversion situation shown in Figure 1. In Parcel 1, located in the lower layer just below the base of the inversion, the temperature (T) equals 26 °C and the dew point (T_d) is 22 °C. In Parcel 2, located in the layer above the base of the inversion, the temperature and dew point are 27 °C and 19 °C, respectively. Now consider what happens between Parcels 1 and 2 if some process lifts the two layers containing the parcels. Both parcels are initially unsaturated, so they cool at the dry adiabatic lapse rate (DALR) of 1 °C per

100 m, and their dew points decrease at 0.2 °C per 100 m. After 500 m of ascent, the temperature of both parcels has fallen by the same amount, so the temperature difference between them is unchanged. However, Parcel 1 is now saturated, so further lifting will cause its temperature to decline at the saturated adiabatic lapse rate (SALR). Meanwhile, Parcel 2 is still unsaturated, so further lifting leads it to cool at the DALR.

Now let's lift the two parcels another 500 m. Assuming an SALR of 0.5 °C per 100 m, the lower parcel cools 2.5 °C to 18.5 °C, while the upper parcel cools at the DALR to 17 °C. We can now see how uplift of the column of air containing the two parcels has affected its stability. Initially, the parcel in the upper layer was warmer than the one below, which meant the layer containing the two parcels was statically stable. After lifting both parcels 1000 m, however, the upper parcel became 1.5 °C cooler than the one

below. This yields a temperature lapse rate between the two parcels of 1.5 °C per 100 m, making the air statically unstable. Thus, the air that is statically stable has the potential to become statically unstable, given sufficient uplift—hence the term *potential instability.*

Both theory and experience show that potential instability is an important factor in the development of severe thunderstorms. During spring and summer, the southern Great Plains region often has warm, humid air near the surface advected from the Gulf of Mexico. In the middle troposphere above the region, westerly winds bring dry air from the southern Rockies. This air in the middle troposphere sinks somewhat after passing the Rockies to form a subsidence inversion, which inhibits the development of air mass thunderstorms. But given sufficient uplift, the surface layer of air can become statically unstable and severe thunderstorms can develop.

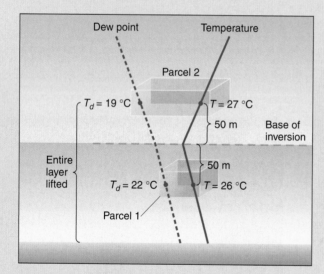

FIGURE 1

Potential instability occurs when a layer of warm, dry air overlies moist air. The temperature (solid line) and dew point (dashed line) profiles show such a situation. Examine what happens to the parcels of air 50 m above and below the base of the inversion. Because the lower parcel is nearly saturated (that is, the dew point is close to the air temperature), lifting causes it to become saturated after 500 m of uplift. The parcel above the inversion does not become saturated until 1000 m of uplift. Thus, after both have been lifted 1000 m, the upper parcel has undergone more cooling than the lower one, and the temperature lapse rate becomes steep enough to produce statically unstable air.

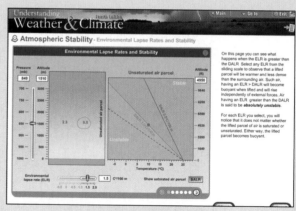

TUTORIAL
Atmospheric Stability

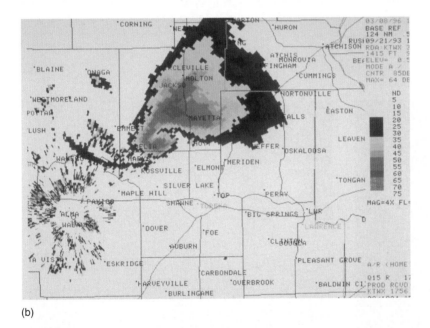

(a)

FIGURE 11-19

The typical organization of a supercell as seen on a radar image (a). The zone of no radar return in the southeast is the vault, which is flanked by a hook echo to the south. An actual radar image showing a hook is seen in (b).

(b)

particular type, called **Doppler radar**, is described in *Box 11–4, Forecasting: Doppler Radar*. Radar can reveal one of the most noteworthy features of a supercell, called a **hook**, which looks like a small appendage attached to the main body of the storm on the radar image (Figure 11–19).[5] Hooks are significant because their appearance usually means tornado formation is imminent.

The zone with no radar return between the main part of the supercell and the hook echo, known as a **vault**, is where the inflow of warm surface air enters the supercell. The air entering the vault rises, and water vapor condenses to form a dense concentration of water droplets. But the newly formed droplets in the vault are too small to effectively reflect radar waves. Thus, this zone does not show up on the radar image despite the dense concentration of water droplets.

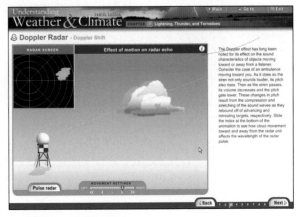

TUTORIAL
Doppler Radar

---

[5]A hook is sometimes called a *hook echo* in reference to the way radar waves reflect off it, as sound waves echo off canyon walls.

Doppler Radar

Just as we are able to distinguish different colors of light by their wavelengths, so can we differentiate sounds by the length of their sound waves. If an object making a sound is moving away from a listener, the sound waves are stretched out and assume a lower pitch. Sound waves are compressed when an object moves toward the listener, making them higher pitched. Unconsciously, we use this principle, called the **Doppler effect**, to determine whether an ambulance siren is coming closer or moving away. If the pitch of the siren seems to become higher, we know the ambulance is getting nearer (of course, the siren would also sound louder). A similar process occurs when electromagnetic waves are reflected by a moving object: The light shifts to shorter wavelengths when reflected by an object moving toward the re-

ceiver and to longer wavelengths as it bounces off an object moving away from the receiver.

Doppler radar is a type of radar system that takes advantage of this principle. It allows the user to observe the movement of raindrops and ice particles (and thus determine wind speed and direction) from the shift in wavelength of the radar waves, as well as the intensity of precipitation. Like any other type of radar, Doppler radar has a transmitter that emits pulses of electromagnetic energy with wavelengths on the order of several centimeters. Depending on the wavelength used, water droplets and snow crystals above certain critical sizes reflect a portion of the radar's electromagnetic energy back to the transmitter/receiver. Doppler radar is special in its ability to observe the motion of the cloud con-

stituents. If a cloud droplet is moving away from the radar unit, the wavelength of the beam is slightly elongated as it bounces off the reflector. Such reflections are normally indicated on the display monitor as reddish-to-yellow. Likewise, a droplet moving toward the radar unit undergoes a shortening of the wavelength. Echoes from these constituents are displayed as blue or green on the radar screen.

A radar unit must rotate 360 degrees to get a complete picture of the weather situation surrounding the transmitter/receiver unit. When the transmitter makes one complete rotation at a fixed angle, it is said to have completed a **sweep**. The angle can then be increased as a second sweep is taken that depicts a higher cloud level. This can be repeated several times so that the

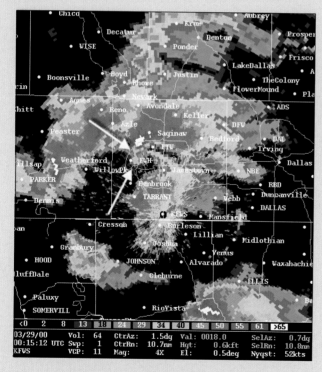

(a)

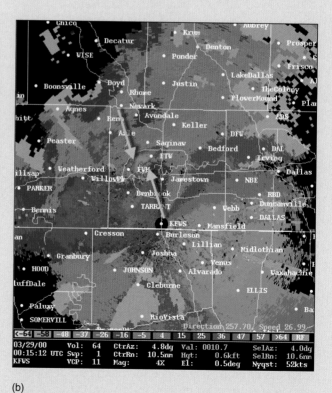

(b)

FIGURE 1

Doppler radar images of a storm near Dallas–Fort Worth, Texas, on March 29, 2000. Part (a) depicts the intensity of precipitation; part (b) shows the storm radial velocity (SRV) pattern, which is the movement of different parts of the storm toward or away from the radar unit.

324

radar can peer into multiple levels of the cloud. The compilation of all the individual sweeps takes approximately five to ten minutes and produces a **volume sweep**.

Figure 1 shows a pair of Doppler radar images of a major storm near Dallas–Fort Worth, Texas, on March 29, 2000. Figure 1a shows the reflectivity of the storm, with redder regions indicating intense cloud cover and green areas representing less intense cloud cover. The white arrows point toward a hook echo (described in the main text of this chapter). Figure 1b displays the *storm radial velocity (SRV) pattern*, which describes the motions taking place within the cloud. SRV displays use redder colors to represent winds blowing away from the radar and green to indicate movement toward the radar. The yellow arrows on this image highlight a region of counterclockwise rotation. As we discuss later in the chapter, this pattern, called a *mesocyclone*, often precedes the formation of a tornado. After the onset of rotation, it takes only 30 minutes or so for the tornado to form, which allows meteorologists to give warning in advance. In this particular case, a tornado did hit the city of Fort Worth.

During the early 1990s, the National Weather Service began replacing the old system of conventional radar with a modern Doppler network called NEXRAD (for NEXt Generation Weather RADar). The first unit, installed at Norman, Oklahoma, became operational in early 1991 and on its very first day in service tracked a tornado that destroyed two houses. Fortunately, the radar allowed forecasters to issue a warning that may have contributed to no one being killed or injured.

A study published in 2005 concluded that the implementation of Doppler radar units across the United States has prevented 79 fatalities and 1050 injuries per year. Doppler radar has increased the average warning time for all tornadoes from 5.3 minutes to 9.5 minutes. Results are even more impressive for F5 tornadoes, the most deadly of all. Warning time for those tornadoes has increased to 16.23 minutes from the previous 11.7 minutes.

Today about 160 Doppler sites are scattered across the United States (Figure 2). The National Weather Service operates 113 of these sites; the rest are owned by the Federal Aviation Administration and the Department of Defense. In part because of budgetary cutbacks, the Atmospheric Environment Service of Canada has just a handful of Doppler radar installations. Because both sides of the border area tend to be heavily populated, Doppler radar from the United States provides extensive coverage of severe storms that could affect many large Canadian urban centers.

NEXRAD is also useful for flood forecasting, providing continual precipitation estimates over large areas. Doppler radar can sometimes observe wind movements even when no clouds exist, as large clusters of flying bugs or heavy dust concentration scatter radar waves toward the transmitter. The resultant echoes are called clear air echoes.

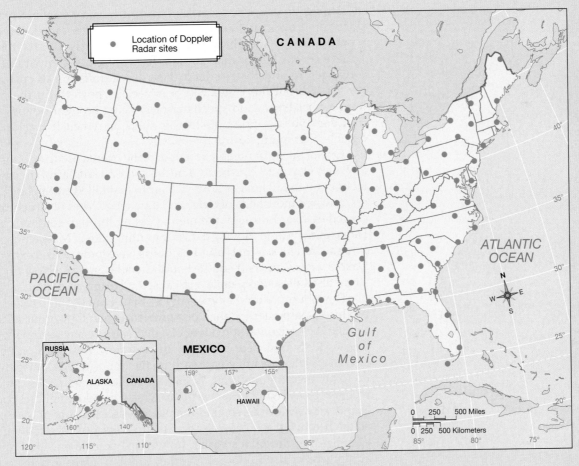

FIGURE 2
Doppler radar sites in the United States.

DRYLINES

Drylines were briefly described in Chapter 9 as boundaries separating mT and cT air masses. They are most likely to form in spring and early summer over Texas and Oklahoma, where they can be the site of severe weather. Drylines develop as moist air near the surface from the Gulf of Mexico flows northward, separated from the dry air to the west. Terrain effects are important, as the elevation of the surface gradually increases westward from east Texas and Oklahoma toward the Rocky Mountains. The higher elevation prevents a westward incursion of the moist air. Dry air flowing eastward out of the high plains overrides the moist air below, creating a situation called *potential instability* (see Box 11–3). Potential instability can lead to severe storm activity if enough lifting occurs below the boundary of the warm and moist air. This can occur due to surface convergence as the mT air from the Gulf of Mexico and the westerly flow of cT air collide along the dryline.

DOWNBURSTS, DERECHOS, AND MICROBURSTS

We have seen how downdrafts are an important feature of thunderstorms, especially in the maintenance of severe thunderstorms. Strong downdrafts may also create **downbursts**, potentially deadly gusts of wind that can reach speeds in excess of 270 km/hr (165 mph). When strong downdrafts reach the surface, they can spread outward in all directions to form intense horizontal winds capable of causing severe damage at the surface. In fact, damage attributed to tornadoes may in some cases be the result of downbursts.

Strong downdrafts associated with mesoscale convective systems can produce very powerful, larger-scale horizontal winds called **derechos** (the Spanish word meaning "straight ahead"). Such winds may last for hours at a time and achieve speeds higher than 200 km/hr (120 mph)—comparable to those of many tornadoes. As already described, these downdrafts spread outward upon reaching the ground. But the winds can be especially powerful if the descending upper-level air has high wind speeds prior to being brought downward. This momentum does not disappear as the air sinks, so when it reaches the surface, it is capable of bringing destructive winds.

One of the most notable recent derecho events occurred shortly after midnight on Labor Day in 1998, when a derecho some 50 km (30 mi) in length brought wind gusts of up to 180 km per hour (115 mph) that killed two people, injured eight others, downed thousands of trees, and damaged scores of structures. The storm proceeded eastward through the night, passing through southern Vermont and New Hampshire prior to moving across the entire state of Massachusetts. The following July saw a derecho that lasted 22 hours, sweeping across extreme eastern North Dakota, across much of south central Canada, and into New England. That windstorm also killed two people and caused widespread damage.

Downbursts with diameters of less than 4 km (2.5 mi), called **microbursts**, can produce a particularly dangerous problem when they occur near airports (Figure 11–20). The horizontal spreading of a microburst creates strong wind shear when it reaches the surface. For example, air may flow westward on one side of the microburst while spreading eastward on the opposite side. Imagine what this might do to an aircraft attempting to land in a microburst. As the plane enters the microburst, a headwind provides lift, to which the pilot might respond by turning the aircraft downward. As soon as the plane passes the core of the downdraft, however, the headwind not only disappears, it is replaced by a tailwind, decreasing lift. Coming after the pilot's earlier downward adjustment, this causes the plane to abruptly drop in altitude. Because the plane is not far above the ground when these events occur, the pilot may not have time to compensate before a deadly crash occurs. Fortunately, such disasters are rare. They are also becoming less likely because the installation of Doppler radar at about 40 U.S. airports has proven highly effective at detecting microbursts, with a detection rate of about 95 percent.

FIGURE 11-20
Microbursts can make aircraft landing perilous. A plane flying into the headwinds of a microburst gets a sudden increase in lift. This lift suddenly disappears and is replaced by a tailwind as it exits the downdraft, thereby reducing the lift. If the pilot overcompensates and guides the plane downward while entering the downdraft, a dangerous drop in altitude may occur.

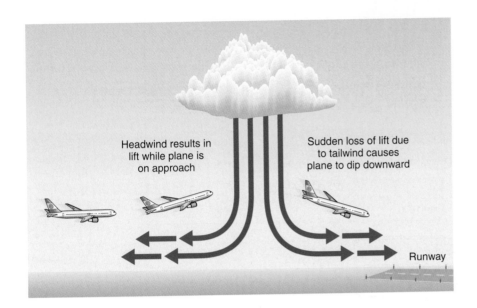

Headwind results in lift while plane is on approach

Sudden loss of lift due to tailwind causes plane to dip downward

Runway

Geographic and Temporal Distribution of Thunderstorms

Thunderstorms are extremely common across much of the globe, numbering some 14.5 million per year. They are most likely to develop where moist air is subject to sustained uplift, and not surprisingly, such conditions occur most commonly in the tropics. Until recently the occurrence of lightning in low-populated or economically underadvantaged areas precluded the gathering of reliable statistics across the globe. Fortunately, satellites have provided high-quality observations of lightning incidence since the mid-1990s (Figure 11–21). Lightning strikes most frequently

DID YOU KNOW?

Thunderstorms account for a large part of the total rainfall that occurs over much of North America. Thunderstorms provide almost half the total rainfall that occurs over the Mississippi River Valley, which covers 41 percent of the conterminous 48-state land area. In the south-central United States they account for as much as 70 percent of total precipitation. Across the western portion of North America the percentages are far lower; generally less than 10 percent of the rainfall in that area falls from thunderstorms.

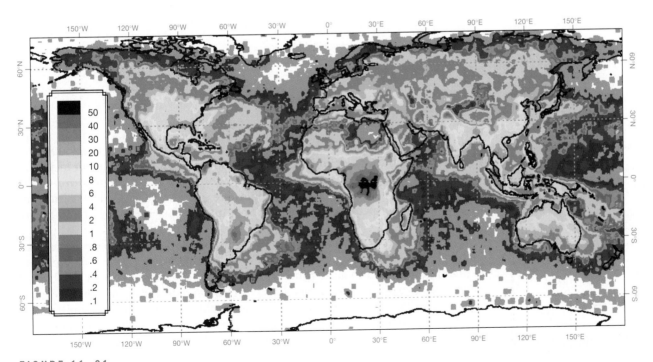

FIGURE 11-21
Data from space-based optical sensors reveal the uneven distribution of worldwide lightning strikes. Units: flashes/km^2/yr.

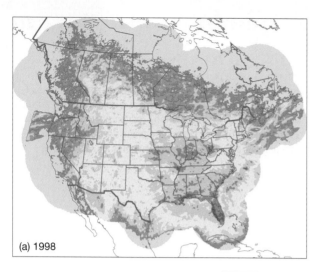

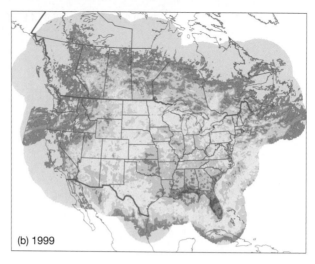

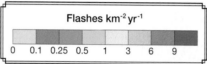

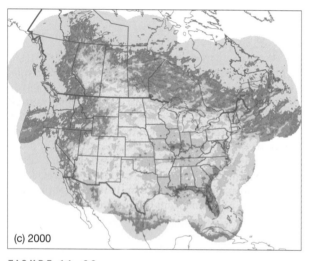

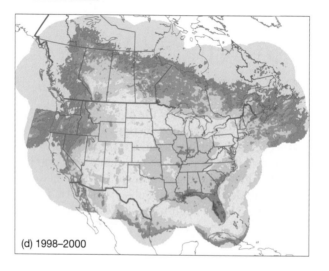

FIGURE 11–22

The annual concentration of cloud-to-ground lightning flashes from U.S. and Canadian detection networks during the years 1998 (a), 1999 (b), and 2000 (c), and average for the three years (d).

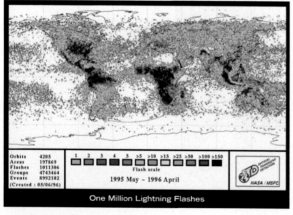

WEATHER IMAGE
Lightning Across the Globe

over the Congo basin of central Africa and occurs frequently over many other regions of the world. Colder and less humid regions typically have a lower incidence of lightning. Outside of the tropical regions, lightning is most common during the summer months when mid-day sun angles are greatest, while equatorial regions usually have their peak activity near the equinoxes (Chapter 2), due to abundant solar radiation.

Detailed maps of cloud-to-ground lightning flashes have been assembled using ground-based data from the Canadian Lightning Detection Network (CLDN) and the National Lightning Detection Network (NLDN) of the United States. The two networks' sensors detect electromagnetic radiation emitted during lightning strikes and automatically relay information on the location, timing, and polarity (positive or negative) of each stroke to a central facility. The networks have recorded an average of 28 million cloud-to-ground lightning flashes and 100,000 thunderstorms (280 flashes per thunderstorm) annually across the continental United States and Canada. Figure 11–22 shows the distribution of these flashes for the years 1998–2000.

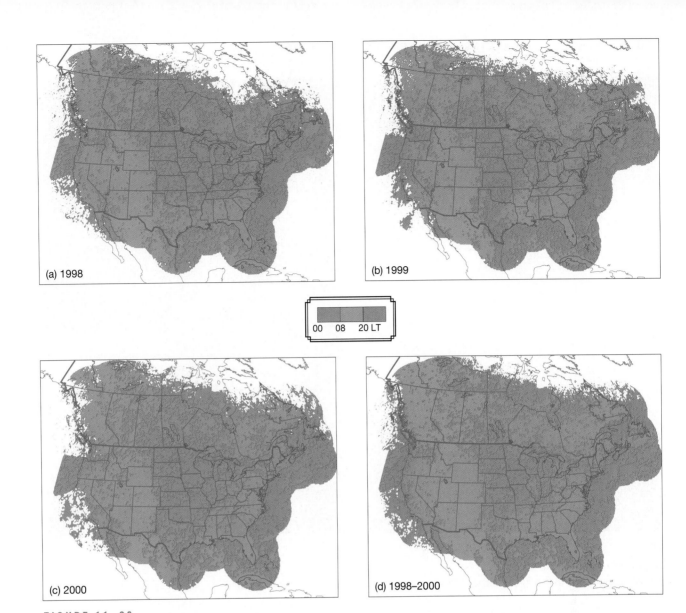

FIGURE 11–23
Locations where most lightning occurs from 8 A.M. to 8 P.M. are shown in red, and those where lightning is more common from 8 P.M. to 8 A.M. are in blue. Information is for the same years as shown in Figure 11–22.

Across the eastern two-thirds of North America there is a general pattern of decreasing thunderstorm activity northward. By far the state with the greatest incidence is Florida, where thunderstorms occur on average more than 100 days per year. Much of the necessary lifting of air results from strong solar heating of the surface. But the situation over the Florida peninsula is unique within the continental United States, because the area is almost completely surrounded by warm water. Thus, air that flows into the interior to replace lifted air has a very high moisture content, which in turn supports heavy precipitation and the development of thunderstorms. The area of the United States and Canada west of the Rocky Mountains experiences considerably fewer thunderstorms than the eastern two-thirds of the continent.

Figure 11–23 illustrates the areas that have the maximum number of lightning strikes during the daytime (8 A.M. to 8 P.M. local standard time), which are shown in red, and those occurring during the 12 hours following 8 P.M. (shown in blue). Much of North America exhibits a daytime maximum in lightning flashes, though much of the mid-continent has a greater likelihood of nighttime events.

(a)

(b)

FIGURE 11–24
Tornadoes come in a wide range of shapes and sizes.

WEATHER IN MOTION
Tornado

Tornadoes

The large hail and strong winds of a severe thunderstorm can bring widespread destruction, but even hailstorms are relatively tame compared to **tornadoes** (Figure 11–24). Tornadoes are zones of extremely rapid, rotating winds beneath the base of cumulonimbus clouds. Though the overwhelming majority of tornadoes rotate cyclonically (counterclockwise in the Northern Hemisphere), a few spin in the opposite direction. Some appear as very thin, rope-shaped columns, while others have the characteristic funnel shape that narrows from the cloud base to the ground. Regardless of their shape or spin, tornadoes are extremely dangerous.

Strong tornadic winds result from extraordinarily large differences in atmospheric pressure over short distances. Over just a few tenths of a kilometer, the pressure difference between the core of a tornado and the area immediately outside the funnel can be as great as 100 mb. To put this in perspective, on a typical day the highest and lowest sea level pressure across all of North America may differ by only about 35 mb—and this difference exists over horizontal distances of up to thousands of kilometers.

TORNADO CHARACTERISTICS AND DIMENSIONS

It is difficult to generalize about tornadoes because they occur in a wide variety of shapes and sizes. While most have diameters about the length of a football field (100 yards or so), some are 15 times as large. Usually they last no longer than a few minutes, but some have lasted for several hours. Tornadoes normally move across the surface at speeds comparable to a car driving down a city street—about 50 km/hr (30 mph). A typical tornado covers about 3 or 4 km (2 to 2.5 mi) from the time it touches the ground to when it dies out.

Estimates of wind speeds within tornadoes are based primarily on the damage they have produced. The weakest have wind speeds as low as 65 km/hr (40 mph); the most severe are in excess of about 450 km/hr (280 mph).

TORNADO FORMATION

Tornadoes can develop in any situation that produces severe weather—frontal boundaries, squall lines, mesoscale convective complexes, supercells, and tropical cyclones (see Chapter 12). The processes that lead to their formation are not very well understood (*Box 11–5, Special Interest: Tornado Research*, discusses some of the equipment that has been used to study the way tornadoes form.) Typically, the most intense and destructive tornadoes arise from supercells.

SUPERCELL TORNADO DEVELOPMENT

In a supercell storm, the first observable step in tornado formation is the slow, horizontal rotation of a large segment of the cloud (up to 10 km—6 mi—in diameter). Such rotation begins deep within the cloud interior, several kilometers above the surface. The resulting large vortices, called **mesocyclones,** often precede the formation of the actual tornado by some 30 minutes or so.

The formation of a mesocyclone depends on the presence of vertical wind shear. Moving upward from the surface, the wind direction in the storm shifts from a southerly to westerly direction as the wind speed increases. This wind shear causes a rolling motion about a horizontal axis, as shown in Figure 11–25a. Under the right conditions, strong updrafts in the storm tilt the horizontally rotating air so that the axis of rotation becomes approximately vertical (Figure 11–25b). This provides the initial rotation within the cloud interior.

Intensification of the mesocyclone requires that the area of rotation decrease, which leads to an increase in

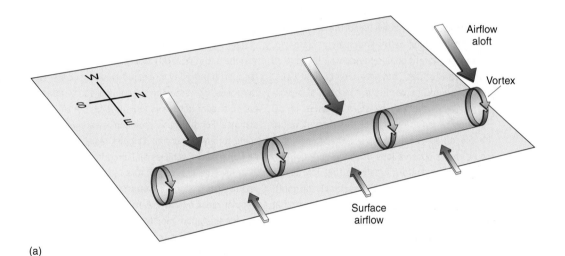

(a)

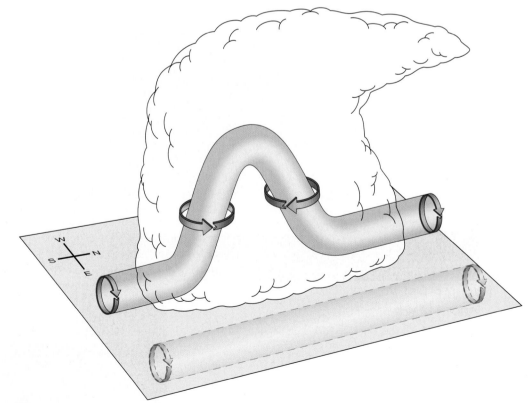

FIGURE 11–25

Mesocyclones can form when a horizontal vortex of air (a) becomes tilted upward (b), yielding vertical counterclockwise and clockwise rotating vortices.

(b)

Tornado Research

Despite the inherent difficulty of studying something as complex, violent, and short-lived as a tornado, scientists can peer into their structure using a number of research tools. For many years, much of the information was obtained by films and pictures of tornadoes. Unfortunately, dust and debris kicked up by a tornado, as well as screening by trees and buildings, often make it impossible to observe the part of the tornado nearest to the ground (and of most concern to people).

In 1980 scientists developed the Totable Tornado Observatory, better known by its acronym, "Toto" (an intentional reference to the dog in *The Wizard of Oz*). Toto was a lightweight package that could be carried in the back of a pickup truck and set up by storm chasers in 30 seconds or less. Ideally, the instrument would be placed in the path of a tornado, where it could make observations of temperature, wind, and pressure. Unfortunately, getting Toto in the path of a tornado proved exceedingly difficult. Moreover, wind tunnel tests have shown that Toto tends to tip over at wind speeds near 50 m/sec (110 mph), well below those that can occur in a tornado. For these reasons storm chasers no longer use Toto.

Following Toto, scientists from the University of Oklahoma began placing a network of smaller instruments called "turtles" in the paths of tornadoes. Though they only recorded temperature and pressure, turtles had an advantage in that they could be laid out in series across a road, crossing a tornado's path. Scientists did not get the results hoped for from turtles and have shifted to more advanced technologies.

The NEXRAD network of Doppler radar, installed primarily for forecasting purposes, provides information on tornadoes that has proven useful to researchers. But NEXRAD radar units are spaced too far apart to provide close scrutiny of most passing tornadoes, so researchers have looked to transportable radar to observe tornadoes from a smaller distance. For this reason, tornado researchers have come to rely on portable Doppler radar units carried on research vehicles and aircraft.

An early version of portable radar was put into use in 1987. Storm chasers transported a small Doppler radar by van toward the site of potential tornado formation, rapidly removed it from the vehicle, and aimed it toward the tornado. Though such radar proved useful, the manually operated radar was supplanted by larger, more powerful

radar in 1995. This Doppler on Wheels (DOW) has a large Doppler radar permanently mounted to a flatbed truck (Figure 1). Storm chasers access current information on storm development from within the vehicle and travel to where they think the greatest likelihood of a tornado exists. DOW has tracked more than 100 tornadoes and provided a wealth of information to researchers. The newest version of the radar was put into use in the spring of 2005. This Rapid-Scanning Doppler on Wheels obtains complete scans of tornadoes every five to ten seconds, a rate much faster than the five minutes per scan of the old DOW. This enables scientists to observe tornado development with a much greater amount of detail.

The primary disadvantage of DOW is the speed at which it can travel to a developing tornado. Storm chasers must make an educated guess about the location of tornado formation and travel considerable distances—often through bad weather—to chase down the storm. Catching a tornado therefore requires some good luck along with skill. Airborne radar, similar to the kind long used to fly through hurricanes, offers great speed in the pursuit of tornadoes and has provided additional information to scientists.

FIGURE 1
Doppler on Wheels.

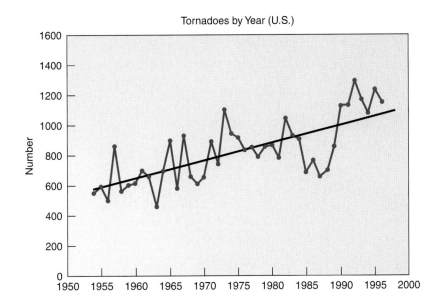

Tornadoes by Year (U.S.)

FIGURE 11–34

FIGURE 11–34
The number of observed tornadoes in the United States has shown an apparent doubling since the 1950s, but this increase may be mostly due to better observation rather than to a genuine increase in tornado activity. The straight line shows the average increase through time based on the plot of actual observations.

TRENDS IN U.S. TORNADO OCCURRENCE

A plot of tornado occurrences in the United States, based on a log of events collected by the National Severe Storms Laboratory of NOAA, shows a near-doubling of the tornado frequency from 1950 to the late 1990s (Figure 11–34). Taken at face value, this would be a very disturbing trend. But there are reasons to believe that this increase might only partially reflect an actual increase in tornado activity, and that the trend may be primarily due to a greater likelihood a given tornado will actually be observed. One possible explanation relates to population increases. As population centers have expanded out into formerly rural areas, there is a greater probability that a tornado will hit a structure or be observed directly. Furthermore, installation of the Doppler radar network has undoubtedly contributed to improved detection and an increase in the number of participants in the national storm spotter network. Changes in the way we classify tornadoes may also have played a role in the apparent increase. Numerous researchers are currently working to explain the relative effects of these factors on the apparent large increase in tornado activity and whether there really has been a significant increase.

The debate on recent global warming and the importance of human activities to that warming has led to concern about a future rise in tornado frequency. At this point in time, however, these ideas are still speculative.

TORNADO DAMAGE

Most structural damage from tornadoes results from their extreme winds. People once believed that homes were destroyed mostly by the pressure differences associated with a tornado's passage, which supposedly caused the interior air to push outward against the walls so violently that the house would explode. For this reason, people were advised to open their windows if they saw an approaching tornado, so that the pressure within the house could be reduced.

We now know that this was not good advice, in part because few homes actually explode. Moreover, though winds are the major factor in tornado damage, flying debris is the primary cause of tornado injuries, and opening a window increases the risk of personal injury from flying debris. (We must also suspect that opening the windows is useless in any case, because they are likely be "opened" anyway by flying objects.)

Although most tornadoes rotate around a single, central core, some of the most violent ones have several relatively small zones of intense rotations (about

FIGURE 11-35
Suction vortices.

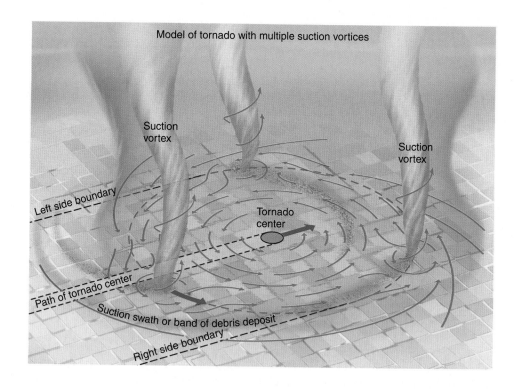

Model of tornado with multiple suction vortices

Suction vortex

Suction vortex

Left side boundary

Tornado center

Path of tornado center

Suction swath or band of debris deposit

Right side boundary

10 m—30 ft—in diameter) called **suction vortices** (Figures 11–35 and 11–36). It is these small vortices that probably cause the familiar phenomenon of one home being totally destroyed while the next one remains relatively unscathed.

Except for those rare times when tornado chasers make firsthand observations of passing tornadoes, it is impossible to get a precise reading on their pressure changes and wind speeds. But it is possible to classify them according to the magnitude of the damage they cause. The **Fujita scale** (named for the eminent tornado specialist Theodore Fujita) provides a widely used system for ranking tornado intensity. As shown in Table 11–2, documented tornadoes fall into seven levels

FIGURE 11-36
A multiple vortex tornado over Friendship, Oklahoma, on May 11, 1982.

of intensity, with each assigned a particular F-value ranging from 0 to 5.[8] In the United States, the majority (69 percent) fall into the *weak* category, which includes F0 and F1 tornadoes. Twenty-nine percent of tornadoes are classified as *strong* (F2 and F3), which makes them capable of causing major structural damage even to well-constructed homes. Fortunately, only 2 percent of tornadoes are *violent* (F4 and F5). Those tornadoes are capable of wreaking incredible destruction. Cars can be picked up and carried tens of meters, pieces of straw can be driven into wooden beams, and freight cars can be carried off their tracks. Indeed, these storms are the true stars in all the movies and videos about tornadoes.

Between 1950 and 1999, 51 F5 tornadoes occurred in the United States and none in Canada. Thus, they happen about once a year in all of North America. Texas holds the lead for the greatest number of F5 tornadoes (6) during that period, while Alabama, Iowa, Kansas, and Oklahoma each had 5.

FATALITIES

Because they are small and last for such a short time, the overwhelming majority of tornadoes kill no one. Consider the fact that between 1950 and 1994, an average of 91 people died in 760 reported tornadoes each year. This means that at least 88 percent of all tornadoes kill no one,

TABLE 11–2	Fujita Intensity Scale		
INTENSITY	**WIND SPEED (km/hr)**	**WIND SPEED (MPH)**	**TYPICAL AMOUNT OF DAMAGE**
F0	<116	<72	**Light:** Broken branches, shallow trees uprooted, damaged signs and chimneys.
F1	116–180	72–112	**Moderate:** Damage to roofs, moving autos swept off road, mobile homes overturned.
F2	181–253	113–157	**Considerable:** Roofs torn off homes, mobile homes completely destroyed, large trees uprooted.
F3	254–332	158–206	**Severe:** Trains overturned, roofs and walls torn off well-constructed houses.
F4	333–419	207–260	**Devastating:** Frame houses completely destroyed, cars picked up and blown downwind.
F5	420–512	261–318	**Incredible:** Steel-reinforced concrete structures badly damaged.
F6	>513	>319	**Inconceivable:** Might possibly occur in small part of an F4 or F5 tornado. It would be difficult to identify the damage done specifically by these winds, as it would be indistinguishable from that of the main body of the tornado.

Note: F0 and F1 tornadoes are collectively called weak, F2 and F3 strong, and F4 and F5 violent.

---

[8]F6 tornadoes are hypothetical and have not been documented in nature.

DID YOU KNOW?

Items lofted by tornadoes to great height can be carried long distances. Following the F3 tornado on August 18, 2005, in Stoughton, Wisconsin (Figure 11–37), birth certificates and other personal papers were found in Milwaukee, more than 100 km (60 mi) away.

FIGURE 11–37
Swath of damage from the tornado that hit near Stoughton, Wisconsin, in 2005.

and the actual proportion of fatality-free tornadoes is even higher than 88 percent, for at least two reasons. First, tornadoes in which no one dies are preferentially undercounted, because a storm that kills is almost certain to be reported. Second, most fatalities result from a few very large storms that kill up to dozens of people. According to the National Severe Storms Laboratory (NSSL), fewer than 5 percent of all U.S. tornado deaths are associated with weak (F0 or F1) tornadoes, nearly 30 percent with strong tornadoes (F2 and F3), and about 70 percent with violent tornadoes (F4 and F5). Thus, only 2 percent of all tornadoes are responsible for more than two-thirds of all fatalities.

If it seems that news reports of killer tornadoes tend to focus on mobile home parks, it's for a reason. A disproportionately large percentage of tornado-related deaths do in fact occur in mobile homes. Trailers offer little protection to their occupants because they are easily blown off their foundations and tossed about by strong winds.

A large percentage of tornado fatalities also occur among passengers and drivers of trucks and automobiles. These victims often panic and make the fatal mistake of leaving the relative safety of their homes to outrun the storm in their vehicles. By far the safest place you can be when a tornado threatens is inside a well-constructed building, preferably in the basement and away from the windows.

A famous video shot in 1991 by a television news crew has led many people to believe that highway overpasses provide good shelter from a tornado. The video shows a film crew along a Kansas highway fleeing an approaching tornado. The reporters stopped at an overpass, where they advised a terrified father and his daughter to take shelter under the girders. Huddled with the man and child, the news crew continued to film the tornado as it passed directly over them—with nobody being hurt. The video then proceeded to show a truck driver who luckily survived after his rig had been blown off the highway, and reported the death of another nearby motorist. The film provided a remarkable view of a tornado from the inside and naturally was repeatedly played on television stations across the country. Unfortunately, it did not issue a disclaimer about the relative lack of safety provided by overpasses in most instances. Though the people in the video survived their ordeal and may have saved their lives by moving under the overpass, this is not usually the best course of action. In this case, the center of the tornado passed just to the south of the overpass so that the road above happened to provide considerable shelter. However, had the tornado taken a slightly different path, the outcome would not have been so fortuitous. Also, this particular tornado was a weak one—either an F0 or F1—and the overpass was a strong one with a narrow opening that allowed the people to hold on tightly as they were buffeted by winds. And finally, the fact that the incident was in a very rural area led to a small amount of debris being thrown toward the people—a situation that would be far different in a more heavily populated area.

People should always obey the following safety rules when a tornado threatens:

1. Stay indoors and seek shelter in a basement.
2. If you are in a building with no basement, move to an interior portion of the lowest floor and crouch to the floor. If possible, cover yourself with a mattress or some other form of padding to protect you from falling or flying debris. Despite what is sometimes said, the southwest corner of a building does not offer additional protection.
3. If you are in a mobile home, evacuate. If no fixed building is nearby, move away from all mobile homes and lie flat on low ground. (Tornado safety experts are now debating whether residents of mobile homes are better off staying in the open or taking shelter in a nearby automobile—or whether it is a good idea to drive that vehicle to a safe shelter. Automobiles offer some protection against flying debris but are subject to being tossed about by the violent winds. People in conventional homes should *never* abandon them for an automobile.)
4. If driving a car or truck, you might be able to avoid a tornado by driving at right angles to its path. Remember that most tornadoes move from southwest

DID YOU KNOW?

Between 1985 and 2003, 40 percent of all tornado fatalities in the United States occurred in mobile homes. This is greater than the 31 percent of fatalities that occurred in permanent homes, and it is very noteworthy in view of the relatively small proportion of people who live in mobile homes.

to northeast. If you cannot avoid the tornado's path, pull off the road and seek shelter. If none is available and strong winds make driving too difficult, run to low ground away from the road.

WATCHES AND WARNINGS

Without question, one of the key responsibilities of the National Weather Service is to issue severe weather advisories. These take two forms, watches and warnings, either of which can be issued for severe storms or tornadoes.[9] The declaration of a watch does not mean that severe weather has developed or is imminent; it simply tells the public that the weather situation is conducive to the formation of such activity. Most watches are issued for a period of 4 to 6 hours, for an area that normally encompasses several counties—about 50,000 to 100,000 sq km (20,000 to 40,000 sq mi).

The Storm Prediction Center (SPC) of the U.S. Weather Service in Norman, Oklahoma, has responsibility for putting out **severe storm and tornado watches** for the entire country. Operating 24 hours a day, every day, the center constantly monitors surface weather station data, information from weather balloons and commercial aircraft, and satellite data for all of the United States. If any particular part of the country appears vulnerable to impending severe storm activity, SPC issues a severe storm or tornado watch. The advisory then goes to the local office of the National Weather Service, which notifies local television and radio stations. The broadcast media then relay the information to the public. Watches for the entire country are also available from a number of sources, many of which are on the World Wide Web. Here is the text of a sample severe thunderstorm watch:

```
BULLETIN - IMMEDIATE BROADCAST REQUESTED

SEVERE THUNDERSTORM WATCH NUMBER 875

STORM PREDICTION CENTER NORMAN OK

319 PM CDT FRI JUL 28 1995

THE STORM PREDICTION CENTER HAS ISSUED A SEVERE THUNDER-
STORM WATCH FOR A LARGE PART OF MISSISSIPPI EFFECTIVE
THIS FRIDAY AFTERNOON AND EVENING UNTIL 1000 PM CDT.

LARGE HAIL..DANGEROUS LIGHTNING AND DAMAGING THUNDER-
STORM WINDS ARE POSSIBLE IN THESE AREAS.

THE SEVERE THUNDERSTORM WATCH AREA IS ALONG AND 75
STATUTE MILES EAST AND WEST OF A LINE FROM 15 MILES SOUTH
SOUTHEAST OF MC COMB MISSISSIPPI TO 5 MILES EAST OF OX-
FORD MISSISSIPPI.

REMEMBER..A SEVERE THUNDERSTORM WATCH MEANS CONDITIONS
ARE FAVORABLE FOR SEVERE THUNDERSTORMS IN AND CLOSE TO
THE WATCH AREA. PERSONS IN THESE AREAS SHOULD BE ON THE
LOOKOUT FOR THREATENING WEATHER CONDITIONS AND LISTEN FOR
LATER STATEMENTS AND POSSIBLE WARNINGS.

OTHER WATCH INFORMATION. THIS SEVERE THUNDERSTORM WATCH
REPLACES SEVERE THUNDERSTORM WATCH NUMBER 874. WATCH NUM-
BER 874 WILL NOT BE IN EFFECT AFTER 345 PM CDT.
```

If a severe thunderstorm has already developed, the public is notified by a **severe thunderstorm warning.** Likewise, **tornado warnings** alert the public to the observation of an actual tornado (usually by a trained weather spotter) or the detection of tornado precursors on Doppler radar. Unlike watches, which are issued

---

[9]Warnings and watches can be issued for other types of threatening weather, such as hurricanes and flash floods.

by the SPC, warnings are given by local weather forecast offices. They warn the public to take immediate safety precautions, such as finding shelter in a basement. The information is broadcast immediately by television and radio stations, and civil defense sirens are sounded.

Sometimes Doppler radar enables meteorologists to give warning of an impending tornado about half an hour before it actually forms. Prior to the advent of Doppler radar, warnings were usually issued only after a funnel cloud had been seen. Many tornado warnings are still based on *in situ* observations, but the use of Doppler radar provides better lead time in many instances. When tornado warnings are based on visual sitings, observers must report their findings to the local Weather Service office, where the meteorologist in charge then passes on the information to the broadcast media. The procedure normally takes several minutes. But most tornadoes have lifespans of about three to four minutes, so in many instances the warnings are issued to the public only after the tornado has died out. Though this might make the procedure seem pointless, that really is not the case. First of all, one tornado is often followed by others within the same storm system. Consequently, the precautions taken for the defunct tornado might still save lives when subsequent ones appear. Second, rare killer tornadoes tend to have considerably longer lifetimes and can remain in existence long after the warning has been issued. Thus the warning goes out while the tornado still presents a serious risk to people.

CONVECTIVE OUTLOOKS

The Storm Prediction Center recently unveiled to the public **convective outlooks** maps readily available from its Web site (listed at the end of this chapter). These maps are issued several times each day, showing the probability of severe weather hitting parts of the United States for the current day, the next day, or the day after. Figure 11–38 shows such a series of maps for May 15, 2003, with (a) plotting the threat of severe weather of any kind across the country. Here the greatest possibility of severe weather occurs around the Texas and Oklahoma panhandles, with much of the Ohio River Valley and Southeast having a slight risk. Panels (b), (c), and (d) give more specific probabilities for the occurrence of F2 or higher tornadoes, damaging winds, and large hail, respectively, occurring within 25 miles of any point within the outlined areas. So, for example, a resident of north Texas or western Oklahoma has about a 25 percent probability that a tornado or dangerous winds will occur within 25 miles of home on that day, with an even greater probability (35 percent) of large hail. Residents of Tennessee and Kentucky, on the other hand, have less than a 2 percent probability of tornadoes hitting, but a much greater likelihood (25 percent) of strong winds or large hail. In addition to these maps, the SPC accompanies the probability maps with narratives further describing the threats.

Despite the excellent warning system, disasters do indeed happen. Sometimes, sheer luck is all that stands in the way of major loss of life. On September 2, 2002, at 4:20 P.M. CDT, an F3 tornado hit the town of Ladysmith, Wisconsin, destroying 26 businesses and 17 homes. This time no tornado warning was issued until 17 minutes *after* the tornado hit the town, although a severe thunderstorm warning had been in effect for more than half an hour. (The fact that the nearby Doppler radar unit at La Crosse, Wisconsin, had been temporarily out of order may have contributed to the absence of a more timely warning.) Although most residents received no warning about the impending tornado, somehow the town escaped with no fatalities.

TORNADO OUTBREAKS

The most devastating of all outbreaks in the United States was the Tri-State tornado that rolled through eastern Missouri, southern Illinois, and southwest Indiana on March 18, 1925. The tornado killed about 700 people (roughly 600 of them

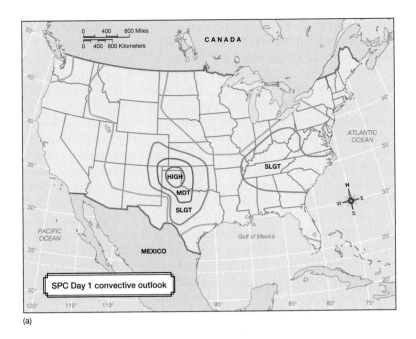

(a)

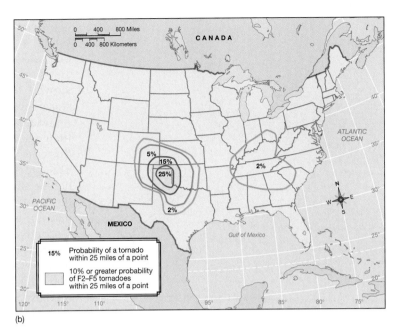

(b)

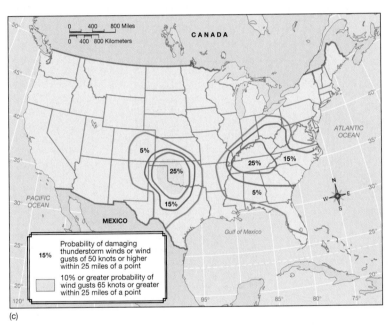

(c)

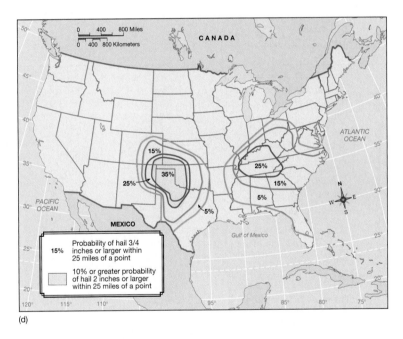

(d)

FIGURE 11–38

Examples of convective outlook maps mapping the threat of any type of severe weather (a), tornadoes (b), strong winds (c) and large hail (d).

in Illinois) and leveled several towns. To this day we do not know for sure if the Tri-State tornado was a single, incredibly large tornado, or part of a larger **tornado outbreak**—an event in which a single weather system produces at least six tornadoes.

Though the Tri-State was the deadliest of all tornado events, it was not the largest. That honor goes to the outbreak of April 3–4, 1974, in which at least 148 tornadoes swept across a wide region extending from northern Alabama and Mississippi to as far north as Windsor, Ontario, in Canada. Thirty of those tornadoes had F4 or F5 intensity and accounted for many of the more than 300 fatalities incurred over a 16-hour period.

Twenty-five years later, in May 1999, another major outbreak occurred. Hardest hit was central Oklahoma, where a series of tornadoes on the first day of the outbreak killed 44 people and destroyed 2600 homes and businesses (Figures 11–39 and 11–40). On the same day, tornadoes killed another 5 people and destroyed 1100 buildings in the Wichita, Kansas, area. And on May 5 and 6 the storms claimed another 5 lives in Texas and Tennessee.

Many factors contribute to the loss of life inflicted by tornadoes, including their maximum wind speed and their paths relative to large populations. In this case, extremely powerful winds descended upon large population centers. The effectiveness of the warning system also affects the death total and, fortunately, the system was at its very best during the 1999 outbreak. Field crews observed and reported the early development of conditions favorable for tornadoes, which complemented the forecasting efforts of meteorologists at the respective Weather Service offices. Meteorologists issued public advisories a full two hours in advance of the tornadoes, television and radio news crews disseminated the information, and the public generally responded wisely. Many residents sought shelter in basements, and others living in homes without basements took cover in interior hallways in time to save their lives. (People living in mobile homes received no protection at all from the storms; in the town of Bridge Creek every one of the 11 people who died were in trailers. Likewise, many fatalities occurred on highways—3 of the fatalities were peo-

FIGURE 11–39
Map of May 3, 1999, tornadoes in Oklahoma.

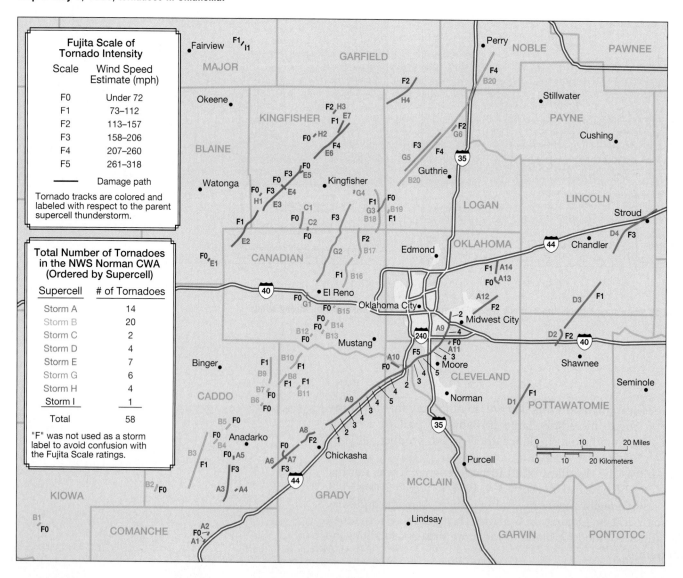

FIGURE 11–40
Tornado damage in central Oklahoma.

ple who took shelter under freeway overpasses!) Experts at the National Oceanic and Atmospheric Administration estimate that the warning system saved as many as 700 lives.

WATERSPOUTS

So far we have discussed tornadoes over land. Similar features, called **waterspouts** (Figure 11–41), occur over warm-water bodies. Waterspouts are typically smaller than tornadoes, having diameters between about 5 and 100 m (17 to 330 ft). Though they are generally weaker than tornadoes, they can have wind speeds of up to 150 km/hr (90 mph), which makes them strong enough to damage boats.

Some waterspouts originate when land-based tornadoes move offshore. Most, however, form over the water itself. These "fair-weather" waterspouts develop as the warm water heats the air from below and causes it to become unstable. As air rises within the unstable atmosphere, adiabatic cooling lowers the air temperature to the dew point, and the resultant condensation gives the waterspout its ropelike appearance. Waterspouts form in conjunction with cumulus congestus clouds, those having strong vertical development but not enough to form the anvil that characterizes cumulonimbus.

Contrary to what we might assume, the visible water in the waterspout is not sucked up from the ocean below; it actually comes from the water vapor in the air. Waterspouts are particularly common in the area around the Florida Keys, where they can occur several times each day during the summer.

Although tornadoes produce the most extreme winds on Earth, they are relatively small and short-lived. Tropical storms and hurricanes, on the other hand, usually produce less intense winds but can wreak more extensive destruction because of their longer lifespans and larger extent. Chapter 12 examines the processes that produce these devastating storms and their consequences.

FIGURE 11–41
A waterspout.

Summary

Some of the most dramatic of all storms are those that produce lightning and thunder. Lightning begins when negative electrical charges build up near the base of a cloud and positive charges gather at the top. In the case of cloud-to-ground lightning, rapidly growing leaders extend downward from the base of the cloud. When they connect with some object at the surface, a visible stroke develops. Most often a rapid sequence of multiple strokes follows the initial one to produce a lightning flash. Extreme heating of the air within the stroke causes the air to expand explosively and create the sound of thunder.

Thunderstorms create downdrafts that can cause the storm to die out or intensify. In air mass thunderstorms, downdrafts eventually destroy the storm. They undergo a sequence from the cumulus to the mature to the dissipative stages in tens of minutes. Severe storms—those that produce large hail, damaging winds, or tornadoes—develop when downdrafts reinforce the storm. Such storms occur within squall lines, in multicellular mesoscale convective complexes, or as supercells.

Tornadoes, among the most fearsome of all natural phenomena, occur more often in the United States than in any other country of the world—and nearly all the states and Canadian provinces experience them sooner or later. They are most likely to form in the spring or early summer, but they can occur at any time of the year. Many tornadoes (especially those that emerge from supercells) follow the formation of large rotating areas within storm clouds called *mesocyclones*. However, other unexplained processes can also lead to their formation.

The majority of tornadoes are classified as weak and cause no fatalities, but relatively rare, extremely large tornadoes can cause tremendous devastation and loss of life. Fortunately, the U.S. Weather Service has a system of tornado watches and warnings that alert the public to threatening weather. In recent years, Doppler radar has shown tremendous value as a research and forecast tool. This tool has also become valuable in the tracking of another weather phenomenon—hurricanes—that we will examine in Chapter 12.

Key Terms

lightning page 308

cloud-to-cloud lightning page 308

sheet lightning page 308

cloud-to-ground lightning page 308

charge separation page 308

runaway discharges page 309

fair-weather electric field page 310

mean electric field page 310

stepped leader page 310

stroke, return stroke page 310

dart leader page 310

flashes page 310

ball lightning page 311

St. Elmo's fire page 312

sprites page 312

blue jets page 312

thunder page 313

heat lightning page 313

air mass thunderstorms page 314

severe thunderstorms page 314

cumulus stage page 315

mature stage page 315

dissipative stage page 316

mesoscale convective systems (MCSs) page 316

squall lines page 316

mesoscale convective complexes (MCCs) page 316

supercells page 316

outflow boundary page 317

potential instability page 318

gust front page 321

shelf cloud page 321

roll cloud page 321

Doppler radar page 323

hook page 323

vault page 323

Doppler effect page 324

sweep page 324

volume sweep page 325

downbursts page 326

derechos page 326

microbursts page 326

tornadoes page 330

mesocyclones page 331

wall clouds page 333

funnel clouds page 333

suction vortices page 340

Fujita scale page 340

severe storm and tornado watches page 343

severe thunderstorm warning page 343

tornado warning page 343

convective outlooks page 344

tornado outbreak page 345

waterspouts page 347

Review Questions

1. How common is cloud-to-ground lightning relative to cloud-to-cloud lightning?

2. Describe the current theories regarding the formation of charge separation.

3. What is the difference between a lightning stroke and a lightning flash?

4. Describe the sequence by which electrical imbalances lead to lightning strokes.

5. Briefly describe the following phenomena:
   a. ball lightning
   b. St. Elmo's fire
   c. sprites
   d. blue jets
6. What causes thunder?
7. Why is the term *heat lightning* misleading?
8. What are the three stages of an air mass thunderstorm?
9. How big are air mass thunderstorms and how long do they usually persist?
10. Explain why some thunderstorms are have short lifespans and yield little damage and others are able to develop into severe thunderstorms.
11. Describe the following types of storm systems:
    a. mesoscale convective systems
    b. squall lines
    c. mesoscale covective complexes
    d. supercells
12. How are outflow boundaries formed, and what effect do they have?

13. Describe the processes that lead to tornado development in supercell and nonsupercell storms.
14. What features of Doppler radar make it an effective tool for severe storm forecasting?
15. What are hook echoes and vaults, and why are they important?
16. Explain how microbursts form and why they present a serious threat to aviation.
17. Describe the location and timing of tornadoes in North America.
18. Describe the process of tornado formation from supercell storms.
19. What are wall clouds, and why is their appearance a cause for concern?
20. What is the leading threat to human safety when tornadoes hit?
21. Describe the Fujita scale for classifying tornadoes. Which category is most common? What is the highest F-value that can actually occur in nature?
22. What is the difference between a tornado watch and a tornado warning?
23. How do waterspouts compare to tornadoes, on average, in terms of intensity?

Critical Thinking

1. Is charge separation necessary for sheet lightning and/or ball lightning?
2. Why does the environmental lapse rate affect the distance at which "heat lightning" can be observed?
3. You are outside on a sunny afternoon and observe a thunderstorm far to the west. An hour later, the storm passes over you. Is this more likely to have been an air mass thunderstorm or some sort of mesoscale convective system?
4. In what fundamental ways are gust fronts different from passing cold fronts?
5. Why is the incidence of thunderstorms much lower near the Pacific Coast than at the Atlantic Coast?

6. What conditions east of the Rocky Mountains promote a much greater incidence of tornadoes than exists in western North America?
7. Why is it not possible for a mesocyclone to occur within an air mass thunderstorm?
8. Why is it extremely unlikely that a tornado will move from east to west?
9. Other than their location with respect to land vs. water, how do waterspouts differ from tornadoes?

Problems and Exercises

1. Compare the map of annual hailstorms (Figure 7–15) to the map depicting the distribution of lightning flashes (Figure 11–22). Identify the regions where both are frequent, and those (if any) where only hail or thunderstorms frequently occur.
2. A tornado has a ring of uniform winds of 200 km/hr (120 mph) around the vortex 20 m (66 ft) away from its center. If the tornado moves to the northeast at 50 km/hr (30 mph), what is the effective wind speed on the northwestern and southeastern portions of the tornado?
3. On a regular basis, go to the Storm Prediction Center (SPC) Web site at **http://www.spc.noaa.gov/** and note the location of all current severe thunderstorm or tornado watches. Then go to one of the Web sites mentioned in previous chapters to observe surface weather maps, satellite images, and radar return maps. Describe the position of the watch area relative to what the maps and images depict.

Quantitative Problems

The Web site for this book, **http://www.prenhall.com/aguado**, offers quantitative problems that can help reinforce your understanding of several of the concepts discussed in this chapter. The problems deal with the timing of the arrival of thunder following a lightning strike, the characteristics of tornadic winds and pressure gradients, and potential instability. We suggest you go to the Web page for Chapter 11 and work out these straightforward problems.

Useful Web Sites

http://www.lightningstorm.com/
Click on the lightning map icon to see where lightning has occurred over the United States during the last 30 minutes. A similar map for Canada exists at **http://weatheroffice.ec.gc.ca/lightning/index_e.html**.

http://www.nssl.noaa.gov/edu/ltg/
This site provides interesting lightning information from the National Severe Storms Laboratory.

http://www.uic.edu/labs/lightninginjury/
Information on the medical aspects of lightning strikes.

http://www.nssl.noaa.gov/headlines/outbreak.shtml and **http://www.srh.noaa.gov/oun/storms/19990503/index.html**
At these two sites, you'll find comprehensive information on the deadly tornado outbreak of May 1999.

http://www.lightningsafety.noaa.gov/
Comprehensive information from the National Weather Service includes medical effects of being struck by lightning, myths, and indoor and outdoor safety tips.

http://www.srh.noaa.gov/oun/papers/overpass.html
This is an outstanding presentation on the dangerous use of highway overpasses as tornado shelters.

Media Enrichment

Tutorial
Doppler Radar

This tutorial describes the use of Doppler radar in weather forecasting, especially as related to severe thunderstorms. It first shows how the Doppler effect allows meteorologists to detect motions within storm clouds and then observe these motions at various levels within the clouds. It also explains how the radar unit can be adjusted to observe different properties of a cloud target. The tutorial concludes with a movie in which a professional meteorologist presents some real examples of Doppler imagery—and how it can sometimes yield surprising results.

Weather in Motion
Lightning

This brief movie shows cloud-to-ground lightning, bolts of cloud-to-cloud lightning, and sheet lightning.

Weather in Motion
Cloud and 3-D Flow Tracers

This is the first of four computer simulations illustrating the typical development of thunderstorm clouds. The white arrows show the movement of air. Initially uplift at the surface causes the development of cumulus clouds. The clouds deepen as rising motions continue well beyond the lifting condensation level. The largest of the thunderstorm clouds penetrates into the stratosphere, where differential wind velocities shear the top of the cloud to produce the characteristic anvil. The scene then rotates 360 degrees to offer a three-dimensional perspective.

Weather in Motion
Cloud and Vertical Velocity

This movie presents the same scene as above but uses color to highlight the speed and direction of vertical motions. Red portions of the clouds indicate rapid updrafts up to 38 m/sec (84 mph), while blue portions represent areas of downdraft. Clearly the clouds are dominated by zones of rapid uplift, with descending winds found only in limited regions, primarily in the downwind margins of the clouds.

Weather in Motion
Vertical Motion Cross Section

This movie, the third in the series, displays a vertical cross section of the winds in the region. Initially, the airflow is primarily horizontal at all but the very lowest level near the surface. Eventually air moves toward the region of convection and rapidly turns upward, where updrafts upset the general pattern of left-to-right winds. Notice that substantial areas of descending air exist outside the margins of the clouds.

Weather in Motion
Cloud and Horizontal Flow

This is the final movie in the sequence showing thunderstorm development. In this case, we see the horizontal wind flow pattern through a progression of altitudes after the thunderstorm is well developed. The airflow consistently turns clockwise with increasing altitude, yielding a pattern of *veering* winds associated with advection of warm air. Also observe that the wind flow is deflected around and over the upper portions of the main storm cell.

Weather in Motion
Squall Line Tornado over Northwestern Louisiana

Severe thunderstorms can develop as part of mesoscale convective complexes, supercells, and squall lines. This movie uses color-enhanced infrared images to depict the movement of a squall line thunderstorm that brought tornadoes to northwestern Louisiana on April 3, 1999. The movie begins with a wide band of cloud cover and storm activity extending nearly north–south from eastern Texas to eastern Oklahoma and western Arkansas. Within the band is a narrow zone of more severe storm activity, indicated by deep red. Scattered regions of somewhat less intense convection exist over northern Louisiana.

As the band approaches northwestern Louisiana, the squall line containing the most intense activity grows longer and wider, and tornadoes occur around Benton. The storm moves from the southwest to the northeast, which is the same direction as the orientation of the squall line. Consequently, the squall line lingers more than an hour before leaving Benton.

Notice that throughout the movie, the leading (eastern) edge of the squall line is not sharply defined. The zone of most intense activity merges with a region mosaiced by cloud cover of varying intensity. In contrast, the western boundary of the storm is very sharp.

Weather in Motion
Squall Line over Kentucky and Indiana

We see a radar loop of the extensive squall line that occurred during the evening of October 24, 2001, over Kentucky and Indiana, ahead of a powerful cold front. Sustained winds of up to 95 km/hr (60 mph) and gusts as strong as 130 km/hr (80 mph) caused considerable damage to the area.

Weather in Motion
Tornado

One of the most recognizable of all weather forms, tornadoes are capable of wreaking widespread death and destruction. Though this is a large and destructive example, tornadoes are sometimes even larger than this one.

Weather Image
Lightning Across the Globe

This image maps the distribution of lightning flashes over a one-year period.

Weather Image
Shelf Cloud over a Gust Front

Strong storms can produce rapid downdrafts. Precipitation causing the downdraft evaporates partially, consuming latent heat and thereby cooling the downdraft. This cooling increases the density of the air in the downdraft. The downdraft spreads outward upon hitting the ground to create a gust front. Because the gust front is denser than the air it displaces (due to evaporative cooling), it lifts the warm air and can produce shelf clouds, as seen in this image.

Weather Image
Mammatus I, II, and III

These pouchlike protrusions are sometimes formed near the anvil of cumulonimbus clouds. Each protrusion forms where air is sinking along the margins of the cloud. The water droplets and ice crystals extend out from the margin of the cloud and evaporate into the surrounding unsaturated air.

Weather Image
Wall Clouds I and II

Wall clouds are important features of supercell storms. Moist air is drawn into this portion of the supercell, causing a lower lifting condensation level and cloud base. Wall clouds are often the site of subsequent tornado development.

Weather Image
Tornado

Tornadoes assume a wide continuum of shapes and sizes. This is one example of a ropelike tornado. Contrast this one to the tornado shown in the *Weather in Motion: Tornado* movie for this chapter.

CHAPTER 2

TROPICAL STORMS AND HURRICANES

FIGURE 12–1
Modified satellite image showing the progression of Hurricane Andrew as it crossed Florida and entered the Gulf of Mexico in August 1992.

Nobody who lived in southern Florida during the August 1992 passage of Hurricane Andrew (Figure 12–1) will ever forget the experience. Andrew was fairly small but remarkably powerful, with wind gusts of up to 280 km/hr (175 mph) that moved rapidly across the peninsula. In one regard Andrew was not as destructive as many other hurricanes; although it did cause some local flooding, it did not yield extremely heavy rainfall. This contrasts strongly with Florida's experience with Hurricane Irene in October 1999.

Irene was typical of the strong tropical storms that form in the latter part of hurricane season: large and slow-moving. Although these storms lack the wind speeds of storms such as Andrew, they can bring heavy rainfall for days and cause extensive flooding. In the case of Irene, much of southern Florida received up to 27 cm (17 in.) of rainfall that produced widespread flooding. Nadia Gorriz of Miami-Dade County was among the many victims. The floodwaters that created a swamp around her house made it uninhabitable for humans but perfectly fine for the snakes and fish that took up residence. She put the cleanup in perspective: "We made it through Andrew. You just threw everything out and there wasn't that much water. This is the worst cleanup. We've gone through two gallons of Clorox II. So far."

Hurricanes do not restrict their fury to coastal and inland regions; they have been the nemesis of mariners for centuries. They have sunk an untold number of ships and even played a role in World War II when a single typhoon (the equivalent of a hurricane over the western Pacific) sank or heavily damaged several U.S. ships, destroyed hundreds of carrier-based aircraft, and killed more than 800 sailors. The death toll exceeded that of most naval battles during the war.

In this chapter we first describe the setting for hurricanes and tropical storms. We then describe their general characteristics, stages of development, and typical patterns of movement, concluding with hurricane monitoring and warning systems.

◄ This NASA image shows the rain accumulation from Hurricane Rita from September 18 through 25, 2005, as determined from data obtained by the Tropical Rainfall Measuring Mission (TRMM) satellite. Areas in green have 30 mm of accumulated rain or less; areas in dark red have 80 mm or more. The satellite-observed cloud pattern of Hurricane Rita on September 25 is depicted in white.

Hurricanes Around the Globe: The Tropical Setting

Extremely strong tropical storms go by a number of different names, depending on where they occur. Over the Atlantic and the eastern Pacific they are known as **hurricanes**. Those over the extreme western Pacific are called **typhoons**; those over the Indian Ocean and Australia, simply **cyclones**. In structure, the three kinds of storms are essentially the same, although typhoons tend to be larger and stronger than the others. We will use the term *hurricane* for the general class of storm, regardless of location.

Most U.S. residents associate hurricanes with storms that form in the Atlantic Ocean or the Gulf of Mexico. Yet other parts of the world have a much greater incidence of hurricanes (Table 12–1 and Figure 12–2). The Atlantic and Gulf of Mexico receive an average of 5.9 hurricanes each year, while the eastern North Pacific off the coast of Mexico has an average of 9.0. Most tropical storms in the east Pacific move westward, away from population centers, and so they receive little public attention. Sometimes, however, they migrate to the northeast and bring severe flooding and loss of life to western Mexico.

The region having the greatest number of these events—by far—is the western part of the North Pacific. In a typical year, nearly 17 typhoons hit the region. Even during the least active season between 1968 and 2004, there were 9 typhoons. At the other extreme, no hurricanes form in the Southern Hemisphere Atlantic, even at tropical latitudes (except for a very unusual event off the coast of Brazil in 2004). As you will see later, hurricanes depend on a large pool of warm water, a condition that does not arise in the relatively small South Atlantic basin.

In Chapter 8 we saw that during much of the year air spirals out of massive high-pressure cells that occupy large parts of the Atlantic and Pacific Oceans. Middle- and upper-level air along the eastern side of these anticyclones sinks as it approaches the west coasts of the adjacent continents. Because the air does not descend all the way to the surface, a subsidence inversion (see Chapter 6) forms above the surface. This particular subsidence inversion is called the **trade wind inversion**. The air below the inversion, called the **marine layer**, is cool and relatively moist.

FIGURE 12–2
Hurricanes around the globe.

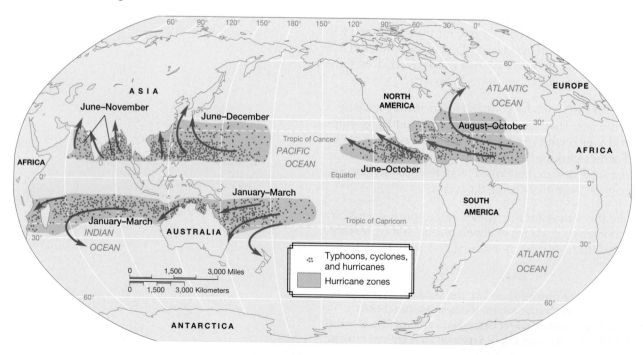

FIGURE 12–7
Some hurricanes develop double eye walls, such as Hurricane Emily in 2005. Usually occurring as the hurricane achieves maximum strength, both eye walls contract, with the innermost eye wall eventually dissipating and giving way to the outer eye wall.

FIGURE 12–7
Some hurricanes develop double eye walls, such as Hurricane Emily in 2005. Usually occurring as the hurricane achieves maximum strength, both eye walls contract, with the innermost eye wall eventually dissipating and giving way to the outer eye wall.

Some intense hurricanes develop **double eye walls**, as surrounding rain bands contract and intensify. As the interior eye wall contracts, it can begin to dissipate and the surrounding band can take its place. Figure 12–7 shows this process occurring in Hurricane Emily in 2005, with the outer band of rainfall (shown as green arcs embedded with red areas of heavier activity) surrounding the small eye wall in the center. The existence of double eye walls often indicates a hurricane is achieving its maximum strength.

NASA scientists have recently uncovered the existence of **hot towers** (Figure 12–8) embedded in some eye walls that last between 30 minutes and 2 hours. Hot towers are localized portions of eye walls that rise to greater heights (up to 12 km, or 7 mi) than the rest of the eye wall. The researchers found that development of hot towers indicates a greater likelihood that the hurricane will intensify within the next 6 hours.

The air temperature at the storm's surface within an eye is several degrees warmer than outside the eye because compression of the sinking air causes it to warm adiabatically. The air is also drier, because warming the unsaturated air lowers its relative humidity. Contrary to common belief, however, the air is not entirely cloud-free within the eye; instead, fair-weather cumulus clouds are scattered throughout the otherwise blue sky.

DID YOU KNOW?

Although hurricane wind speeds are, by definition, extremely fast, it still takes a parcel of air an average of eight days to flow into and out of a hurricane.

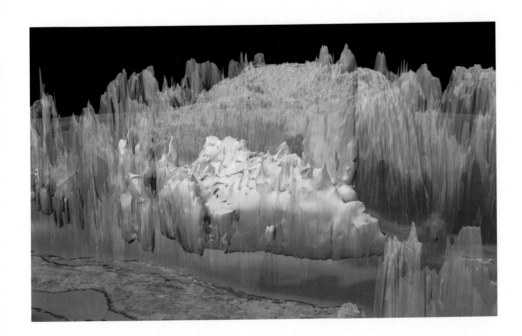

FIGURE 12–8
This hot tower in Hurricane Rita in 2005 was observed by NASA's Tropical Rainfall Measuring Mission (TRMM) spacecraft. These localized areas of deep cloud cover and intense precipitation often precede hurricane intensification.

Hurricane Formation

Hurricanes do not suddenly appear out of nowhere; they begin as much weaker systems that often migrate large distances before turning into hurricanes. In this section we examine hurricane development, with particular emphasis on how it occurs in the Atlantic.

STEPS IN THE FORMATION OF HURRICANES

Although most tropical storms attain hurricane status in the western portions of the oceans, their earliest origins often lie far to the east as small clusters of small thunderstorms called **tropical disturbances**. Tropical disturbances are disorganized groups of thunderstorms having weak pressure gradients and little or no rotation.

Tropical disturbances can form in several different environments. Some form when midlatitude troughs migrate into the tropics; others develop as part of the normal convection associated with the ITCZ. But most tropical disturbances that enter the western Atlantic and become hurricanes originate in **easterly waves**, large undulations or ripples in the normal trade wind pattern. Figure 12–9 illustrates a typical easterly wave. Because pressure gradients in the tropics are normally weaker than those of the extratropical regions, the easterly waves are better shown by plotting lines of wind direction (called *streamlines*) rather than isobars. The air in the wave initially flows westward, turns poleward, and then flows back toward the equator, with the entire wave pattern extending 2000 to 3000 km (1200 to 1800 mi) in length. On the upwind (eastern) side of the axis, the streamlines become progressively closer together, indicating that surface motions are convergent. With convergence there is rising motion (see Chapter 6), thus the tropical disturbance is located upwind of the wave axis (the dashed line) of the easterly wave. Surface divergence downwind of the wave axis leads to clear skies. (An explanation for why the streamlines are convergent and divergent is somewhat complicated; the main factor involves changes in relative vorticity that occur as the air moves poleward and equatorward.)

FIGURE 12–9
Easterly waves have surface convergence and cloud cover east of the axis and divergence to the west.

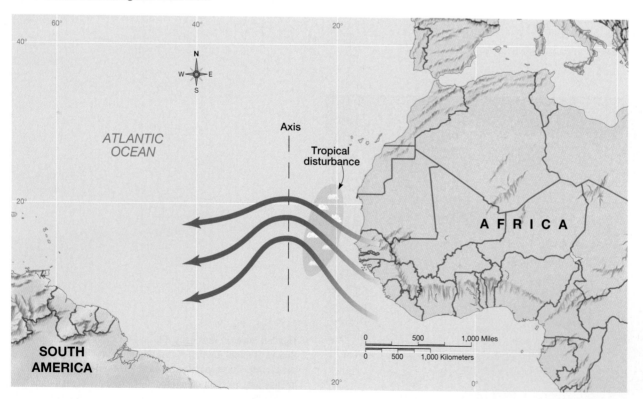

The tropical disturbances that affect the Atlantic Ocean, Caribbean, and the Americas mostly form over western Africa, south of the Saharan desert. Being in the zone of the trade winds, these storms tend to migrate westward. When they reach the west coast of Africa, they weaken as they pass over the cold Canary current over the eastern Atlantic. There the low water temperatures chill the air near the water surface and cause the air to become statically stable. If the disturbances migrate beyond the coastal zone of surface upwelling, however, warmer waters farther offshore raise the temperature and humidity of the lower atmosphere and cause the air to become unstable. Then, as the storms continue westward, a small percentage develop into more intense and organized thunderstorm systems. Easterly waves move westward at about 15 to 35 km/hr (10 to 20 mph), and so it typically takes about a week to ten days for an embedded tropical disturbance to migrate across the Atlantic.

The vast majority (probably more than 90 percent) of tropical disturbances die out without ever organizing into more powerful systems. But some undergo a lowering of pressure and begin to rotate cyclonically. When a tropical disturbance develops to the point where there is at least one closed isobar on a weather map, the disturbance is classified as a **tropical depression**. If the depression intensifies further and maintains wind speeds above 60 km/hr (37 mph), it becomes a **tropical storm**. (At this point the system is named. See *Box 12–1, Special Interest: Naming Hurricanes*, for more information on this practice.) A further increase in sustained wind speeds to 120 km/hr (74 mph) creates a true hurricane. While only a small fraction of tropical disturbances ever become tropical depressions, a larger proportion of depressions become tropical storms, and an even greater percentage of tropical storms ultimately become hurricanes.

The location at which hurricanes are most likely to form varies seasonally. Early in the Atlantic season, weak fronts in the western ocean extend southward over warm tropical water. Wind shear across the fronts provides the circulation necessary for cyclone development. Later in the season, fronts are confined to higher latitudes and no longer play a role in cyclogenesis. Instead, warm waters are found progressively farther to the east, so that disturbances leaving the African continent can grow into full-scale cyclones. (Systems that become tropical storms in the tropical waters just off western Africa and become hurricanes before reaching the Caribbean are often referred to as *Cape Verde hurricanes*, so named for the islands near which they originate.) The net effect is that the birthplace of tropical cyclones moves from west to east across the tropical ocean during the first half of the season. In the late fall, the breeding ground moves westward as frontal activity again emerges as a primary agent of cyclone genesis.

As with their Atlantic counterparts, Pacific hurricanes move westward during their formative stages. Many come near Hawaii, but most bypass the islands or die out before reaching them. Unfortunately, this is not always the case. In September 1992, Hurricane Iniki battered the island of Kauai with wind gusts up to 258 km/hr (160 mph) and brought heavy flooding to the beach resort areas. The hurricane destroyed or severely damaged half of the homes on the island and devastated most of the tourist industry.

CONDITIONS NECESSARY FOR HURRICANE FORMATION

Although the dynamics of hurricanes are extremely complex, meteorologists have long recognized the conditions that favor their development. Great amounts of heat are needed to fuel hurricanes, and the primary source of this energy is the release of latent heat supplied by evaporation from the ocean surface. Because high evaporation rates depend on the presence of warm water, hurricanes form only where the ocean has a deep surface layer (several tens of meters in depth) with temperatures above 27 °C (81 °F). The need for warm water precludes hurricane formation poleward of about 20 degrees; sea surface temperatures are usually too low there. Hurricanes develop most often in late summer and early fall, when tropical waters are warmest.

Naming Hurricanes

12-1 SPECIAL INTEREST

During hurricane season, several tropical storms or hurricanes can arise simultaneously over various oceans. Meteorologists identify these systems by assigning names when they reach tropical storm status. The World Meteorological Organization (WMO) has created several lists of names for tropical storms over each ocean. The names on each list are ordered alphabetically, starting with the letter *A* and continuing up to the letter *W*. When a depression attains tropical storm status, it is assigned the next unused name on that year's list. At the beginning of the following season, names are taken from the next list, regardless of how many names were unused in the previous season. Six lists have been compiled for the Atlantic Ocean, and the names on each list are used again at the end of each 6-year cycle. English, Spanish, and French names are used for Atlantic hurricanes. If all the available names on a season's list are used, subsequent storms will be the given the names of the letters of the Greek alphabet. Thus, the twenty-second named storm for a season would be Alpha. This is exactly what happened in October 2005 when Tropical Storm Alpha appeared in the west-

TABLE 1	Western Atlantic Tropical Storm and Hurricane Names				
2006	**2007**	**2008**	**2009**	**2010**	**2011***
Alberto	Andrea	Arthur	Ana	Alex	Arlene
Beryl	Barry	Bertha	Bill	Bonnie	Bret
Chris	Chantal	Cristobal	Claudette	Colin	Cindy
Debby	Dean	Dolly	Danny	Danielle	Dennis
Ernesto	Erin	Edouard	Erika	Earl	Emily
Florence	Felix	Fay	Fred	Fiona	Franklin
Gordon	Gabrielle	Gustav	Grace	Gaston	Gert
Helene	Humberto	Hanna	Henri	Hermine	Harvey
Isaac	Ingrid	Ike	Ida	Igor	Irene
Joyce	Jerry	Josephine	Joaquin	Julia	Jose
Kirk	Karen	Kyle	Kate	Karl	Katrina
Leslie	Lorenzo	Laura	Larry	Lisa	Lee
Michael	Melissa	Marco	Mindy	Matthew	Maria
Nadine	Noel	Nana	Nicholas	Nicole	Nate
Oscar	Olga	Omar	Odette	Otto	Ophelia
Patty	Pablo	Paloma	Peter	Paula	Philippe
Rafael	Rebekah	Rene	Rose	Richard	Rita
Sandy	Sebastien	Sally	Sam	Shary	Stan
Tony	Tanya	Teddy	Teresa	Tomas	Tammy
Valerie	Van	Vicky	Victor	Virginie	Vince
William	Wendy	Wilfred	Wanda	Walter	Wilma

*Some of the scheduled 2011 storm names have been retired. Details were not known at the time this book went to press.

Hurricane formation also depends on the Coriolis force, which must be strong enough to prevent filling of the central low pressure. The absence of a Coriolis effect at the equator prohibits hurricane formation between 0° and 5° latitude. This factor and the need for high water temperatures explain the pattern shown in Figure 12–2, in which tropical storms attain hurricane status between the latitudes of 5° and 20°.

ern Atlantic—the first time ever that the list of names was unable to accommodate all of the tropical storms in a single season.

Particularly notable hurricanes can have their names "retired" by the WMO if an affected nation requests the removal of that name from the list. All replacements are made with names of the same gender and language. As of the end of 2004, 62 names had been retired from the Atlantic hurricane list.

The practice of naming hurricanes appears to have begun during World War II when meteorologists in the Pacific assigned female names (possibly after wives and girlfriends) to tropical storms and typhoons. This practice was adopted by the U.S. National Weather Service (then called the Weather Bureau) in 1953 and maintained until 1979, when male names were added to the lists. The names for Atlantic and east Pacific hurricanes for the years 2006–11 are presented in Tables 1 and 2.

TABLE 2	Eastern Pacific Tropical Storm and Hurricane Names				
2006	2007	2008	2009	2010	2011*
Aletta	Alvin	Alma	Andres	Agatha	Adrian
Bud	Barbara	Boris	Blanca	Blas	Beatriz
Carlotta	Cosme	Cristina	Carlos	Celia	Calvin
Daniel	Dalila	Douglas	Dolores	Darby	Dora
Emilia	Erick	Elida	Enrique	Estelle	Eugene
Fabio	Flossie	Fausto	Felicia	Frank	Fernanda
Gilma	Gil	Genevieve	Guillermo	Georgette	Greg
Hector	Henriette	Hernan	Hilda	Howard	Hilary
Ileana	Ivo	Iselle	Ignacio	Isis	Irwin
John	Juliette	Julio	Jimena	Javier	Jova
Kristy	Kiko	Karina	Kevin	Kay	Kenneth
Lane	Lorena	Lowell	Linda	Lester	Lidia
Miriam	Manuel	Marie	Marty	Madeline	Max
Norman	Narda	Norbert	Nora	Newton	Norma
Olivia	Octave	Odile	Olaf	Orlene	Otis
Paul	Priscilla	Polo	Patricia	Paine	Pilar
Rosa	Raymond	Rachel	Rick	Roslyn	Ramon
Sergio	Sonia	Simon	Sandra	Seymour	Selma
Tara	Tico	Trudy	Terry	Tina	Todd
Vicente	Velma	Vance	Vivian	Virgil	Veronica
Willa	Wallis	Winnie	Waldo	Winifred	Wiley
Xavier	Xina	Xavier	Xina	Xavier	Xina
Yolanda	York	Yolanda	York	Yolanda	York
Zeke	Zelda	Zeke	Zelda	Zeke	Zelda

*Some of the scheduled 2011 storm names have been retired. Details were not known at the time this book went to press.

Stability is also important in hurricane development, with unstable conditions throughout the troposphere an absolute necessity. Along the eastern margins of the oceans, cold currents and upwelling cause the lower troposphere to be statically stable, inhibiting uplift. Moreover, the trade wind inversion puts a cap on any mixing that might otherwise occur. Moving westward, water temperatures typically increase and the trade wind inversion increases with height or disappears altogether, and so hurricanes become more prevalent. Finally, hurricane formation requires an absence of strong vertical wind shear, which disrupts the vertical transport of latent heat.

Once formed, hurricanes are self-propagating (just as severe storms outside the tropics are self-maintaining). That is, the release of latent heat within the cumulus clouds causes the air to warm and expand upward. The expansion of the air supports upper-level divergence, which draws air upward and promotes low pressure and convergence at the surface. This leads to continued uplift, condensation, and the release of latent heat.

So, if hurricanes are self-propagating, can they intensify indefinitely, until they attain supersonic speeds? No, because they are ultimately limited by the supply of latent heat, which in turn is constrained by the temperature of the ocean below and by the processes underlying evaporation and convection. The importance of ocean temperature suggests that if the oceans were to become warmer, hurricanes would theoretically become more intense. This topic has received considerable attention lately because of the possibility of climatic warming, which could be accompanied by higher ocean temperatures.

Hurricane Movement and Dissipation

The movement of tropical systems is related to the stage in their development. Tropical disturbances and depressions are guided mainly by the trade winds and, therefore, tend to migrate westward. The influence of the trade winds often diminishes after the depressions intensify into tropical storms. Then the upper-level winds and the spatial distribution of water temperature more strongly determine their speed and direction (with the storms tending to move toward warmer seas). Once fully developed, tropical storms become more likely to move poleward, as shown in Figures 12–2 and 12–10. Figure 12–11 shows the locations of all hurricane landfalls from 1950 through 2004. (*Box 12–2, Special Interest: 2004 and 2005: Two Historic Hurricane Seasons*, describes the paths and damage done by several notable hurricanes of these two remarkable hurricane years.)

HURRICANE PATHS

Hurricanes and tropical storms often move in wildly erratic ways—for example, moving in a constant direction for a time, then suddenly changing speed and direction, and even backtracking along its previous path. Figure 12–10 plots some particularly erratic paths along the east coast of North America.

While hurricanes can hit any part of the Gulf of Mexico or Atlantic coasts at any time during hurricane season, they have a greater likelihood of taking particular paths during different months. Figure 12–12 shows the likelihood of hurricane passage for August (a), September (b), and October (c). In August, the most likely path of hurricanes tracks over the West Indies. From there, hurricanes are about equally likely to track toward the Texas coast or along the Atlantic coast from Florida to North Carolina. Two prominent paths dominate in September. One goes from between the Yucatan Peninsula and western Cuba, northward toward the central Gulf of Mexico coast; the other moves northward from around Haiti, the Dominican Republic, and Puerto Rico into the western Atlantic. Hurricanes taking the more easterly track are most likely to hit the middle Atlantic states if they make landfall. October paths exhibit a greater tendency to track from eastern Mexico, northward to Florida and the rest of the southeastern United States.

Atlantic tropical storms and hurricanes can travel great distances along the North American east coast, but they usually weaken considerably as they approach the northeastern United States and the Maritime Provinces of Canada. These storms usually lack the strong winds that characterize hurricanes in the low

FIGURE 12–10

Tropical storms and hurricanes have a tendency to move north or northeast out of the tropics along the southeast coast of North America. Their paths are often erratic, as seen in these examples.

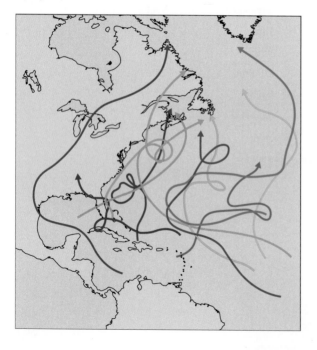

latitudes but can still bring intense rainfall and flooding. On rare occasions, the storms can maintain their strong winds even as they move considerable distances from the subtropics. For example, an intense September 1938 hurricane brought 200 km/hr (120 mph) winds to Long Island, New York, as it moved toward New England. Its estimated 600 fatalities made it the fourth deadliest of all U.S. hurricanes. More recently, in September 1985, Hurricane Gloria brought considerable wind and flood damage to Long Island and Connecticut.

Although hurricanes and tropical storms can move into the northeastern United States, along the West Coast they do not migrate nearly as far north without weakening to tropical depressions. The reason for this is the difference in water temperatures along the two coasts. The Pacific Coast is dominated by upwelling and the cold California current, while the warm Gulf Stream flowing along the East Coast provides a greater supply of latent heat. Sometimes, storms off the coast of Mexico move to the northeast across Baja California and into southern California. These storms lose their supply of latent heat and lose their intensity as they move inland, but they can still bring heavy rains and flooding. In 1976 Hurricane Kathleen caused massive flooding in the desert of southern California that wiped out part of Interstate Highway 8.

EFFECT OF LANDFALL

After making landfall, a tropical storm may die out completely within a few days. Even as the storm weakens, though, it can still import huge amounts of water vapor and bring very heavy rainfall hundreds of kilometers inland. This is especially true when the remnant of a hurricane moving poleward joins with a midlatitude cyclone moving eastward. Exactly this happened in 1969 when one of North

DID YOU KNOW?

A quick glimpse of Figure 12–2 reveals that hurricanes never occur in the South Atlantic. Well, almost never. On March 27, 2004, for the first time in recorded history a hurricane hit the coast of Brazil. Tropical cyclones have hit the Brazilian coast twice before, but "Catarina" (so named for the Brazilian state of Santa Catarina, which was hit hardest) packed 147 km/hour (90 mph) winds—making it a true hurricane. This storm did not develop in exactly the same manner as do most of those in the North Atlantic. It appears to have been a hybrid between a tropical system and a midlatitude cyclone, having initially formed along a cold front in the South Atlantic. Considerable damage occurred along the coastal region, but effective warnings to the public minimized the loss of life.

FIGURE 12–11

The locations of all hurricane landfalls over the continental United States from 1950 through 2004. Saffir-Simpson categories are those at the time of landfall.

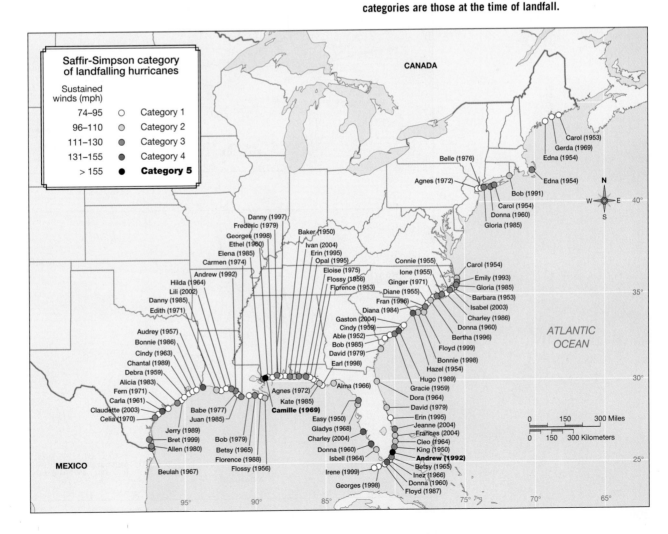

FIGURE 12–12

The predominant paths of Atlantic hurricanes in August (a), September (b), and October (c).

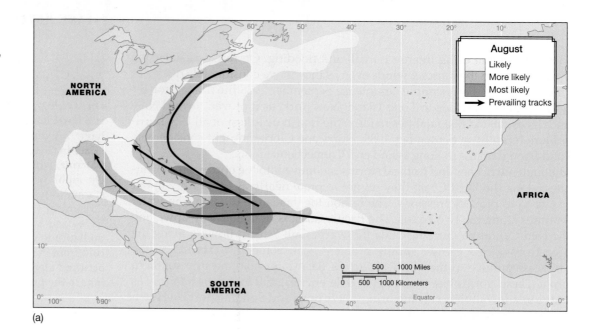

(a)

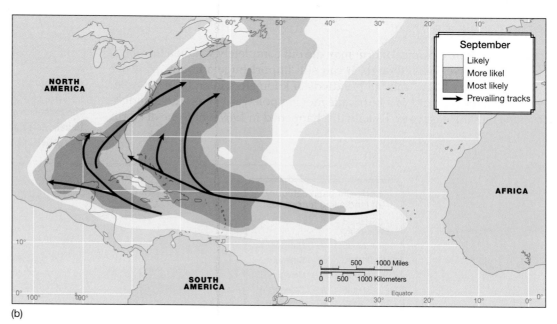

(b)

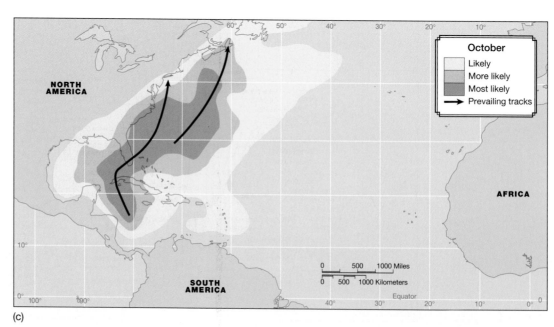

(c)

http://www.usatoday.com/weather/hurricane/whur0.htm
Some basic information from *USA Today.*

http://www.ncdc.noaa.gov/oa/climate/severeweather/hurricanes.html
Climatological information and data on past hurricanes from the National Climate Data Center.

http://tropical.atmos.colostate.edu/
Official Web site of the Tropical Meteorology Project at Colorado State University. Includes much information, such as the seasonal tropical storm forecast and a description of factors used to predict the upcoming season.

http://www2.sunysuffolk.edu/mandias/38hurricane/
A comprehensive and interesting review of the hurricane of 1938 and what it tells us about the vulnerability of the New York area to future events.

http://www.ncdc.noaa.gov/oa/reports/tech-report-200501z.pdf
An excellent analysis of Hurricane Katrina.

Media Enrichment

Weather in Motion
The 2005 Hurricane Season
Satellite movie showing all tropical storms and hurricanes in the Atlantic from June 1 to November 1, 2005.

Weather in Motion
Hurricane Katrina 1
A satellite movie showing Hurricane Katrina from the time it approaches Florida to its second landfall in Louisiana and Mississippi.

Hurricane Katrina 2
Radar loop of Hurricane Katrina as it hits Florida.

Hurricane Katrina 3
Radar loop of Hurricane Katrina prior to hitting the Gulf coast.

Weather in Motion
Hurricane Rita
View the movement of Category 5 Hurricane Rita as it moves over the Gulf of Mexico.

Weather in Motion
Hurricane Wilma
Hurricane Wilma as it moves toward Florida.

Weather in Motion
A Fly-Through of Hurricane Mitch
This movie opens with a view of this devastating hurricane obtained by satellite sensors. As the movie zooms in on the hurricane, the overlying cirrus clouds are stripped away so that the interior structure of the storm is evident. This movie was compiled using microwave, visible, and infrared imagery obtained by the Tropical Rainfall Measuring Mission (TRMM).

Weather in Motion
Hurricane Slice
This brief movie shows the pinwheel structure of a hurricane after clearing away the overlying cirrus.

Weather in Motion
Hurricane Eye Wall
A view of a hurricane eye wall as seen from a hurricane chase plane. Note the abrupt transition from clear air within the hurricane eye and the massive wall of cloud encircling the eye.

Weather in Motion
Hurricane Dennis
A computer-enhanced movie showing the movement of Hurricane Dennis along the Atlantic Coast.

Weather in Motion
Storm Surge
The combination of low pressure and strong onshore winds can bring elevated sea levels and strong waves ashore. This scene shows what a moderate storm surge can look like.

Weather in Motion
Hurricane Damage
A view of severe inland damage in the wake of a major hurricane, as seen from a helicopter.

Weather in Motion
Interview with Chase Plane Pilot
Have you ever wondered what it must be like to fly into a hurricane? Take a look at this film clip and hear firsthand from an expert.

PART FIVE
Human Activities

Greenwich Village, New York, during the 1996 blizzard.

Forecasters for the National Weather Service (NWS) spend a large portion of each workday fielding telephone inquiries about upcoming weather. Sometimes people want to know if their round of golf is likely to be rained out or if their outdoor plants might be vulnerable to overnight frost damage. But the most important aspects of a forecaster's job do not involve matters of simple convenience but questions that deal with life and death.

Such was the case for Mark Moede, a meteorologist with the NWS Forecast Office in San Diego. In late August 1998 much of southern California was threatened by brush fires triggered by unusually heavy thunderstorm activity. Firefighters battling the blazes received constant updates on weather conditions that could either suppress or enhance the spread of the fire. On September 2 the situation reached its most critical stage. Pete Curran of the Orange County Fire Department maintained close contact with Moede for constant updates on a line of approaching thunderstorms. At issue was whether the thunderstorms would continue to strengthen and whether they would pass through the areas where the firefighters were working. If the storms passed that way, then strong winds, deadly lightning, and blinding rains would place Curran's crews in jeopardy. With the aid of Doppler radar and information from automated weather stations, Moede made the right call. He advised Curran to evacuate his crews from the eventual path of the storm, where winds in excess of 95 km/hr (60 mph) created an uncontrollable firestorm. Afterward, firefighters reported that the fire line they had evacuated had been completely burned over. If there had been no call to evacuate, in all likelihood a number of firefighters would have been killed.

This chapter discusses the methods by which forecasters perform their job. We first look at some important issues regarding the general concept of weather forecasting and then discuss ways in which necessary data are obtained and processed. We then study the various types of weather maps and how they are used in weather analysis.

Weather Forecasting— Both Art and Science

Weather forecasting is a highly sophisticated and data-intensive endeavor. Forecasters routinely employ state-of-the-art computer hardware and software systems that perform millions of calculations, based on a huge amount of input data and displayed at sophisticated work stations. The meteorologists who work with the information come armed with rigorous university training that includes intensive coursework in mathematics, physics, chemistry, and, of course, meteorology. But the scientists who work at the public and private forecasting agencies also rely heavily on their ability to subjectively analyze the current weather situation, thus making their task a blending of scientific principle and art.

Forecasters for the **National Weather Service (NWS)** in the United States and the **Meteorological Service of Canada (MSC)** work 8-hour shifts in offices that are open 24 hours a day, 7 days a week. All forecasters apply the same scientific underpinnings to their work and make use of the same types of resources, but there is no precisely defined routine by which a forecast is made. Every meteorologist has his or her own preferred set of procedures for analyzing the weather situation—and those procedures may vary considerably under differing conditions. For example, a forecaster concerned with the possibility of near-term tornado development might spend much of his or her time looking at radar images, while one concerned with the possibility of an advancing snowstorm might spend more time assessing the changing pattern shown by satellite images. Furthermore, meteorologists employ

◀ A wildfire near Lake Piru, California. The intense heat induced convection that formed the cumulus cloud above the plume of smoke.

FIGURE 13–1

A typical National Weather Service office (a) and workstation (b). Two AWIPS graphical display monitors (c).

(a)

(b)

(c)

numerous rules of thumb that might help them assess the likelihood of a given storm producing rain instead of snow, and these rules of thumb often vary by geographic region.

Upon starting his or her shift, a forecaster will probably receive a verbal briefing about the weather of the last 24 hours from the meteorologist whose shift is about to end. The incoming meteorologist will also examine a large amount of textual and graphic information. Many professional meteorologists can still remember a time when most of the maps and information came in via a facsimile machine that produced black-and-white output. Each weather map or satellite image would take several minutes to print on a damp sheet of paper, and then it would be hung up on a wall atop the previously printed maps or images. If the meteorologist wanted to examine the changes that occurred in the weather over a certain period, he or she would have to manually flip through a sequence of maps. It would take a large section of wall to display the various types of maps and images to which the forecaster might want to refer. This situation has changed radically over the last several decades.

Figure 13–1 shows a typical National Weather Service office (a) and workstation (b) and (c), at which the forecaster will perform most of the job. Each workstation employs the **Advanced Weather Interactive Processing System (AWIPS),** which allows forecasters to display maps of current weather conditions, output of computer forecast models displayed in map form, satellite and radar images, forecast advisories and discussions from other weather facilities, and a great deal more. Each AWIPS station has two graphical display monitors and a separate alphanumeric display. The graphical display monitor is typically set up to present one large panel and four smaller panels off to the left side. Forecasters can zoom in on or out of any image and even superimpose several different types of information simultaneously. For example, a forecaster might set the large panel to show a satellite image of North America, with superimposed arrows depicting the upper-level winds. At the same time, the four smaller panels on the left side of the screen might display the current temperature, humidity, surface winds, and air pressure distributions. With just a few clicks of a mouse, the forecaster can call up just about any information that might help decipher the current weather patterns and help produce a successful forecast. AWIPS can also display all maps and images as movie loops to show the recent movement of significant weather features.

The forecaster goes through a sequence of procedures using AWIPS, depending on what the current conditions are and what type of weather may be developing. The first step is often to determine what type of cloud or precipitation conditions might be in the offing. This is of primary concern not only because of their intrinsic importance but also because such knowledge is necessary before any kind of temperature forecast can be made. (Obviously, a heavy overcast will reduce the amount of sunlight that can reach the surface, which will retard daytime warming.) If cloudy conditions currently exist, the question at hand is whether the clouds will move on and be replaced by clear skies. Forecasters rely heavily on visual inspection of satellite loops in these cases. Morning cloud cover can also "burn off" later in the day as the sun gets higher above the horizon. Meteorologists would want to look at the temperature and moisture profiles in these instances to determine just how thick the cloud layers are; if clouds are relatively thin, they are more likely to give way to clear skies. Clear days also include the possibility that clouds will move in from other regions or develop later on and create local rain showers. Vertical profiles showing the trend in temperatures and humidity from the surface to the upper stratosphere can be particularly useful in these situations.

Changes in the weather can occur even if there is little movement of major weather systems. Assume, for example, that a forecaster starting the morning shift at Charleston, South Carolina, determines that no major systems are moving toward or away from the southeastern United States. The meteorologist still has to look for a change in temperatures over the next few days. A good start in such instances

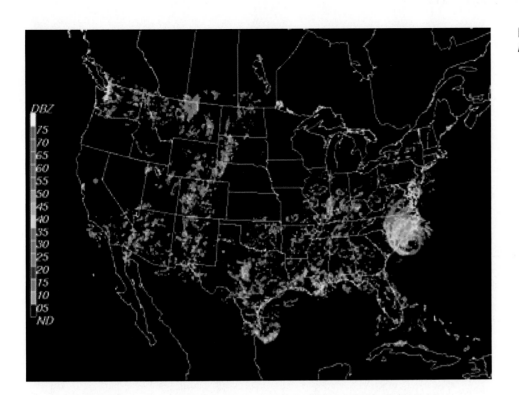

FIGURE 13–21
A radar composite map.

instance, the curvature of Earth's surface causes a horizontally emitted beam of radiation to assume greater heights above the surface as it moves far from the transmitter. Also, radar waves are refracted (bent) somewhat as they travel through the atmosphere. The extent to which the refraction occurs depends on the stability. Thus, the meteorologist not accounting for this effect can obtain false readings of cloud top height.

Thermodynamic Diagrams

The maps and images previously described provide two-dimensional views of atmospheric conditions, but they fail to provide detailed vertical information. Vertical profiles of temperature and dew point observed by radiosondes are plotted on **thermodynamic diagrams** (also called *pseudo-adiabatic charts*). The simplest thermodynamic diagram is the Stuve chart (Figure 13–22), on which the air temperature is scaled along the horizontal axis and the pressure on a nearly logarithmic

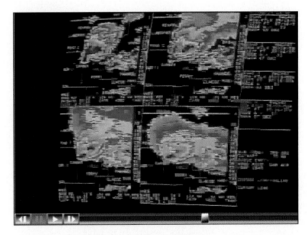

WEATHER IN MOTION
**Simulated Doppler I and II
Doppler Radar I, II, and III**

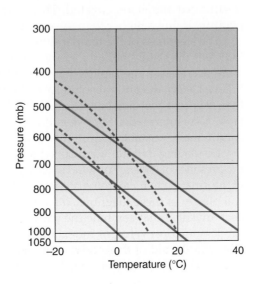

FIGURE 13–22

A Stuve thermodynamic diagram. The vertical lines across the horizontal axis represent temperature (°C); the horizontal lines, pressure. The solid and dashed lines sloping to the left with height are dry and saturated adiabats, respectively.

FIGURE 13–23

An example of a sounding on a Stuve diagram. The heavy line on the right plots the temperature; the one on the left shows the dew point profile.

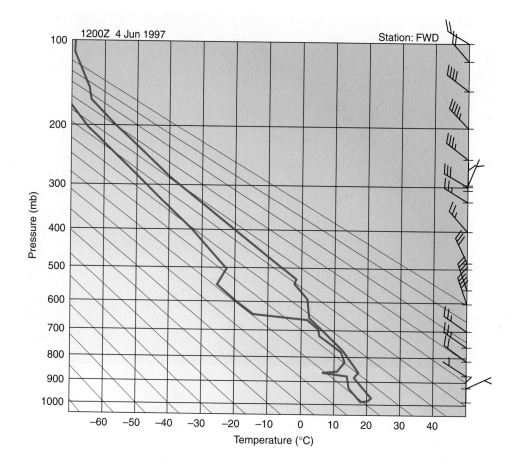

FIGURE 13–24

Stability, as indicated by the temperature profile on a Stuve diagram. Profile A is steeper than the dry adiabat, so the air is absolutely unstable. Profile B is less steep than the wet adiabat, so the air is absolutely stable. Profiles with slopes intermediate between the dry and wet adiabats (such as C) indicate conditionally unstable air.

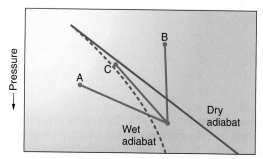

vertical axis. The straight, solid lines that slant upward to the left are called *dry adiabats.* These show the rate of temperature change for an unsaturated parcel of air that is lifted or lowered (in other words, they plot the dry adiabatic lapse rate). The dashed, slightly curved lines are called *wet adiabats,* showing temperature changes experienced by a rising saturated parcel.

Soundings are plotted on the charts by marking the temperature and dew point data at numerous pressure levels and connecting the dots. Figure 13–23 shows an actual sounding for the Dallas–Ft. Worth airport on the morning of June 4, 1997. The heavy lines represent the temperature and dew point profiles (the temperature profile is the one on the right—which must be the case because the temperature is always equal to or greater than the dew point). Notice that the two profile lines nearly merge between about 800 and 660 mb, indicating that the layer is saturated and cloud-covered. Throughout the rest of the atmosphere, the temperature is much higher than the dew point and the air is unsaturated.

The steepness of the temperature profiles relative to the dry and wet adiabats indicates the static stability for any portion of the atmosphere. Where the temperature decreases more rapidly with height than does the DALR (as in profile A in Figure 13–24), the air is absolutely unstable. Where it slants less steeply to the left than adjacent wet adiabats (profile B), the air is absolutely stable. Temperature lapse rates intermediate between the two adiabatic lapse rates (profile C) indicate conditionally unstable conditions, as we discussed in Chapter 5.

Displaying data on wind speed and direction at numerous heights (and thus the presence or absence of wind shear), thermodynamic diagrams provide a wealth of information to meteorologists, especially in predicting severe weather. Given the complexity of these diagrams, it's not surprising that a number of measures have been invented that combine the various elements related to severe weather into a single numerical value, or index. In the past these values were determined manually with the aid of thermodynamic diagrams; today they are calculated by computers. The lifted index and the K-index are two such summary measures.

Television Weather Segments

Television weather broadcasters are often local or even national celebrities. In many cases people choose a particular news broadcast each morning or evening based on the person doing the weather segment. So who exactly are these people and how do they do their jobs? Some weather anchors have little formal training in the subject, but many others have professional backgrounds in meteorology. These backgrounds range from a few college courses to formal degrees in the subject.

Weather anchors ultimately base their forecasts on the same information used by National Weather Service meteorologists, but the level of external support they use varies. Some of them depend entirely on forecasts issued by the NWS; others rely on forecasts provided by private weather companies; and some with strong meteorological training rely solely on their own analyses.

Regardless of the level of meteorological training, weather broadcasters can use an impressive array of graphics. The weather anchors will often use specialized software that allows them to set up a sequence of text pages, maps, and satellite and radar loops. These images and loops appear to viewers to be behind the meteorologist, but in fact he or she is standing in front of a blank screen (Figure 1). So how does the presenter manage to point to spots on a weather image that aren't really shown on the screen? This is done by looking at one of several strategically placed television monitors. Whichever direction the weather anchor is facing, he or she can see a monitor.

Since 1957 the American Meteorological Society has given the AMS Seal of Approval to television meteorologists meeting established criteria. Holders of the seal must have a certain level of scientific understanding, as demonstrated by the completion of at least 20 semester units of meteorology and the submission of three examples of their on-air work. Since its inception, more than 1400 broadcasters have been awarded the seal of approval.

In January 2005 the AMS initiated the Certified Broadcast Meteorologist Program (CBM). This program requires completing a college or university degree in meteorology or atmospheric science from an accredited institution, passing a written test, and submitting samples of televised work that are judged for their methodological soundness and the broadcaster's communication skills. Until 2008, the AMS will issue both the seal of approval and the CBM certification. In 2008 the AMS will stop issuing new seals of approval and will grant only the CBM certification (holders of the seal of approval will be able to retain their certifications after that date with annual renewals and the completion of professional development requirements every five years).

FIGURE 1
Television meteorologist Christina Russo doing a live, on-air weather segment.

LIFTED INDEX

The **lifted index** was developed in the mid-1950s. It combines the average humidity in the lowest kilometer of the atmosphere, the predicted maximum temperature for the day, and the temperature at the 500 mb level into a single number. The magnitude and sign of the values together indicate the potential for thunderstorms; negative values indicate sufficient water vapor and instability to trigger thunderstorms. More specifically, lifted index values between −2 and −6 indicate a high potential for thunderstorms, whereas values lower than −6 suggest a threat of severe thunderstorms.

K-INDEX

The **K-index** is similar to the lifted index but works better for predicting air mass thunderstorms and heavy rain than for severe weather. The index uses values of temperature and dew point at the surface and the 850, 700, and 500 mb levels. Various rules of thumb for different geographic regions and times of year translate K-values to the probability of heavy rains and thunderstorms. In general, K-values less than 15 indicate no potential for thunderstorms; values above 40 suggest that they are highly likely.

The National Centers for Environmental Prediction compile maps of K-index and lifted index values across the United States and Canada each day, giving forecasters a quick reference for the possibility of thunderstorms. Though the indices are very useful as a first step in predicting thunderstorms, a meteorologist would never use them alone in making a forecast. Rather, the indices help identify regions in which the forecaster should make more detailed analyses.

Summary

Weather analysis and forecasting depend on the cooperative efforts of nations belonging to the World Meteorological Organization. The weather services of these nations obtain surface data from land and sea-based stations, radiosondes, and satellites. For decades, weather forecasting involved subjective analysis of the available data. Although today's meteorologists still rely heavily on their own interpretation of current weather patterns, they also make wide use of output from numerical models. There is a wide variety of these models, and they undergo constant evolution, but all attempt to forecast the atmosphere by applying known physical laws. Short-term forecasts (up to two days) have improved considerably in recent years in response to improved data acquisition and processing, as well as model improvements. Medium-range forecasts offer some degree of skill out to about a week. On the other hand, long-term forecasts for 30- and 90-day periods have not displayed a high degree of accuracy or skill.

While information is available in many different forms, the weather map continues to be one of the most important tools for forecasting. The NWS updates current surface maps for the contiguous 48 states and southern Canada every 3 hours, and forecast maps are published at 12-hour increments. Maps of observed conditions plot the distribution of pressure with isobars and the location of fronts with heavy lines. Station models provide detailed information on temperature, dew point, pressure, wind, and other conditions at large number of stations.

Surface weather maps provide only part of the story and must be augmented by maps of atmospheric conditions at higher levels. In the United States, maps are routinely produced for the standard levels: the 850, 700, 500, 300, and 200 mb pressure levels. Each of these has its own set of advantages. For example, at the 850 mb level, analysis of air flow relative to the temperature distribution helps identify areas of low-level uplift. The 700 mb map has many of the same uses as the 850 mb map. In addition, it is particularly useful in predicting the movement of air mass thunderstorms. Of all the upper-atmosphere maps, 500 mb charts are probably the most widely used. Maps of the 300 and 200 mb levels are best for defining the number and position of Rossby waves and for analyzing jet stream patterns.

While weather maps provide two-dimensional views of the atmosphere at a number of levels, they do not provide much information on the vertical structure of the atmosphere at particular locations. This is where thermodynamic charts make a contribution. Profiles of temperature and dew point identify moist and dry layers within the atmosphere along with the location of clouds. The steepness of temperature profiles relative to dry and wet adiabats also indicate the stability of the atmosphere. Over the years, meteorologists have formulated several numerical indices that condense the information from these diagrams into single numerical values. Values above or below certain critical numbers indicate the need for further analysis of the possibility of heavy precipitation or severe weather.

The material in this chapter has relied heavily on the principles discussed in the previous 12 chapters and has applied these principles to an activity of great interest to society, forecasting. The next chapter looks at another human aspect of meteorology—the effects of human activity on weather.

Key Terms

National Weather Service
(NWS)   page 389

Meteorological Service of
Canada (MSC)   page 389

Advanced Weather
Interactive Processing
System (AWIPS)   page 390

zone forecast   page 391

National Oceanic and
Atmospheric
Administration (NOAA)
page 394

regional weather centre
page 394

climatological forecasts
page 394

persistence forecasts
page 395

analog approach   page 395

numerical weather
forecasting   page 395

quantitative forecasts
page 395

qualitative forecasts
page 395

probability forecasts
page 396

forecast quality   page 396

forecast value   page 396

forecast accuracy   page 396

forecast bias   page 396

forecast skill   page 397

World Meteorological
Organization (WMO)
page 397

World Meteorological
Centers   page 397

National Centers for
Environmental Prediction
(NCEP)   page 397

Canadian Meteorological
Centre   page 397

Atmospheric Environment
Service (AES)   page 397

Automated Surface
Observing System (ASOS)
page 398

Automated Weather
Observation Systems
(AWOS)   page 398

radiosondes   page 398

wind profilers   page 399

Weather Forecast Offices
(WFOs)   page 399

rawinsondes   page 399

analysis phase   page 400

prediction phase   page 401

post-processing phase
page 401

medium-range forecasts
(MRFs)   page 403

European Center for
Medium-Range Weather
Forecasting (ECMWF)
page 403

ensemble forecasting
page 404

chaos   page 404

long-range forecasts
page 405

Climate Prediction Center
(CPC)   page 405

seasonal outlooks   page 405

surface maps   page 407

station models   page 408

850 mb map   page 408

700 mb map   page 411

500 mb map   page 411

300 mb map   page 413

200 mb map   page 413

isotachs   page 413

visible images   page 414

infrared images   page 414

water vapor images
page 414

radar images   page 414

thermodynamic diagrams
page 415

lifted index   page 417

K-index   page 418

Review Questions

1. Briefly describe some of the variables that complicate weather forecasting.

2. Describe the basic characteristics of climatological forecasts, persistence forecasts, the analog approach, and numerical forecasting.

3. What are the distinguishing characteristics of quantitative, qualitative, and probability forecasts?

4. Explain how weather data are obtained and disseminated to agencies across the globe.

5. What are radiosondes and rawinsondes? What other sources of upper-atmosphere information are available to forecasters?

6. Describe the analysis, prediction, and post-processing phases in numerical forecasting.

7. What are model output statistics?

8. What are the primary characteristics of short-range, medium-range, and long-range forecasts? What types of information are needed to prepare the individual forecasts?

9. What is ensemble forecasting?

10. Describe the station model used for surface weather maps. How is the information presented on the station model? What measures must be used to convert the numerical data on the station model to real values?

11. Describe the characteristics of the 850 mb, 700 mb, 500 mb, 300 mb, and 200 mb maps that make each of them useful to forecasting.

12. What is the significance of cold air advection? Which weather map is most useful for locating it?

13. Which upper-level weather map would you use to locate short waves?

14. Why are omega highs significant? Which map is best for identifying them?

15. What is an isotach?

16. Which maps are most useful for locating the polar jet stream?

17. Describe the three types of satellite images discussed in this chapter. What characteristics make them useful?

18. Describe how radar works and how its information is presented.

19. Explain what a thermodynamic diagram does and how it is constructed.

20. Describe the lifted index and the K-index. How are they valuable to forecasters?

Critical Thinking

1. Why is it that the climatological, persistence, and analog approaches will never be entirely eliminated from the process of weather prediction?

2. If a forecast calls for a 70 percent chance of rain and no precipitation occurs, was the forecast actually wrong? What if this happens two days in a row? Ten days?

3. What would have to happen to the data acquisition network for the analysis phase of forecasting to be bypassed?

4. Are further improvements in weather forecasting more likely to occur for large-scale phenomena or for smaller-scale events? Explain your answer.

Problems and Exercises

1. Go to the Web site **http://www.wrh.noaa.gov/wrhq/nwspage. html (http://weatheroffice.ec.gc.ca/ canada_e. html** in Canada), and click on the National Weather Service office nearest to you. Make note of the 24-hour forecast and follow up the next day to see if the forecast was correct. Do this for extended forecasts as well. In general, do you find the forecasts to be accurate?

2. During times of unusual or inclement weather, visit the Web sites listed below to obtain weather-map, satellite, radar, and thermodynamic-diagram information. Is the weather you are experiencing consistent with what you would have expected, based on the information you obtained?

3. On a daily basis, make use of the available information from the Web sites listed below and make your own forecast (before reading the official forecast for your area). Then compare your forecast to that of the local weather office. Are your forecasts generally consistent with those of professional meteorologists?

4. Read the forecast discussion on the Web page of your local weather service office each day. These discussions explain the meteorologists' reasons for making their particular forecasts. Determine how this elaborates on the zone forecast for the same area.

Quantitative Problems

Weather forecasting is now strongly based on numerical techniques. This chapter is followed by an appendix that discusses some detailed aspects of numerical forecasting. The book's Web site, **www.prenhall.com/aguado,** offers several quantitative problems to help you better understand some nuances of numerical techniques.

Useful Web Sites

http://ww2010.atmos.uiuc.edu/(Gh)/guides/mtr/fcst/home.rxml
An excellent primer on forecasting from the University of Illinois.

http://www.crh.noaa.gov/lmk/soo/
Excellent information on forecasting tools available to meteorologists, courtesy of the Louisville, Kentucky, office of NWS. Includes pages describing AWIPS, Doppler radar, and numerous training modules.

http://meted.ucar.edu/nwp/course/index.htm
Tutorials on using numerical weather prediction.

http://www.theweatherprediction.com/habyhints/index.html
An extensive array of forecasting tips and tidbits.

http://www.nrlmry.navy.mil/~kuciausk/esis/frames/image_ interp_basic_frame.html
Excellent introduction to the use of satellite imagery in forecasting. Use the panel on the left side of the screen for navigation.

http://www.wrh.noaa.gov/wrhq/nwspage.html
Provides links to local NWS forecast offices and various support centers.

http://weatheroffice.ec.gc.ca/canada_e.html
A great first stop for forecast information in Canada.

http://weather.uwyo.edu/
Contains just about everything you need to make your own forecasts.

Media Enrichment

Weather in Motion
Simulated Doppler I and II

The first movie shows a horizontal slice of the thunderstorm simulation presented in Chapter 11. North–south winds are colored according to wind speed, with deep blue representing fast southward-moving wind and deep red used for fast northward-moving wind. (Greens indicate that air is not moving north or south.) As you view the movie, note the relationship between the streamlines and the colors.

While viewing the second movie, imagine a Doppler radar situated far to the south of the storm shown in the first movie. Southward-moving air would approach the radar and be rendered in blue. Similarly, northward-moving air would be shown in shades of red. This movie depicts what such a radar would observe. (It's the same sequence, viewed from above.) Areas where deep red adjoins deep blue correspond to areas of strong shear and would be interpreted as places for possible tornado formation.

Weather in Motion
Doppler Radar I, II, and III

In these three movies, a professional meteorologist gives a lesson on the use of Doppler radar. In the first movie, the radar displays four separate slices of a supercell storm, each at a different height. The top left panel displays a cross-section of the lower portion of the supercell; moving clockwise, the other three panels show a progressively higher cut. The most significant feature of the radar display is the presence of a hook echo. Hook echoes frequently signal the presence or imminent formation of tornadoes. This was indeed the case with this storm, which spawned an F4 tornado.

The second movie shows a different use of Doppler radar—one that displays the direction and speed of movement of the air with-

in the same storm as the first movie. A rotational signature is clearly evident in this radar return display.

Forecasters undergo considerable training in the use and interpretation of radar imagery. This knowledge helps them distinguish between significant features on the radar display and other less meteorologically important objects. The third movie shows a surprising feature on a radar display from Dodge City, Kansas.

Weather in Motion
Ensemble Forecasts I and II

These two movies show global forecasts of the 500 mb height surfaces for a two-week period from April 15, 2000, to April 30, 2000. NCEP produces a set of 17 global forecasts each day. These forecasts are run with slightly different initial conditions.

The first movie shows all 17 individual forecasts, contoured at two heights: 5490 m (light blue) and 5730 m (red). Climatological values are shown in green, and the medium-range forecast (MRF) control run (no perturbations in the initial conditions) is shown in white. "Spaghetti" plots like this may help forecasters determine model skill. Similarities between ensemble members help forecasters assess the confidence they can place in the model forecast. In addition, differences between the ensemble and observations may help expose model errors. Notice that as the animation runs, the various forecasts become increasingly different from one another, especially after one week or so (180 hours). Because the atmosphere displays chaotic behavior, the small differences in the initial conditions of each ensemble member grow with time, causing the forecasts to diverge.

The second animation shows the mean of all 17 ensemble members. The ensemble mean usually has greater skill than the individual model runs because averaging filters out the inconsistencies among ensemble members.

Chapter 13 Appendix
Numerical Forecast Models

Over the last four decades, weather forecasters have relied on several generations of computers and a variety of different models for guidance. As computers have increased in speed and capacity, the models have become increasingly complex, always straining the limits of computer power in an effort to achieve greater realism and accuracy. Even with today's computers, numerous compromises and approximations are necessary. In fact, it might be said that the spectrum of models arises not so much from differences in purpose or theory, but rather from differing approaches to the basic problems of abstraction and simplification. The result is tremendous variety in both the details and gross features of numerical models. In this appendix we first discuss the major features of numerical models, using the primary NCEP operational models as examples. We then describe some methods for assessing forecast quality.

MODEL CHARACTERISTICS

Today there are three numerical forecast models (ignoring variants): the **Global Spectral Model**, the **Nested Grid Model (NGM)**, and the **North American Mesoscale (NAM) Model** (formerly called the eta model).

SCALE CONSIDERATIONS

Thinking first about gross features, perhaps the most fundamental difference among models concerns the model **domain,** the region of the globe to be represented. Of course, if one wants to forecast for the entire globe, no decision is necessary—the computational domain must be the entire globe. But what if the goal is a European or a North American forecast? Where should the edges be? The boundaries of the domain require special treatment, usually in the form of strong assumptions about mass and energy transfer. To minimize their effect, one wants as large a domain as possible, with boundaries far outside the forecast region. Just how far outside depends in part on the forecast lead time. As lead time increases, locations farther removed from the forecast region begin to affect the forecast. Here, then, is a classic trade-off. A larger domain is preferable but requires more computer resources, which means some other aspect of the model must be compromised. Of the three NCEP models mentioned, only the Global Spectral Model has a global domain; the others have domains centered on the United States and cover only part of the Northern Hemisphere.

Another issue is spatial resolution. The fundamental equations governing atmospheric behavior are continuous in space, meaning that they describe the evolution of the atmosphere everywhere. If the governing equations were solved directly, they would yield a solution for an infinite number of points (we could find forecast values at every location within the domain). But the equations are far too complex to solve directly by analytical means; no such solution exists. Instead, the equations are written and solved in approximate form, with the result that forecast values are available only at widely spaced locations.

The approximation is accomplished in a number of different ways, but the result is always the same: There is some minimum size below which explicit representation is impossible. Current models are limited to a few tens of kilometers and larger, so, for example, nothing as small as an individual thunderstorm cloud can possibly appear. If such "sub-scale" phenomena are to be considered, they can only be expressed in terms of features that *are* resolved, an error-prone process called *parameterization*. A cloud parameterization would typically use simple relations that compute ice and liquid water fractions on the basis of relative humidity and air temperature, without modeling the details of cloud growth. Obviously, one wants high resolution so that small-scale processes and phenomena can be modeled and appear in the forecast, but this can come only at the cost of more computation. (Roughly speaking, doubling the resolution leads to eight times the computation.) Increases in resolution can be subsidized by reducing the domain, but that creates its own problems, as previously described.

The resolution issue applies to vertical coordinates as well as horizontal coordinates. Modelers face difficult questions about how many vertical levels to include, as well as how they should be arranged (at what altitudes). As a point of comparison, the horizontal resolution of the Global Spectral Model is about 1° in latitude and longitude (about 60 mi). It has 28 levels in the vertical, ranging from the surface to the 2.7 mb level. These are far from equally spaced—there are 8 levels below 800 mb, with increasing spacing at lower pressures (higher altitudes). Operational versions of the NAM model have been run with increasing resolution since its inception, ranging from 80 km initially to 29 km (17 mi) at the time of this writing. There are now 50 levels in the vertical. The NGM has 16 layers and two grids. The smaller inner grid, centered over the United States and Canada, has a resolution of about 80 km (48 mi). It lies completely within a larger, coarser outer grid that extends the domain to much of the Northern Hemisphere. The advantage of this scheme is that most of the computing is confined to the region of interest—computer resources are not wasted on detailed features far outside the forecast area. But at the same time, those areas are not ignored, as they would be with a smaller domain.

HORIZONTAL REPRESENTATION

Yet another major difference among models is the horizontal representation. Many models adopt a **grid representation**, in which the domain is subdivided into a lattice of grid points. (The grids at various levels need not coincide, nor need they be the same for all variables.) The equations are solved only at the grid points. As we've said, the finer the grid, the higher the model's resolution. Implicit in this is the idea that the grid captures horizontal variation in the atmosphere and that intermediate values can be inferred knowing values at nearby grid nodes.

The alternative is called a **spectral representation**. (Can you guess which representation is used in the Global Spectral Model?) Variables are represented as a series of "waves" in space, each having a characteristic wavelength. For example, there are waves in the model corresponding to Rossby waves, repeating just a few times around a parallel of latitude. Superimposed on these are other waves representing smaller-scale and larger-scale variations. To obtain the value for a particular variable at a point, its various wave functions are summed over all the wavelengths (the Global Spectral Model uses 126).

Spectral models have some advantages, especially for a global domain (the North and South Poles are easily handled in a spectral representation). There are other quite technical advantages as well, which we won't discuss. Instead we'll make a few general points about spectral modes. First, horizontal resolution is determined by the smallest of the "harmonics" present (smallest wavelength), so there is no escaping the problem of how to represent sub-scale processes. Second, not all of the variables can be represented in spectral terms. Only quantities subject to advection, such as heat and moisture, are treated this way. Other variables, such as radiation, must be computed on a point-by-point basis. Thus, the Global Spectral Model uses a grid for these "physical" quantities, and there is constant transformation of information from spectral to grid representation. (Fortunately, there are fast computational methods for this.) Finally, the spectral representation applies only to the horizontal—spectral models are layered in the vertical.

PHYSICAL PROCESSES

Numerical models also differ greatly in their "physics," which basically includes all processes other than those related to motion ("dynamics"). A model's physics package includes purely atmospheric processes (such as condensation), atmosphere–surface interactions (such as friction between the atmosphere and ground), and purely surface processes (such as soil moisture or depth of snow). Consider, for example, just some of the physical processes included in the Global Spectral Model:

- **Radiation:** Shortwave absorption and scattering in three wavelength bands (including the effects of ozone, water vapor, and carbon dioxide), with multiple reflections between clouds and ground. Effects of clouds, water vapor, CO_2, and O_3 on longwave absorption and emission are explicitly modeled, including overlap in absorption bands. Radiative properties of clouds depend on cloud thickness, temperature, and moisture content.

- **Convection, Clouds, and Precipitation:** Stratiform (nonconvective) clouds, as might be found with fronts and tropical disturbances, are determined from relative humidity using a statistical relationship. Convective clouds are either precipitating (deep convection) or nonprecipitating (shallow convection). Temperature and moisture profiles are used to determine which occurs. Downdrafts and evaporation of precipitation are simulated, as are entrainment and detrainment of updraft and downdrafts. Precipitation arises from both deep convection and from large-scale condensation (when the air at a point becomes saturated, regardless of convection). Precipitation is evaporated into unsaturated air below the cloud; only that which survives the descent is deposited on the surface.

- **Surface Properties and Processes:** Sea surface temperature and sea-ice distributions are fixed for the model run. Sea-ice temperature is computed from heat exchange between the atmosphere and ocean below the ice. Surface albedo depends on zenith angle, snow cover, and vegetation type. Snow cover is determined by accumulation from falling snow, snow melt, and sublimation. Land surface evaporation consists of evaporation from the ground and plant canopy, as well as transpiration by the canopy. Precipitation not intercepted by the canopy is partitioned into soil moisture recharge and runoff. (Melting snow also contributes to soil moisture.)

Notice that many of the preceding are sub-scale processes; thus, the physics package uses parameterization heavily. Different models not only include different processes but also employ different parameterizations for the same processes. Sometimes this is by necessity rather than preference, as parameterizations appropriate at one scale are not necessarily useful at another scale.

As we mentioned, there is a continual evolution of models (and no doubt many of the specifics in this appendix will have changed by the time you read this). For example, when it was introduced in 1973, the Global Spectral Model was the state of the art at NCEP. Today, the NAM model is generally considered to occupy that position and is itself evolving. Is this effort at improving models justified—has forecasting skill improved? It definitely has, as we will see later in this appendix.

The future of numerical prediction calls for ever more realistic models. For example, NCEP plans a 5-km/150-level NAM-like model in the near future. This is a daunting task: Such a model will require 1000 times the computing power and 100 times the memory of today's model. Even a 5-km model will not permit explicit calculation of convection and other extremely important processes. Estimates suggest that this will require about 10,000 times the power of today's supercomputers. Whether or not this is possible depends on both the public's willingness to pay and the development of computer technologies not yet on the drawing board.

MEASURES OF FORECAST ACCURACY AND SKILL

For some variables, such as temperature or precipitation amount, accuracy measures, such as bias and mean absolute error (MAE), practically suggest themselves. Bias is most easily defined as the difference between the average forecast value and the average observed value. It reveals any tendency for the method to give forecasts values above or below the true value (for example, too hot on average, too much rain on average). A mathematically equivalent definition of bias is simply the average error. For each forecast, we find the departure from observed and compute the average of those errors.

Figure 1 illustrates how bias can be used as a means to assess the accuracy of models. The figure plots the bias for 1-in. (2.5-cm) precipitation events predicted by NCEP for the last few decades. Considering the Day 1 forecasts, the figure shows that in the early 1960s and again in the mid-1970s errors were large, as forecasters overpredicted rain areas (as indicated by values >1). Otherwise, from 1980 onward, there is remarkably little improvement in bias over the entire period. Bias for Day 2 forecasts are slightly below unity from about 1980 onward, indicating a slight tendency for underprediction.

Does the absence of bias mean a forecast method is perfect? Certainly not—it is always possible for overprediction (positive errors) to nearly balance underprediction (negative errors). As a matter of fact, most statistical forecast methods are totally unbiased, yet give far from perfect forecasts. To avoid this problem, one can use MAE, defined as the average of the absolute errors. Because absolute values are always positive, there is no cancellation of errors. MAE therefore provides information about how far an individual forecast is likely to be from true value, without regard to sign (positive or negative). If the MAE is 2 °C, we expect an individual forecast to be 2 °C away from the true value.

Note that in MAE, we averaged the absolute values to remove the sign of each error. We could accomplish the same thing in another way: by squaring each error before taking the average. This is done during the calculation of the **root-mean-square error (RMSE)**, which has some statistical advantages over MAE and is therefore more common. Obviously, as we sum only positive values, there is no cancellation of error. The root-mean-square error is just the square root of average squared error. Like MAE, it has the same units as the forecast variable (°C, millimeters of snow, and so on) and provides an estimate of the error expected for a single forecast.

For a qualitative variable, the most common accuracy measure is the proportion of correct forecasts, known as the **hit rate**. In the case of a binary variable (just two classes), we would ask, of all the forecasts made, whether "yes" or "no," what percentage turned out correct? For example, suppose rain/no rain forecasts have been made for a 100 km² area, with the results shown in Table 1.

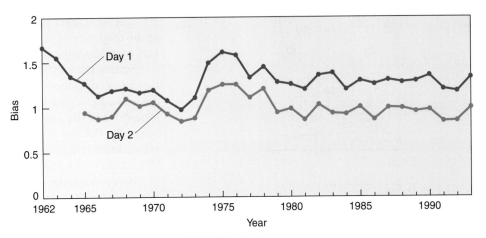

FIGURE 1

Changes in accuracy (bias) for 1-in. (2.5-cm) precipitation forecasts for Day 1 and Day 2. Day 1 forecasts are for the period 12 to 36 hours ahead, whereas Day 2 represents the period 36 to 60 hours ahead. A bias value of unity occurs when the size of the forecast and observed precipitation areas are equal.

TABLE 1	Rain Forecasting Accuracy		
	RAIN OBSERVED	**NO RAIN OBSERVED**	**TOTAL**
Rain Forecast	10 km² (Correct)	30 km² (Error)	40 km²
No Rain Forecast	20 km² (Error)	40 km² (Correct)	60 km²
Total	30 km²	70 km²	100 km² Total Area

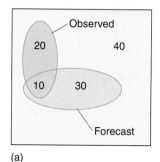

(a)

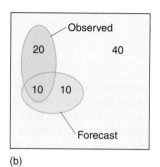

(b)

FIGURE 2

Hypothetical distributions of observed and forecasted precipitation. Numbers in single ovals refer to percentages of times that precipitation was observed but not forecast, or vice versa. Numbers contained in both ovals indicate the percentage of times precipitation occurred in an area forecasted to have precipitation. Numbers outside the ovals are for areas in which precipitation was not observed or forecasted.

The same data are shown in map form in Figure 2a. Looking at either the table or the map, we see rain was correctly forecast for 10 percent of the region, and no rain was correctly forecast for 40 percent of the area, giving a hit rate of 0.5 (50 percent).

Somewhat similar is the **probability of detection (PoD)**, which is the proportion of occurrences that were correctly forecast. The difference is that nonevents are excluded in the PoD. Thus, for example, in Figure 2a, the PoD is 10/30 = 0.33 (one-third of the area receiving rain was correctly forecast). Because one can maximize the PoD by always forecasting "yes," the PoD is often accompanied by the **false alarm rate**, the proportion of "yes" forecasts that were wrong. In Figure 2a, 75 percent of the rain forecasts are false alarms (30/40).

Bias can also be computed for a binary variable; in this case, bias is simply the ratio of forecast to observed occurrences. Here, however, a perfect method has a bias value of 1. Values larger than unity imply a tendency to overpredict, whereas values less than 1 suggest underprediction. For our example (Figure 2a), the rain forecast area is too large by one-third: the bias is 133 = (40/30). As before, a perfect bias value does not indicate a perfect method. For example, if the rain forecast area were sliced down to 20 km², the size of forecast and observed areas would agree perfectly, even though the forecast would be wrong for a large part of the region.

Another widely used measure for binary variables is the **threat score**. This measure is similar to PoD in that it is concerned only with occurrences. However, it tries to penalize false alarms as well as missed occurrences (PoD responds only to the latter). Looking at Figure 2a, you can see the 10 km² area of correctly forecast occurrences is flanked by large areas of false alarm as well as missed occurrences. Given the size of those areas, the 33 percent PoD seems a rather inflated measure of success. The threat score accounts for this, and is given by

$$TS = \frac{\text{Correct}}{\text{Forecast} + \text{Observed} - \text{Correct}}$$

where the variables on the right hand side are the area (or number of places) correct, area forecast, and area observed. The term *threat* is used because the denominator is the area "threatened" by an occurrence—either forecast, observed, or both. For Figure 2a, the threat score is 0.17 = 10/(40 + 30 − 10), half as large as PoD for the same map. Note that a perfect forecast will have a threat score of unity, because the forecast and observed areas will coincide perfectly.

Figure 3 illustrates the improvement in NCEP forecast model performance by plotting annual threat scores for 1 in. or more of precipitation. Unlike bias, threat scores over time have shown considerable improvement, reflecting better performance for the numerical models for both one- and two-day forecasts. Some of the variability is weather-related (wet years have higher scores), but significant upturns can be tied to new models coming online. Particularly striking is a narrowing of the gap between Day 1 and Day 2 forecasts, implying that Day 2 forecasts have improved more. In fact, by the early 1990s the two-day forecasts were about as good as the one-day forecasts of the early 1960s. In that sense, predictive

FIGURE 3

Change in skill (threat score) for 1-in. (2.5-cm) precipitation forecasts for Day 1 and Day 2. A perfect forecast has a skill score of unity.

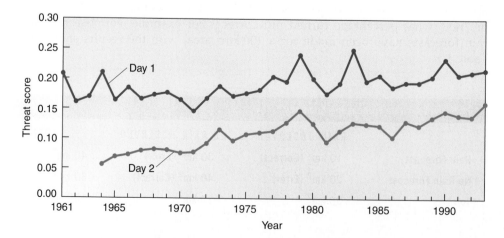

ability doubled over the period. Looking at changes in the two-day score alone, there is a threefold improvement.

It's a little more complicated to measure accuracy for probability forecasts. As we mentioned in the text, the most common probability forecast is a probability-of-precipitation, or PoP forecast. The PoP refers to the chance that a random place in the forecast area will receive 0.01 in. or more of precipitation in the forecast period. Equivalently, considering just a single location, a 60 percent chance of rain means that out of ten forecast days, six will be rainy. A forecast might say, for instance, "the rain chance is 70 percent." How can we assess the accuracy of this statement? The forecast itself means that there is a 70 percent probability of rain for a randomly chosen location in the forecast area. Obviously, there is no way to assign accuracy on the basis of just a single precipitation measurement. (Failing to observe rain on a single day does not mean the forecast was wrong.) But if we were to count the number of times rain is recorded on many of these "70 percent" days, we could see if the actual frequency is above or below the forecast frequency, and thereby assess the method's accuracy. To fully verify the forecast method, we would compare all the forecast classes, not just the 70 percent class. In effect, we compare the forecast probability distribution with the observed probability distribution.

As we mentioned, forecast skill refers to the improvement provided by a forecast over and above some reference accuracy. The reference, or "no-skill," system is often taken to be persistence, or alternatively, climatology. But these aren't the only choices for the reference system; we could compare one numerical model with another, or with purely random values. For example, the threat score is often adjusted by the number of hits expected by random assignment of forecasts. In Figure 2a, rain was observed over 30 percent of the area. If we were to randomly forecast rain for various places, we would expect considerable success simply because so much of the area is wet. (If you forecast a single point by throwing a dart at the diagram, you have a 30 percent chance of being correct—assuming you can hit the diagram.) As the rain forecast area grows with no change in observed rain, the threat score increases. To account for this, we first find the expected area of correct forecasts, assuming purely random forecasting. This is given by

$$EC = \text{Forecast}\left(\frac{\text{Observed}}{\text{Total}}\right)$$

The term in parentheses is the proportion of the total area where rain fell—multiplying this by the forecast area gives the area expected to be correctly forecast. With this, the so-called **equitable threat score** is

$$ETS = \frac{\text{Correct} - EC}{\text{Forecast} + \text{Observed} - \text{Correct} - EC}$$

For Figure 2a, $EC = 40(30/100)$, or 12 km^2. That is, knowing nothing about the atmosphere, we'd expect rain to be correctly forecast for 12 km^2, given that we are forecasting rain to occur over 40 km^2. For this example, ETS is $(10 - 12)/(40 + 30 - 10 - 12)$ or -0.04, a negative number! Does this make sense? Well, as our forecast method generated fewer successes than would occur by chance, it certainly seems reasonable to treat this as "negative" skill.

Figure 2b shows a similar situation, but with a smaller rain forecast area. The PoD is exactly the same as before (0.33), and the threat score is only slightly larger (0.25 vs. 0.17). However, where in Figure 2a we saw negative skill compared to chance, here the equitable threat score is 0.12. Although there are no more successful rain forecasts than before, this situation has fewer false rain forecasts, and therefore shows positive skill.

We see that there is no single measure of forecast quality. Rather, various measures provide different kinds of information, so each has its own uses. The other side of this is that whether or not a forecast model shows improvement over time, or whether it is better than some other competing model, depends in part on the way we measure accuracy and skill. Careful assessment always requires that we compare a number of measures.

With 18 million people and 3.5 million cars, most of which are not equipped with modern emission-control devices, it is little wonder that Mexico City is one of the smoggiest places on Earth. Surrounded by mountains that confine the polluted air and subject to frequent temperature inversions that inhibit vertical dispersion of pollutants, Mexico City has all the right ingredients for a serious smog problem. But the spring of 1998 brought prolonged periods of unhealthful air quality that were extreme even for this city. The smog crisis was brought on primarily by a rash of forest fires in southern Mexico. To make matters worse, Popocatepetl volcano, 50 km (30 mi) southeast of the city, spewed tons of smoke and ash into the region. Wind transported the pollution south to Honduras and north all the way to Florida and Texas, where people were advised to stay indoors to mitigate health hazards.

Normally, Mexico City's worst smog occurs in January and February, when stagnant air traps automobile and other urban pollutants near the surface. Smog levels decline somewhat by springtime, although the slash-and-burn methods of field clearance used by farmers makes the air quality worse than it would be otherwise. But in 1998 a major drought, believed by many to have resulted from the strong El Niño, created particularly dry conditions that caused the fires to cover three times their normal area. By May the polluted air forced automobile drivers to use their headlights in the middle of the day in the state capital of Chiapas, and in Mexico City extreme ozone concentrations brought a dramatic surge in the number of people seeking medical care for respiratory problems. In response, the government exercised emergency powers that restricted automobile traffic and closed down many factories. However, that was not enough to make conditions tolerable for Pedro Chavez, who said, "You notice it in your eyes, in the tiredness you feel. Our children are getting sick more often. . . . If we could, we would leave, but this is where our business is."

The effects of human activities are not restricted to air quality degradation. We change the atmosphere in more subtle ways as well. For example, the construction of cities influences the way energy and water are exchanged near the surface. Every time a subdivision is laid out, natural soil and vegetation are replaced by concrete or asphalt. This greatly reduces the amount of water that can evaporate into the air and thereby increases the sensible heat flux (Chapter 3) to the atmosphere. We also build structures with vertical walls that receive sunlight at a more direct angle than the original absorbing surface. These processes work to increase the temperature of urban areas relative to their rural counterparts, creating the heat islands that we describe later in the chapter.

Atmospheric Pollutants

Nowhere is the air entirely pristine. Small suspended solids and liquids (called *particulates*) enter the atmosphere from natural and human sources. Likewise, many gases that are considered pollutants also arise naturally from processes such as lightning-induced forest fires and volcanic eruptions. Nonetheless, natural dilution and removal of these gases and particulates makes them relatively unimportant to the air quality experienced by most people. More important are the effects of human activity, especially in and around urban and industrial centers where anthropogenic emissions are concentrated into much smaller areas. In this chapter all references to **air pollution** will concern the introduction of undesirable gases and particulates by humans. The varying sources of particulates and other pollutants in the United States and their relative concentrations are shown in

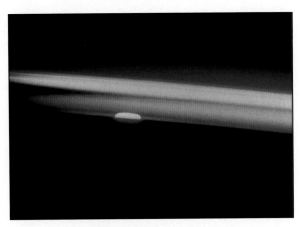

WEATHER IMAGE
Atmospheric Pollution as Seen from Space Shuttle

Pollution over Amazon Basin

Kuwait Oil Field Fires

◄ Massive fires over Indonesia in the summer and fall of 1997 caused extreme air pollution that spread into Malaysia and Thailand.

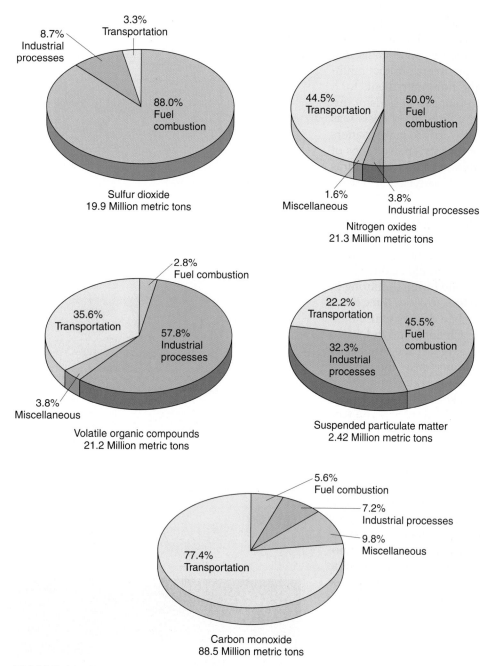

FIGURE 14-1
The sources of various pollutants in the United States.

Figure 14–1. (We present some background on major pollution episodes of the past in *Box 14–1, Focus on the Environment: Severe Pollution Episodes.*)

In the most general sense, pollutants can be divided into two categories. Some, called **primary pollutants,** are emitted directly into the atmosphere. Others, called **secondary pollutants,** do not go directly into the atmosphere but result from one or more chemical transformations. Thus, a chemical emitted into the atmosphere may be innocuous in its original state but becomes a noxious gas or particulate after combining with other emissions or naturally occurring compounds. Several primary and secondary pollutants figure most prominently in the degradation of air quality.

PARTICULATES

Particulates (also called *aerosols*) are solid and liquid materials in the air that are of natural or anthropogenic (human-made) origin. Though always small, particulates come in a wide range of sizes ranging from about 0.1 to 100 μm. Some of the particulates are primary pollutants put directly into the atmosphere, while others are secondary pollutants formed by the transformation of preexisting gases or from the growth of smaller particulates into larger ones by coagulation.

SOURCES OF PARTICULATES

Particulates introduced directly into the air can originate from natural fires, volcanic eruptions, the ejection of salt crystals by breaking ocean waves—and as any sufferer of hay fever can tell you, by the entrainment of pollen by wind. Human activities, especially those involving combustion, produce primary and secondary particulates.

Some secondary particulates form by the coagulation of gases. This process is most rapid when the humidity is high, which creates an interesting situation. Recall from Chapter 5 that water droplets in nature always form on condensation nuclei—with large, hygroscopic aerosols being particularly effective at attracting water and lowering the relative humidity needed for droplet formation. Thus, the introduction of particulates, especially large ones, promotes the formation of fog or cloud droplets. At the same time, high humidities favor the conversion of certain gases into secondary particulates, which in turn promote the condensation of water vapor into liquid droplets. As a result, humid areas with a high concentration of industrial activities can become foggy at relative humidities considerably below 100%. This symbiotic relationship helped make the London type of smog ubiquitous in eastern, industrial North American cities in previous years.

REMOVAL OF PARTICULATES

Though particulates are always present in the air, no individual particulate stays in the atmosphere forever. As we have seen in Chapter 7, terminal velocities increase with the size of falling objects. Thus, particulates, which are always small, can remain suspended in the atmosphere for considerable lengths of time. Larger

FIGURE 14-4
Smokestacks on manufacturing and power plants are designed to keep emissions away from the ground near the source. Unfortunately, pollutants are carried downwind for hundreds of kilometers, where they can exacerbate acid deposition.

greater use of coal and heating oil. A huge proportion of the sulfur dioxide contributing to the acid rain originates from a relatively small number of sources. It is estimated that the 50 largest sulfur emitters in the region (all of which are power-generating plants) account for half of the acid deposition.

Interestingly, one of the measures undertaken to improve air quality near sulfur-emitting industries and power utilities may have exacerbated the acid deposition problem farther downwind. To help in the dispersion of sulfur oxides from industrial plants, many industries and utilities have built large smokestacks to release the pollutants well above ground level (Figure 14–4). The idea behind the stacks is that, by releasing the smoke far above the surface, sulfur compounds will be transported considerable distances downwind before settling near the ground. While the stacks have successfully reduced sulfur concentrations near their source, they have had the unintended consequence of allowing the sulfur compounds to be transported hundreds of kilometers downwind, where they react to form acid deposition. Thus, the acid problem over much of the eastern United States and Canada is due to transported, rather than locally generated, pollutants. This has led to many years of litigation between states in the Midwest and Northeast and between the United States and Canada.

While most acid deposition in eastern North America is associated with sulfur compounds, this is not always the case in other areas. Some acid deposition, especially in the western United States and Canada, is related to compounds made of nitrogen and oxygen.

NITROGEN OXIDES (NO$_X$)

Nitrogen oxides (also called *oxides of nitrogen*) are compounds consisting of nitrogen and oxygen atoms. The two most important of these from an air pollution viewpoint are **nitric oxide** (NO) and **nitrogen dioxide** (NO$_2$). Together the two gases are commonly referred to as NO$_X$. Nitric oxide is a nontoxic, colorless, and odorless gas that forms naturally by biological processes in soil and water. While millions of tons of the material are introduced into the atmosphere each year, it is highly reactive and tends to break down very quickly. Nitric oxide also forms as a byproduct of high-temperature combustion associated with automobile engines, industrial manufacturing, and electric power generation. Its primary importance from an air quality perspective is that it oxidizes to form nitrogen dioxide, a major component of smog in many places.

FIGURE 14–5
Nitrogen dioxide gives polluted air a yellowish to reddish brown color, as in this photo of Hong Kong.

Nitrogen dioxide is a toxic gas that gives polluted air its familiar yellow to reddish brown color (Figure 14–5), as well as a pungent odor. It is an important component in air pollution in that it is relatively toxic and corrosive and undergoes transformations that contribute to acid deposition and other secondary pollutants. As with nitric oxide, nitrogen dioxide breaks down very readily and, as a result, NO_2 concentrations in urban areas tend to rise and fall in accordance with vehicular traffic patterns. Furthermore, the rapid decay of nitrogen dioxide prevents large concentrations from occurring in rural areas surrounding source areas.

Like sulfur compounds, nitrogen oxides can cause serious pulmonary health problems. Clinical studies have shown that NO_2 easily passes through bronchial passages and irritates tissue deep within the lungs. Laboratory tests have shown animals to experience severe lung damage and reduced immunity to infection when exposed to high levels of NO_2.

VOLATILE ORGANIC COMPOUNDS (HYDROCARBONS)

Volatile organic compounds (VOC), also called *hydrocarbons*, are materials made entirely of carbon and hydrogen atoms. These compounds, including methane, butane, propane, and octane, occur in both gaseous and particulate forms. Globally, the vast majority of hydrocarbons arrive in the atmosphere via natural processes, including plant and animal emissions and decomposition. In the United States, industrial activities account for the greatest proportion of anthropogenic hydrocarbons, with automobiles also contributing a major share. The emissions associated with automobiles arise primarily from incomplete fuel combustion and the evaporation of gasoline (often while filling gas tanks).

Even in cities that have high VOC concentrations, there is little evidence that these chemicals have any direct adverse health impacts. Nonetheless, they are extremely important because in the presence of sunlight they recombine with nitrogen oxides and oxygen to produce photochemical smog.

PHOTOCHEMICAL SMOG

If you have ever visited Los Angeles in the summer, you have probably heard the term **photochemical smog** and know what it feels like. Burning eyes, sore lungs, and a subtle but unpleasant odor accompany an atmosphere with poor visibility. Photochemical smog consists of secondary pollutants that include ozone (O_3), NO_2, peroxyacyl nitrate (PAN), formaldehyde, and other gases that occur in very minute quantities. As the name implies, this type of smog forms when sunlight triggers numerous reactions and transformations of gases and aerosols. Unlike the **London-type smog** found in many places where smoke combines with damp air (the word *smog*, in fact, originally derived from the terms *smoke* and *fog*), this **Los Angeles-type smog** usually involves dry air.

Ozone has been designated by the Environmental Protection Agency as the most important agent of photochemical smog. It can cause serious physical and environmental harm at surprisingly low concentrations, so low that the EPA established a concentration of only 0.12 ppm averaged over a 1-hour period as the maximum allowable without exceeding federal standards.

Exposure to ozone causes inflammation of air passages that can reduce lung capacity by as much as 20 percent. The EPA estimates that perhaps 20 percent of all respiratory-related hospital visits in the northeastern United States during the

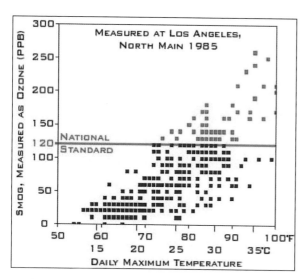

WEATHER IMAGE
Ozone Concentration and Air Temperature in Los Angeles

summer result from exposure to ozone. Although acute symptoms usually subside fairly quickly after ozone concentrations decline, research has shown that long-term exposure to ozone can cause permanent damage to lung tissue and impairment of the body's ability to resist bronchitis, pneumonia, and other diseases. Of course, ozone causes even greater problems for people with asthma and other preexisting pulmonary problems. For those people, ozone constricts lung passages to the point where breathing becomes nearly impossible, and the gas may contribute considerably to the number of asthma-related fatalities that occur in the United States annually. Currently, an average of 5000 Americans die each year during acute asthma attacks.

Not only are people directly harmed by ozone, but high levels also result in serious environmental degradation. Damage to agricultural crops by photochemical smog (mainly ozone) was identified in southern California grape fields in the late 1950s. Since that time, plant damage from oxidants (and to a lesser extent, PAN) has been widespread across North America, with conifers being particularly vulnerable. (*Box 14–2, Focus on the Environment: The Counteroffensive on Air Pollution,* summarizes recent efforts to combat air pollution.)

Atmospheric Controls on Air Pollution

As we have seen, a large portion of the chemicals that we consider pollutants occur naturally in the environment. These emissions do not create high concentrations, however, because their release into the atmosphere is over such a large area that they are immediately diluted. Urban emissions, on the other hand, are concentrated over much smaller areas and can thereby lead to significant pollution episodes.

Atmospheric conditions play a major role in determining pollution concentrations in several ways. Atmospheric stability and wind conditions control the vertical and horizontal dispersion of pollutants, and cloud conditions can influence the rate of photochemical reactions taking place. Furthermore, unusually cold or warm conditions encourage the increased use of heaters or air conditioners, which can increase emissions.

EFFECT OF WINDS ON HORIZONTAL TRANSPORT

Strong winds aid in the dispersal of pollutants two ways. First, they rapidly transport emissions from their source and spread them over a wide horizontal extent. Figure 14–6 illustrates how the concentration of pollution is inversely proportional to the wind speed (to make this easier to visualize, the figure depicts puffs of smoke being released from a stationary source every second, though in reality pollutants would be released continuously). In (a), the wind blows at 5 m/sec (11 mph) so that each puff of smoke travels 5 m before the next one is released. In (b), the wind flows twice as fast as in (a), and the distance between successive puffs of smoke likewise doubles. Thus, the greater wind speed in (b) causes the same amount of pollution to be diluted within twice as large a volume of air.

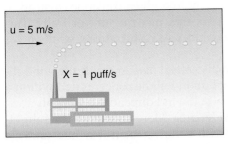

(a)

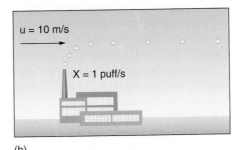

(b)

FIGURE 14–6

The effect of wind speed on pollutant dispersal. In (a), the 5 m/sec (11 mph) wind moves individual puffs of smoke downwind slowly, so each successive puff is only 5 m behind the previous one. In (b), the 10 m/sec (22 mph) wind causes twice as great a distance between puffs, and thus only one-half the concentration of smoke.

The Counteroffensive on Air Pollution

Regulations designed to improve air quality have made a substantial impact on the lives of people in the United States and Canada. Although federal regulations regarding air pollution in the United States did not come into being until the 1950s, certain cities and states have had laws on the books regulating smoke emissions since the late nineteenth century. In some cases, such as in Pittsburgh, fairly stringent controls were in effect by the 1940s.

The first major U.S. initiative to clean up the nation's air was the 1963 Clean Air Act, which, among other things, expanded the role of the federal government in interstate air pollution control and authorized increased research and technical development initiatives. Major extensions of government involvement were subsequently added to the Clean Air Act by the adoption of numerous amendments in 1970, 1977, and 1990.

The original Clean Air Act and its amendments through 1977 established air pollution standards and created government agencies to ensure that those standards were being met. Maximum concentrations were established for PM_{10}, sulfur dioxide, carbon monoxide, nitrogen oxides, ozone, and lead (formerly an ingredient in gasoline but now outlawed by federal regulations). Individual states were required to establish agencies to ensure enforcement of the standards, and regions where air quality fell short of the standards were designated nonattainment areas and required to take appropriate actions.

The act and its amendments also required automobile manufacturers to install emission reduction devices, such as the catalytic converter, that have lowered individual vehicle emissions by about 95 percent since the 1960s. The law also ordered the phaseout of open burning of refuse, the installation of filters on industrial smoke-stacks, and other emission-reducing measures. Overall, the act and its amendments have resulted in substantial reductions in pollutant levels in urban areas, despite a large increase in motor vehicle miles driven each year. Figure 1 shows the reduction in pollution levels.

Though the original act and the earliest amendments were monumental in their scope and effect, dozens of metropolitan areas still failed to meet the standards by 1990. In response, the U.S. Congress passed the most sweeping set of changes to the act—the amendments of 1990. Consisting of 11 sets of provisions, the 1990 amendments established specific dates by which cities must meet ozone, PM_{10}, and CO standards, mandated corrective measures for areas of noncompliance, determined new emission standards for vehicles and stationary sources, and created new provisions to reduce acid deposition. Figure 2 shows the

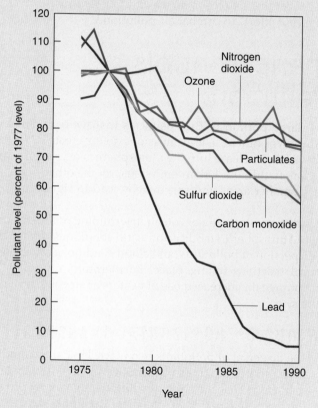

FIGURE 1

Trends in pollutant levels in the United States relative to 1977.

Greater wind speeds also lower the pollution concentration indirectly. Recall from Chapter 3 that air does not flow uniformly in a given direction. Instead, it contains small, swirling motions, called *eddies*, that mix the air vertically. This forced convection increases with wind speed and, as a result, strong winds favor greater vertical dispersion.

Short-term variations in wind direction also affect dispersion. If the wind direction varies only slightly through time, pollution will be concentrated within a relatively narrow area downwind of the source. If wind directions are highly variable, the pollutants will spread out over a wider area. More people will be subjected to the pollutants, but the concentration will be lower than it would under a more constant wind regime.

counties that have failed to meet ozone standards as of October 2002.

Recent research has shown that the standards adopted for various pollutants were not always based on the appropriate targets. For example, scientists now know that 1-hour exposures to ozone do not have as great an effect on human health as do longer-term exposures at lower levels. Thus, in July 1997 the EPA adopted the phasing-in of a new standard in which compliance will be based on 8-hour concentrations above 0.08 ppm, as opposed to the current 1-hour standard of 0.12 ppm. The EPA believes that the enactment of the new standards will annually prevent 15,000 premature deaths, 350,000 cases of aggravated asthma, and 1 million cases of decreased lung capacity in children. But the new standards did not get implemented without opposition. In May 1999 the District of Columbia Circuit Court of Appeals ruled against the EPA, which in turn appealed the decision to the U.S. Supreme Court. In February 2001 the Supreme Court reversed the lower court decision and determined that the EPA did not exceed its authority in issuing the new guidelines. However, it also declared that certain details of the regulations would have to be reviewed by the lower court. In March 2002, the issue was brought to rest when the appellate court ruled in favor of the EPA and declared the regulations binding.

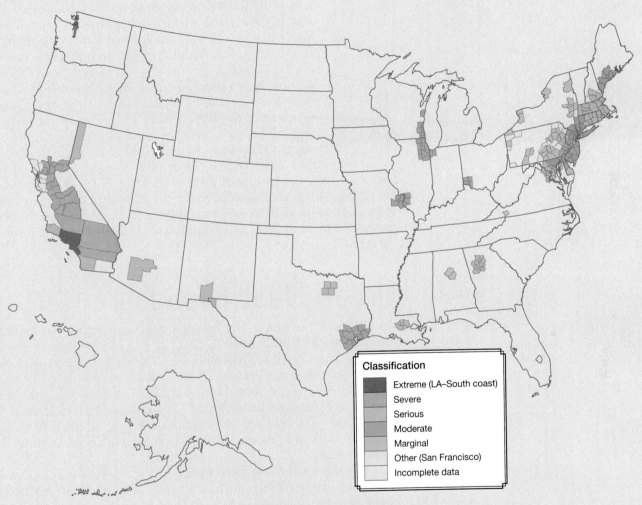

FIGURE 2
Map of counties that have failed to meet ozone standards as of October 2002.

EFFECT OF ATMOSPHERIC STABILITY

Just as the stability of the air (Chapter 6) influences lifting and cloud formation, it also affects the vertical movement of pollutants. Recall that when the air temperature decreases slowly with height (or if it increases with height), the air is said to be stable. Stable air resists vertical displacement and leads to higher pollutant concentrations near the ground. Unstable air, on the other hand, enhances vertical mixing, and any material introduced near the surface is easily displaced upward. This reduces pollution concentrations near the surface.

Inversions, the situation in which air temperature increases with height, make the air extremely stable and impose the greatest restraint on vertical mixing

FIGURE 14–7
The base of an inversion is clearly evident above the smog layer in Los Angeles.

(Figure 14–7). Radiation inversions (described in Chapter 6) originate at the surface in response to cooling of the lower atmosphere. These inversions usually dissipate in the late morning after the Sun has warmed the surface and lower atmosphere. As a result, they tend to have the greatest impact on pollution concentrations in the early morning. These inversions are most important in areas subject to a London-type smog.

Subsidence inversions are often important where photochemical smog is the major problem. The base of a subsidence inversion marks the maximum height to which the air below can be easily mixed. An inversion with a base at 1000 m (0.6 mi) above the surface will result in a *mixing depth* of 1000 m, doubling the concentration of pollutants that would accompany a mixing depth of 2000 m (1.2 mi).

Just as the semipermanent Hawaiian high-pressure system accounts for the mostly dry summers of southern California, subsidence from the same system also plays an important role in the region's poor air quality. As air rotates clockwise out of the eastern margin of the high, air in the middle troposphere descends and creates an inversion. During the summer, the base of the inversion level over Los Angeles typically occurs at about 700 m (2300 ft) above sea level, but the base of the inversion can also occur at lower levels and lead to particularly bad smog events. (*Box 14–3, Focus on the Environment: Smog in Southern California*, presents further information on atmospheric and other controls on air pollution in the region.)

Urban Heat Islands

Not all human impacts on the atmosphere are as dramatic as the pollution of the atmosphere. The well-known **urban heat island** is an excellent case in point. For centuries it has been known that urbanized areas often have higher temperatures than do adjacent countrysides. These differences can be quite dramatic, with temperatures in major metropolitan areas sometimes exceeding those of their hinterlands by as much as 12 °C (22 °F). Although the exact nature of the urban heat island varies from one city to another, in general the urban–rural temperature differences are greatest during the late evening and night and during the winter months.

Urban heat islands occur because of modifications to the energy balance (Chapter 3) that result when natural surfaces are paved and built upon, and when human activities release heat into the local environment. Though it is not possible to generalize the relative importance of these processes for every city, all of them probably play some role in causing the phenomenon to occur.

Several variables influence the magnitude of the heat island effect. Some are related to the local setting, others to the activities undertaken within it. But the most important are the size of the city itself and the density of its population, with large, densely populated cities having the largest heat island effect. Figure 14–8 illustrates the relationship between the population of North American cities and the maximum urban and rural temperature differences (note that the horizontal axis is plotted on a logarithmic scale).

The intensity of an urban heat island varies spatially across a city, with highest temperatures normally found within the city core. Figure 14–9 illustrates this effect by plotting temperatures over Vancouver, British Columbia, on a July evening. The downtown area is located on the southeastern part of the peninsula that juts into Burrard Inlet. Immediately to the northwest of downtown (on the northwestern part of the peninsula) lies Stanley Park, a wooded area with few buildings. As expected, temperatures are greatest over the downtown region and decrease substantially over less-populated areas. Temperatures on the peninsula

WEATHER IMAGE
Atlanta Urban Heat Island I, II, and III

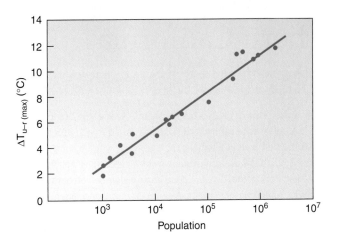

FIGURE 14–8
Urban heat islands vary with city population. The vertical axis plots temperature differences between urban centers and surrounding areas in °C against city population. Note that the horizontal axis (population) is on a logarithmic scale.

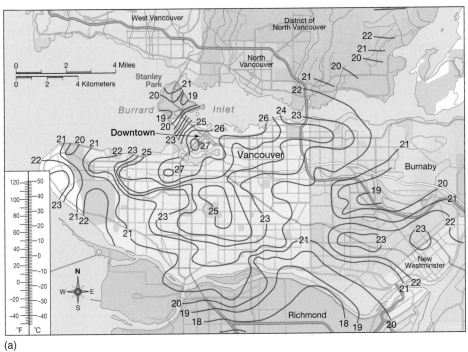

(a)

(b)

FIGURE 14–9
(a) Temperatures in Vancouver, British Columbia, at 9 P.M. on July 4, 1972. Note the sizable temperature gradient between downtown and Stanley Park. (b) Photo showing the Vancouver downtown area with Stanley Park in the background.

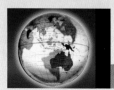

Smog in Southern California

Los Angeles has long had a reputation for extremely bad air quality—and for good reason. Of all the cities in the United States, Los Angeles is the only one classified by the Environmental Protection Agency as an "extreme area" of noncompliance of ozone standards. A number of factors work together to make the air quality bad enough to earn this dubious distinction. As shown in Figure 1, Los Angeles occupies part of a basin bounded by mountains to the north and east that block the free movement of pollutants by the sea breeze, while the presence of a subsidence inversion during the warmer months restricts vertical dispersion. Add to that the typically cloud-free conditions during the mid-day period that trigger photochemical reactions. And finally there is the city's well-known love affair with the automobile, which contributes much of the estimated 2 million kilograms (2200 tons) of hydrocarbons and 1 million kilograms (1200 tons) of NO_x released each day into the four-county South Coast Air Basin.

Fortunately, a number of new regulations have improved the situation. For example, beginning in 1984 all automobiles were required to undergo biannual smog checks. More recently, regulations have been enacted requiring gas pump nozzles to have rubber sleeves to capture fumes that would otherwise escape into the air as peo-

ple fill their tanks. Also, a cleaner-burning type of gasoline that releases fewer hydrocarbons has been phased in at all area gas stations. To illustrate how much progress has been made, consider the fact that during the 5-year period from 1976 to 1980, there was an average of 112 stage-1 smog alerts (ozone $\geq$ 0.20 ppm) each

year in the South Coast Air Basin. During the period from 1995 to 1999, the average dropped to 7, with no smog alerts at all in 1999! In fact, in 1999 Los Angeles did not top the nation in the number of days in which ozone concentrations exceeded federal standards—that honor went to Houston, Texas.

FIGURE 1
The topography of the Los Angeles basin.

decrease dramatically between downtown and the middle of the park, with a difference of about 9 °C (16 °F) occurring over a distance of only about 1.5 km (1 mi). Wind speeds also play an important role. During windy conditions, cooler air from the surrounding countryside displaces the warmer urban air and thus reduces the magnitude of the urban heat island.

RADIATION EFFECTS

Urban particulates can also affect the intensity of the urban heat island through their effect on the radiation balance. Increased particulates associated with urban activity can absorb and scatter incoming solar radiation and also increase the amount of absorption and reradiation of longwave energy in the atmosphere. Although it is hard to generalize for all cities, it is believed that the particulates tend to decrease the amount of incoming solar radiation at the urban surface, but the increase in net longwave radiation probably offsets the reduction in absorbed solar energy. Thus, the direct effect of particulates on urban temperatures is probably negligible.

Particulates also affect the radiation balance indirectly. Recall that water droplets in the atmosphere form onto condensation nuclei. The increase in particulates due to human activity can increase cloud cover, as in the case of London,

During the summer, daily concentrations of photochemical smog vary on a regular basis in the course of a day. Prior to the morning rush hour, residual primary and secondary pollutants from the previous day leave a background level of air pollution. As traffic increases during the morning, emissions increase substantially. Early morning winds are usually weak, which leads to little movement of pollutants. At the same time, the low Sun angle and common presence of early morning fog and low clouds inhibit photochemical activity. The situation normally changes by late morning. A sea breeze usually develops along the coast and moves pollutants inland, while clearing skies and increasing Sun angles increase photochemical conversions.

As the sea breeze develops, a boundary called a **sea breeze front** separates the relatively clean marine air from the more polluted, drier air ahead. As the sea breeze front moves inland, it pushes the emissions eastward or northeastward. This often creates a strong gradient in ozone concentrations near the sea breeze front, with relatively clean air behind it and increasing concentrations to the east or northeast (Figure 2). By late afternoon, the cities in the eastern portion of the basin get the full onslaught of the advected pollutants, while local commuters add their own contribution to the photochemical smog. As a result, pollution levels can become extraordinarily high in the areas downwind of Los Angeles.

The area to the south of Los Angeles—including Orange and San Diego counties—also has a significant air pollution problem. Usually the air pollution in San Diego, some 150 km (100 mi) to the south, is of local origin. During severe episodes, however, most of the pollution originates over Los Angeles and Orange counties and gets carried into San Diego by the wind. These episodes often occur as Santa Ana winds die out. During the Santa Ana, the easterly winds force the basin's pollutants offshore. As the Santa Ana begins to weaken, a thermally induced low-pressure system over the eastern desert stretches into the San Diego area. This creates a northwesterly flow that transports pollution originating over Los Angeles and Orange counties. Not surprisingly, the trend toward decreasing pollution in Los Angeles has also occurred in San Diego.

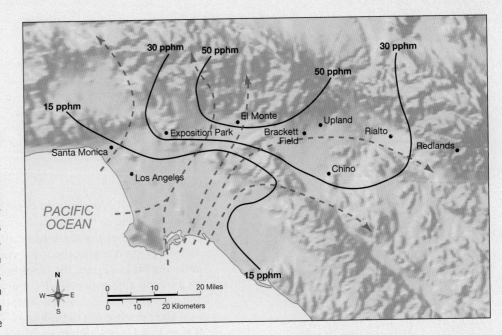

FIGURE 2

The distribution of ozone between 4 and 5 P.M. on July 25, 1973. Highest concentrations (shown by solid lines) occur in the northeast, ahead of the sea breeze front. Dashed lines show the wind direction.

England, which has been shown to receive 270 fewer hours of bright sunlight annually than the surrounding area. Particulates have long been known also to increase precipitation downwind of urban regions. Interestingly, studies have also shown that precipitation can decrease downwind of major industrial centers, perhaps because cloud water is spread over many condensation nuclei, which lessens the chance of growth to precipitation size.

More important than the effect of increasing particulate concentrations is the impact that buildings have on the radiation balance. Consider the impacts that the construction of buildings with vertical walls might have on the receipt of solar radiation. When the sun is low in the sky—near sunrise and sunset, and during much of the day at high latitudes in the winter—direct sunlight that would otherwise reach the horizontal surface hits the vertical walls of buildings. This causes the angle of incidence to become closer to perpendicular and increases surface heating, which leads to a higher temperature.

The presence of buildings also affects the rate of heating by changing the surface albedo. Darker buildings, of course, absorb more sunlight than lighter ones, and, in general, urban surfaces (asphalt streets, roofing materials) have lower albedos than the natural surfaces they replace. The presence of buildings also affects the amount of absorption by causing multiple reflections to occur, as shown

FIGURE 14–10

Effect of buildings on solar radiation receipt. As incoming radiation contacts a building, some is scattered off in all directions and some is absorbed. The scattered radiation may in turn hit an adjacent building, where further absorption can take place. This lowers the urban albedo.

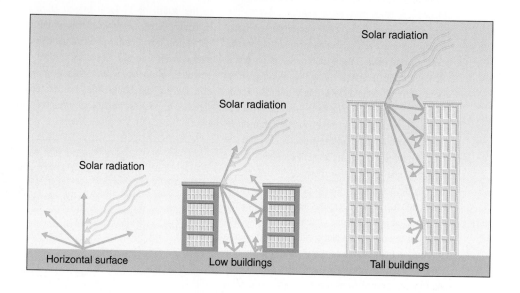

in Figure 14–10. As sunlight penetrates the urban landscape and hits the side of a building, some of the energy is absorbed and some is scattered back as diffuse radiation. Some of the scattered radiation strikes an adjacent building, where once again a portion is absorbed. This process goes on repeatedly, with each successive reflection at least partially absorbed upon contact with another wall. This increases the total absorption so that the albedo of the urban area is actually lower than the albedo of the individual surfaces.

The presence of tall buildings also affects the transfer of longwave radiation in a way that favors higher nighttime temperatures. Essentially, the process is very similar to the multiple reflection of solar radiation just discussed. Longwave radiation emitted from an open, rural surface travels upward without being impeded by buildings. Urban areas, in contrast, reduce the amount of longwave radiation that freely escapes to space because walls absorb a portion of the outgoing radiation. The resultant reduction in longwave radiation loss slows the rate of nocturnal cooling and promotes higher daily minimum temperatures.

CHANGES IN HEAT STORAGE

As explained in Chapter 3, radiation is not the only mechanism that transfers energy from one place to another; conduction and convection are also important heat transfer mechanisms. In the middle of a sunny day, absorption of solar radiation warms the surface of the ground. Conduction within a very thin layer of the atmosphere and convection transfer much of this energy to the air. At the same time, a gradient develops in which soil temperature decreases with depth, and heat is conducted downward.

During the late afternoon, the surface begins to cool when energy losses by radiation and convection exceed the absorption of shortwave and longwave radiation. The soil temperature profile eventually reverses, with temperature increasing with depth. Thus, during the evening hours, heat that has been stored within the soil is transferred to the surface.

The same processes just described occur in the walls and roofs of urban buildings. As the surface is warmed during the day, a temperature gradient develops that conducts heat toward the building interior. When cooling occurs in late afternoon, the stored heat is released to the surface. What is different from the rural setting is that materials used in building construction have a much greater capacity to store heat than most natural surfaces. As a result, more stored heat is available for transfer to the lower atmosphere during the evening and nighttime than for natural surfaces, and nocturnal temperatures are increased.

The release of heat from buildings just described is supplemented by the anthropogenic heat produced for comfort (e.g., space heating) or as a byproduct of other activities (e.g., waste heat from a hot car engine). Anthropogenic heating is greatest during the winter, which partially explains why heat islands are most pronounced during the low sun season. Anthropogenic heat can be a surprisingly large component of the urban energy balance. In Vancouver, British Columbia (49° N), for example, the amount of anthropogenic heat released in the winter has been estimated to be nearly four times that received as net radiation. That estimate, based on 1970 data, probably understates the importance of anthropogenic heat currently released because the city has grown significantly since then.

SENSIBLE AND LATENT HEAT TRANSFER

In Chapter 3 it was shown that most of the global surface has a surplus of net all-wave radiation on an annual basis. That surplus is transferred to the atmosphere as sensible and latent heat. When moisture is available near the surface, the transfer of energy as latent heat can exceed sensible heat transfer, indicating that most of the surplus is consumed by evaporation. On the other hand, if the surface is completely dry, surplus energy raises the surface temperature far above the air temperature, and sensible heat dominates. All other things being equal, the higher the ratio of sensible to latent heat, the greater the temperature.

Urbanization affects the routing of precipitation in a way that favors increased sensible heat transfer. Unlike natural surfaces that allow rainfall or snow melt to permeate the soil and be retained below ground, city streets and sidewalks are almost impervious to water. So when precipitation occurs, most of the water runs off the surface and ultimately flows out through the flood-control system. The reduction in available water increases the input of sensible heat to the atmosphere at the expense of latent heat and helps increase the temperature of the urban environment.

URBAN HEAT ISLANDS AND THE DETECTION OF CLIMATE CHANGE

By now everyone has heard much discussion about the possibility of impending climatic change resulting from the anthropogenic emission of greenhouse gases. Many atmospheric scientists believe that this change has already begun to take place. To support this notion, they point to an overall warming of 0.3 to 0.6 °C (0.5 to 1 °F) occurring since the late nineteenth century for weather stations having records going back more than a century. However, we cannot use these recorded temperature changes at face value, because many of the data come from urban weather stations, and most cities having long-running weather stations have undergone considerable growth over the last century. Thus, we must contend with the problem of enhanced urban heat islands influencing the data, which means that records from large urban areas are not representative of the surrounding region. Atmospheric scientists are well aware of this source of bias in temperature records and routinely account for its effect, either by discarding contaminated data or by adjusting values downward for affected stations.

DID YOU KNOW?

It has long been known that cities create their own urban heat islands. But cities can also affect precipitation patterns—in conflicting ways. The urban areas around the San Francisco Bay area, Los Angeles, and San Diego increase the amount of small particulates in the air that are transported downwind. The increase in particulates causes clouds over the foothills of the mountains to the east to have their liquid or ice content distributed among a larger number of droplets or crystals. The smaller size of the cloud constituents makes it harder for precipitation to develop and may reduce mountain snowfalls by as much as 20 percent in some place.

But a separate NASA study shows that in cities that are more prone to thunderstorm activity, the urban heat island effect can *enhance* precipitation downwind by as much as 51 percent. Summertime precipitation increases have been observed downwind of numerous cities, ranging from Atlanta, Georgia, in the Southeast, to Dallas, Texas, in the South, and as far north as Chicago, Illinois. These increases can be traced to enhanced free convection (Chapter 3) due to urban heat islands.

Summary

As the human population has grown in size and become more industrialized, societies have increased their impact on the atmospheric environment. The most dramatic effects result from the release of numerous gases and particulates. Although many of the emissions that we consider pollutants result from natural processes, rapid dispersion of these materials prevents them from causing negative impacts. In industrial and urban areas, on the other hand, these emissions are concentrated into smaller areas and often lead to serious problems.

Some atmospheric pollutants are released directly into the atmosphere (primary pollutants), while others result from transformations of other gases (secondary pollutants). Particulates are solid and liquid aerosols that can be produced as either primary or secondary pollutants. Recent research has shown that the health effects of the smallest particulates are most critical because they are easily lodged in lung tissue. Pollutant gases include carbon oxides, sulfur compounds, nitrogen oxides, volatile organic compounds (also called *hydrocarbons*), and photochemically formed gases (the most notable of which is ozone). Each gas presents its own set of health problems, ranging from reduced immunity to permanent lung damage to cardiovascular problems.

Up until the middle part of the twentieth century, efforts to control these emissions were instituted on a local scale. Beginning in 1955, the U.S. government began to enact laws designed to improve air quality across the nation and reduce the number of major health problems that result from air pollution. The original Clean Air Act and its amendments have had a dramatic impact on air quality. This legislation has required the formation of local agencies to monitor pollution levels and ensure compliance with federal standards. It has also required the automobile industry, power utilities, ore smelting plants, and manufacturing industries to reduce the amount of their emissions. While reductions in many pollutants have been dramatic, numerous cities in the United States still have not met clean air goals.

The amount of air pollution does not depend entirely on the activities of people; atmospheric conditions also affect the dispersal of pollutants. If winds are strong and constantly shift direction, pollutants are distributed over a larger area and concentrations decrease. Statically unstable air also favors the dilution of gases and particulates by enhancing vertical mixing. On the other hand, stable conditions and, in particular, the presence of an inversion, can greatly restrict vertical motions and concentrate pollutants near the ground.

Human impacts on the atmosphere are not restricted to pollution. The urban heat island is a well-known phenomenon in which changes in the surface (such as the replacement of vegetated surfaces with concrete and asphalt), the existence of buildings with vertical walls, and the release of heat as a byproduct of human activity combine to increase temperatures. These increases are most notable during the evening and nighttime hours and in the winter season.

We have now looked at the natural processes that make up daily weather, and the ways humans analyze, predict, and alter the resultant patterns. The next two chapters of this book concern the longer-term state of the atmosphere—the climate. Chapter 15 looks at general patterns across the globe, and Chapter 16 examines past and possible future changes in climate.

Key Terms

air pollution   page 429
primary pollutants   page 430
secondary pollutants
  page 430
particulates   page 430
PM_{10} page 431
$PM_{2.5}$ page 431

carbon oxides   page 431
carbon monoxide   page 431
carbon dioxide   page 431
sulfur dioxide   page 433
sulfur trioxide   page 433
acid fog   page 434
acid rain   page 434

nitrogen oxides   page 435
nitric oxide   page 435
nitrogen dioxide   page 435
volatile organic compounds
  (hydrocarbons)   page 436
photochemical smog
  page 436

London-type smog
  page 436
Los Angeles-type smog
  page 436
urban heat island   page 440
sea breeze front   page 443

Review Questions

1. Explain the distinction between primary and secondary pollutants.

2. What are particulates and how are they introduced into the atmosphere?

3. What are the two processes most responsible for removing particulates from the atmosphere?

4. What are PM_{10} and $PM_{2.5}$? Does one pose a greater health risk than the other?

5. List the most important gases that contribute to air pollution.

6. What are the primary sources of carbon monoxide in the atmosphere? If these sources are nonanthropogenic, why is it that CO is considered a pollutant?

7. In what way does carbon monoxide harm the human body?

8. What are the primary sources of sulfur dioxide and sulfur trioxide in the atmosphere?

9. Would a person be more likely to notice the presence of high CO or high SO_2 contents in the ambient air?

10. Which primary pollutant is most likely to promote the formation of acid fog or acid rain?

11. Why is it that nitric oxide is much less common in the atmosphere than nitrogen dioxide?

12. Describe the general composition of volatile organic compounds.

13. How do London-type and Los Angeles-type smog differ from each other?

14. Which pollutant gases cause a noticeable coloration of the atmosphere? Which have a characteristic odor?

15. Describe the various atmospheric controls that affect the concentration of air pollutants.

16. How does the construction of buildings in cities alter the radiation exchange near the surface and contribute to the urban heat island effect?

17. Describe the way in which heat storage in cities differs from that of surrounding rural areas.

18. What effect does urbanization have on the exchange of sensible and latent heat?

19. Does the urban heat island manifest itself equally during the day and night? Are there seasonal differences in the magnitude of the heat island effect?

Critical Thinking

1. There has been much improvement in air quality for North America as a whole. Further improvements will only arise from measures that may be costly, both directly through the application of technology and by reductions in certain economic activities. Do you personally believe that further improvements can be brought about at a cost that people are willing to bear?

2. Visit the Web page at **http://www.gsfc.nasa.gov/gsfc/earth/ terra/co.htm**, and form your own conclusions about whether air pollution is now primarily a local or a global process.

3. It is believed that global warming over the last few decades might have been even greater, were it not for the effect of certain types of air pollution. Explain how this could come about.

Problems and Exercises

1. Examine the map in *Box 14–2, Focus on the Environment: The Counteroffensive on Air Pollution.* How is the air pollution situation in your area? Is the information on the map consistent with your perception of the local air quality? What factors do you think lead to the type of air quality that your area has?

2. Visit the Web page at **http://www.epa.gov/air/partners.html** for a listing of EPA, state, and local agencies that provide data on air quality. Refer to one of the agencies for a daily report on the air quality in your area. How do the changes in air quality fluctuate with different weather patterns where you live?

the air quality in your area. How do the changes in air quality fluctuate with different weather patterns where you live?

3. If you live in or near an urban area, make note of the daily maximum and minimum temperatures in your region. Do you detect a significant urban heat island? How does the magnitude of this heat island compare for maximum and minimum temperatures? Are there seasonal differences?

Quantitative Problems

The companion Web site to this textbook, **http://www.prenhall. com/aguado**, has some interesting problems related to the concentration of pollutants and precipitation pH values. We suggest you solve those problems to further your understanding of air pollution.

Useful Web Sites

http://www.epa.gov/air/EPA
Environmental Protection Agency Web page. Includes a wide range of information on various types of air pollution.

http://www.cdc.gov/nceh/airpollution/default.htm
National Center for Environmental Health, an agency of the Centers for Disease Control. Includes numerous links to reports on the effect of air pollutants on human health.

http://www.tc.gc.ca/programs/Environment/AirPollution/menu.htm
Information on air pollution in Canada.

http://www.ces.ncsu.edu/depts/pp/notes/Ozone/ozone.html
Describes the effects of ozone on plants.

http://www.epa.gov/oar/oaq_caa.html
Presents comprehensive information on the Clean Air Act.

http://www.nrdc.org/air/pollution/default.asp
Web page of the Natural Resources Defense Council, a nonprofit public interest group that advocates environmental causes.

http://www.gsfc.nasa.gov/gsfc/earth/terra/co.htm
Reports on how NASA has been monitoring global air pollution. Includes some very interesting movies.

Media Enrichment

Weather Image

Atmospheric Pollution as Seen from Space Shuttle

This cross-section of Earth's atmosphere at sunset and earth limb (24.5° S, 43.5° E) displays an unusual layering believed to be caused by temperature inversions that concentrate aerosols into narrow, horizontal layers. The white layer, possibly containing volcanic particles, represents the top of the stratosphere. Smoke created by biomass burning is found in the troposphere (the purple layer).

Weather Image

Pollution over Amazon Basin

This NASA space shuttle image, taken September 18, 1991, is centered over the Bolivian Altiplano (20.0° S, 65.0° W), looking northeast into the lower elevations of Bolivia and Brazil. Thick haze is present over the southern Amazon Basin due to seasonal biomass burning. This haze is trapped in the lower atmosphere by a temperature inversion.

Weather Image
Kuwait Oil Field Fires

This Landsat-5 Thematic Mapper image shows oil wells burning in Kuwait at the end of the Persian Gulf War. Fires appear as red dots, and smoke plumes stretch along the Persian Gulf coast from Kuwait to Saudi Arabia. The smoke plumes were relatively narrow—25 to 60 km—and extended to a height of 3 to 4 km in the atmosphere. Although it was believed that the smoke from these fires would have some effect on global climate, effects were limited to the Gulf region.

Weather Image
Industrial Pollution in Siberia

This photograph, taken April 2, 1992, shows industrial pollution (soot-blackened snow) around the Siberian city of Troisk (54.0° N, 61.0° E), an industrial city located east of the Ural Mountains. Troisk is considered one of the most polluted cities in the region. Due to the large amount of soot in the atmosphere, respiratory diseases among citizens of Troisk are chronic.

Weather Image
Ozone Concentration and Air Temperature

This graph shows the relationship between photochemical smog (measured as ozone in parts per billion) and temperature. Photochemical smog is created when sunlight triggers numerous reactions and transformations of gases and aerosols. These reactions are more intense and occur more often at high temperatures. As this graph for 1985 shows, when the daily maximum temperature was below 70 °F in Los Angeles, ozone amounts seldom exceeded the national standard of 120 ppb. When temperatures increase, however, ozone concentrations tend to increase as well. Note that ozone alerts were common at temperatures above 95 °F.

Weather Image
Atlanta Urban Heat Island I, II, and III

These satellite images were taken of Atlanta in 1997 as part of a study on the effects of Atlanta's urban expansion on the area weather and air quality. Researchers found that urban Atlanta experienced temperatures that were warmer by 4.5 to 5.5 °C (8 to 10 °F) than outlying rural areas, making a good example of a heat island.

The first scene, obtained on May 11, 1997, is a black-and-white image showing daytime thermal infrared data (oriented with north at the top). Surfaces that are "hot" or "warm" appear in varying shades of white to light gray, whereas cooler surfaces are darker. Most of the buildings and other urban surfaces are hotter than the surrounding suburban and rural areas during the daytime.

The second picture, a false-color image, was obtained using a multispectral sensor flown on an aircraft. Trees and other vegetation appear red. Buildings, streets, and other urban land covers appear white, blue-green, or black.

The third picture is a GOES 8 satellite image showing a regional perspective of the urban heat island. The thermal channels of the GOES 8/10 geostationary satellites measure outgoing longwave radiation. In the absence of clouds, the satellite data can be used to monitor hourly changes in thermal characteristics of the surface. This image was taken at 10:30 P.M. EDT on May 6, 1997, and shows surface temperatures in degrees Fahrenheit. Atlanta is warmer than the surrounding area by several degrees.

PART SIX

Current, Past, and Future Climates

Stonehenge in Salisbury Plain, England. About 4000 years old, the arrangement of the stones probably served to track the seasonal changes in Earth–Sun geometry.

EARTH'S CLIMATES

Although we don't always know in advance what type of weather will occur at any particular place on a particular day, we do have some idea of what type of weather is considered normal for a given location. For example, Florida typically has hot, humid summers and mild winters—although cold spells do occur from time to time. But there are times when truly anomalous weather events occur in some particular location, events that defy our expectations. These may take the form of unusual heat or cold, but they may also appear as severe weather events uncommon to a particular region. Such an event occurred in Salt Lake City, Utah, on August 11, 1999, when an F2 tornado roared through the downtown area.

Tornadoes are not unknown in Utah; they occur on average about twice a year, but they almost always are classified as "weak" tornadoes, F0 or F1. The Salt Lake City tornado was an F2 that covered a 3-km (2-mi) path. The tornado brought winds in excess of 180 km/hr (110 mph) to the downtown area, toppling trees, knocking over tractor trailers, and damaging more than 120 homes. The Delta Center (the home of the Utah Jazz basketball team) and the Salt Palace Convention Center were among the more notable structures damaged. David Gross, who was inside the Convention Hall, reported that, "The roof opened and it [the wind] ripped off a door. It was over in 15 or 20 seconds, but it seemed like a lot longer than that."

One man was killed by flying debris in a large tent set up for an outdoor retailers convention. The victim was believed to be the first fatality from a tornado in Utah's recorded history. One hundred other people were injured, and 40 of them required hospitalization.

Of course, the climate we associate with an area involves more than just severe weather; it concerns the entire range of weather conditions in the area. Climate deals with long-range conditions, as opposed to the passage of daily weather occurrences. In this chapter we examine the issues involved in classifying climate, and we describe the broad climatic regions across the globe.

Defining Climate

We define **climate** formally as the statistical properties of the atmosphere. This is consistent with the notion that the details of an individual event or moment in time are not of interest—instead, climate is concerned with the long-term behavior, or expected (typical) conditions. Thus, for example, the mean (average) temperature is a climatological value, and by comparing mean temperatures one gains information about differences in climate. But climate is more than average values. For example, two places might have similar long-term average rainfall. But if one place typically has both very high and very low values, while the other tends to receive nearly the same total year in and year out, we would surely say they have different climates. This year-to-year variability is another statistical property and is thus a measure of climate.

In addition to variability, we might be interested in the degree to which above-average or below-average values tend to come in runs or be clustered in successive years. For example, do periods of drought tend to be followed by wet periods? Or are extended episodes of extreme heat and cold a common occurrence? These are questions about another statistical property, namely the correlation between values in successive years. Again, this is part of climate. The point we want to make is that climate consists of all statistical properties. For our purposes, a consideration of average values will be sufficient, but a complete description of climate would include much more.

Although the delineation of distinct climates may seem like a very straightforward endeavor, establishing the criteria by which climates are delineated requires considerable subjectivity. Consider what you would do if called on to devise a

◀ **Rain forest in Venezuela.**

scheme by which Earth's surface would be covered by distinct climatic zones, each having properties that set them apart from the others. The job would require you to set distinct boundaries that separate one climate zone from another. Yet in nature such clear boundaries are rare. Thus, there is a considerable difference in temperature and precipitation along the east coast of North America from Florida to the Maritime Provinces of Canada, and you certainly would not want to put St. John's, Newfoundland, in the same climatic zone as Tallahassee, Florida. But exactly where would you draw the lines that separate the various climates? And how would you decide how many climates? Too many would make the system too complex; too few would fail to capture the patterns you would want to identify.

Climatologists over the years have made numerous attempts to establish useful climatic classification schemes. Some are based on the obvious properties of temperature and precipitation. Others use the frequency with which air mass types occupy various regions, differences in energy budget components, or seasonal characteristics of the water balance at the surface. Each has its own advantages, depending on the purpose of the classification. Agriculturalists, for example, would probably be most interested in using a classification that yields information on water availability relative to plant needs, reflecting gains and losses of water in the soil column (precipitation, evapotranspiration,[1] runoff, and losses to groundwater). For more on this topic, see *Box 15–1, Special Interest: Different Climate Classification Schemes for Different Purposes.*

The Koeppen System

For many people, a climatic classification scheme based on temperature and precipitation is useful because it yields information on the two meteorological variables of greatest general interest. The most widely used systems based on these variables have followed on the work of Vladimir Koeppen, a German citizen of Russian ancestry. The **Koeppen system** was developed over the period from 1918 to 1936 in a process of almost continual revision and refinement. Koeppen looked at the world distribution of natural vegetation types, located the boundaries that separated them, and determined what combinations of monthly mean temperature and precipitation were associated with those boundaries. Thus, although its climatic types are determined by temperature and precipitation, Koeppen's system is inherently tied to natural vegetation. Contrary to what one might assume, it does not begin with the idea of "natural" temperature/precipitation regimes. The boundaries are associated with plant associations, which may or may not coincide with what seem to be obvious or striking gradients in temperature and precipitation.

This chapter will use a version of the Koeppen system as modified by Trewartha. Various versions of the Koeppen system apply different names for each of the climates and may have slightly varying criteria for distinguishing them. Thus, our maps and descriptions are necessarily somewhat different than some other portrayals of the Koeppen scheme. Moreover, this scheme is totally descriptive and does not attempt to explain the causes of the various climates.

Koeppen used a multi-tiered classification system that delineated primary climates by capital letters ranging from A through E. These five broad categories tend to arrange themselves across Earth's surface in response to the latitude, degree of continentality, and location relative to major topographic features. In addition to these climates, the version we are using includes an additional one for mountain environments, designated by H. The main climate groups (designated by first letter only) can be briefly described as follows:

- A—**Tropical.** Climates in which the average temperature for all months is greater than 18 °C (64 °F). Almost entirely confined to the region between the equator and the tropics of Cancer and Capricorn.

---

[1]Evapotranspiration is the combination of water directly evaporated at the ground surface and that absorbed by plants and evaporated through their leaves.

- B—**Dry.** Potential evaporation exceeds precipitation.
- C—**Mild Midlatitude.** The coldest month of the year has an average temperature higher than −3 °C (27 °F) but below 18 °C (64 °F). Summers can be hot.
- D—**Severe Midlatitude.** Winters have at least occasional snow cover, with the coldest month having a mean temperature below −3 °C (27 °F). Summers are typically mild.
- E—**Polar.** All months have mean temperatures below 10 °C (50 °F).

Climatic zones A, C, D, and E are based on temperature characteristics. The A climates (tropical) tend to straddle the equatorial regions; C, D, and E climates usually occur sequentially further from the tropics and toward the polar regions. The sole primary climate zone that considers precipitation is the B climate, designating deserts and semi-deserts.

The A through E climates are subdivided into smaller zones represented by a second letter, which are further subdivided. We will use subdivisions only through the third letter; thus, each individual climate is represented by a three-letter combination describing its temperature and precipitation characteristics. The descriptions for each of the three-letter climates are presented in Table 15–1 and

DID YOU KNOW?

Climate classifications have been derived based on some unconventional but nonetheless significant climatological elements. For example, scientists have calculated and mapped the average apparent temperatures (an index of temperature discomfort due to a combination of high temperature and humidity, which was discussed in Chapter 5) for the hottest 12 days of the year for weather stations across the United States. The maps indicate that typically the most severe heat occurs in the region extending from south Texas northeastward to about St. Louis, Missouri. There the heat index on the hottest days of the year exceed those found in much of southern Arizona or the Florida panhandle.

TABLE 15–1   Climate Types According to Koeppen

TYPE	SUBTYPE	LETTER CODE	CHARACTERISTICS
A—Tropical	Tropical wet	Af	No dry season
	Tropical monsoonal	Am	Short dry season
	Tropical wet and dry	Aw	Winter dry season
B—Dry	Subtropical desert	BWh	Low-latitude dry
	Subtropical steppe	BSh	Low-latitude semi-dry
	Midlatitude desert	BWk	Midlatitude dry
	Midlatitude steppe	BSk	Midlatitude semi-dry
C—Mild	Mediterranean	Csa	Dry, hot summer
	Midlatitude	Csb	Dry, warm summer
	Humid subtropical	Cfa	Hot summer, no dry season
		Cwa	Hot summer, brief winter dry season
	Marine west coast	Cfb	Mild throughout year, no dry season, warm summer
		Cfc	Mild throughout year, no dry season, cool summer
D—Severe Midlatitude	Humid continental	Dfa	Severe winter, no dry season, hot summer
		Dfb	Severe winter, no dry season, warm summer
		Dwa	Severe winter, winter dry season, hot summer
		Dwb	Severe winter, winter dry season, warm summer
	Subarctic	Dfc	Severe winter, no dry season, cool summer
		Dfd	Extremely severe winter, no dry season, cool summer
		Dwc	Severe winter, winter dry season, cool summer
		Dwd	Extremely severe winter, winter dry season, cool summer
E—Polar	Tundra	ET	No true summer
	Polar ice cap	EF	Perennial ice
H—Highland	Highland	H	Highland

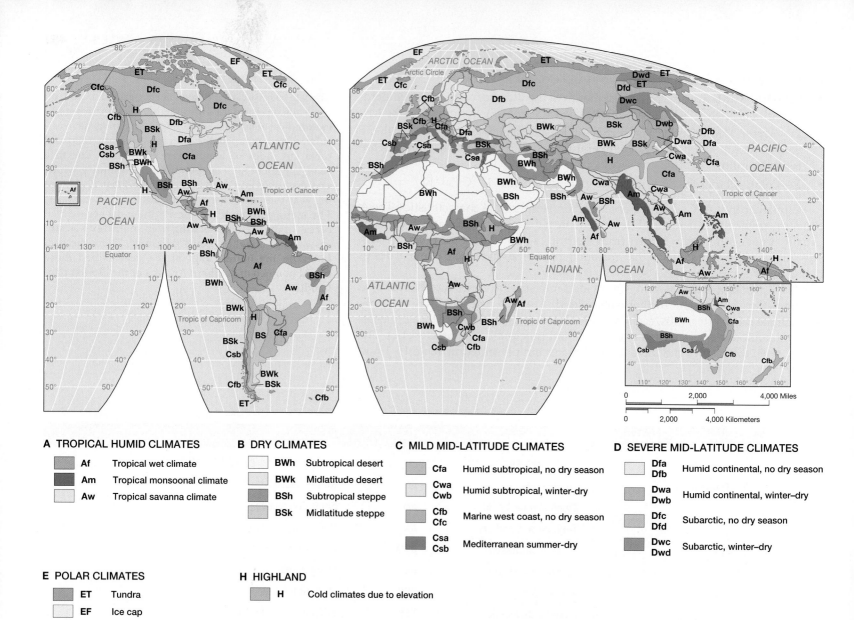

A TROPICAL HUMID CLIMATES

	Af	Tropical wet climate
	Am	Tropical monsoonal climate
	Aw	Tropical savanna climate

B DRY CLIMATES

	BWh	Subtropical desert
	BWk	Midlatitude desert
	BSh	Subtropical steppe
	BSk	Midlatitude steppe

C MILD MID-LATITUDE CLIMATES

	Cfa	Humid subtropical, no dry season
	Cwa Cwb	Humid subtropical, winter-dry
	Cfb Cfc	Marine west coast, no dry season
	Csa Csb	Mediterranean summer-dry

D SEVERE MID-LATITUDE CLIMATES

	Dfa Dfb	Humid continental, no dry season
	Dwa Dwb	Humid continental, winter-dry
	Dfc Dfd	Subarctic, no dry season
	Dwc Dwd	Subarctic, winter-dry

E POLAR CLIMATES

	ET	Tundra
	EF	Ice cap

H HIGHLAND

	H	Cold climates due to elevation

FIGURE 15–1
World map of Koeppen climates.

their distributions are depicted in Figure 15–1. While this system has been taught to countless numbers of students over the years, there is little insight to be gained by memorizing the exact temperature and precipitation values that define the climate boundaries. Thus, the critical values are omitted from the table.

For the A climates, the second letters *f, m,* and *w* indicate if and when a dry season occurs. **Af** climates have no dry season at all. **Am** climates are the monsoonal climates in which a short dry season is normally experienced while the rest of the year is rainy. **Aw** climates have a distinct dry season, usually coinciding with the seasonal presence of the subtropical high pressure of the Hadley circulation. The dry climates are divided into two classes: the true deserts (BW) and the semi-deserts (BS). The second letter of the C and D climates signifies the timing of the dry season. The letter *f* indicates no dry season at all (as with the A climates), while *s* and *w* represent dry summers and winters, respectively. The second letter of the E climates (capitalized) distinguishes polar tundra (ET) regions from areas covered by glaciers (EF). The third letter of each climate represents the temperature regime. As you can see, the meaning of the third symbol varies among the major groups.

Tropical Climates

The name of this climatic group could not be more straightforward or accurate. Tropical climates exist almost entirely between the tropics of Cancer and Capricorn. The tropical group consists of three climates, each of which is warm year-round, with only minor—and in some cases, minimal—variation in temperature throughout the year. The three climates are distinguished by their different degrees of precipitation seasonality. The **tropical wet** climate has significant rainfall every month of the year, the **tropical wet and dry** climate has a pronounced dry season, and the **monsoonal** climate undergoes relative dryness for 1 to 3 months but receives sufficient moisture that vegetation need not be adapted to seasonal drought. All three tropical climates are dominated by the seasonal movement of the Hadley cells, described in Chapter 8.

TROPICAL WET (Af)

The three largest tropical wet climates are found in the Amazon Basin of South America, over western equatorial Africa, and on the islands of the East Indies (see Figure 15-1). As the map shows, the majority of these locations exist within about 10° on either side of the equator, though some are found as far poleward as 20°. Tropical wet climates have no dry period because their position near the equator puts them under the constant influence of the intertropical convergence zone. For this reason, precipitation is almost always convectional, with strong solar heating of the surface triggering brief but heavy thundershowers in the mid- to late afternoon. In the more poleward areas in which Af climates occur, rain often results from orographic uplift of the predominant trade winds. The tropical wet climate of the Atlantic coast of Central America serves as an excellent example of this phenomenon.

Figure 15–2 presents two **climographs** (a climograph depicts monthly mean temperatures and precipitation, with line and bar graphs plotted simultaneously) for Singapore and for Belém, Brazil, typical tropical wet climate stations. Notice that the rainfall is distributed nearly uniformly throughout the year for most Af locations (although Belém has greater seasonality than Singapore), and that all months average at least 22 cm (5 in.) of precipitation. Even more striking is the uniformity of average monthly temperatures, which vary in these examples by only about 2 °C (4 °F).

While temperatures are often high throughout the year, these climates are not among the hottest on Earth. The ever-present moisture availability at the surface allows a large portion of the incoming solar radiation to be expended on evaporation rather than increasing the surface temperature. Furthermore, the convection of humid air promotes the formation of cumulus clouds that scatter much of the incoming solar radiation back to space. Thus, maximum temperatures never come close to those found in the subtropical deserts. On the other hand, the high humidity retards nighttime cooling, so diurnal temperature ranges are low compared to drier climates. Unlike most climates, the diurnal range in tropical wet climates often exceeds the annual range. Minimum and maximum temperatures normally range from about the low 20s Celsius (low 70s Fahrenheit) in the morning to the low 30s Celsius (high 80s Fahrenheit) in the afternoon.

In addition to the unsurpassed consistency in temperature and precipitation, areas having tropical wet climates tend to have the same type of natural vegetation—the tropical rain forests, which house a very dense canopy of tree cover and a tremendous amount of species diversity.

MONSOONAL (Am)

The monsoonal climate can be thought of as a transition between the tropical wet to the tropical wet and dry climates. Monsoonal climates usually occur along tropical, coastal areas subjected to predominant onshore winds that supply warm, moist air to the region throughout most of the year. Such areas are found along northeastern South America,

FIGURE 15–2

Climographs for Belém, Brazil, and Singapore, representative of tropical wet climates. The bars plot the monthly mean precipitation (scaled on the right vertical axis). The lines represent mean monthly temperature (scaled on the left).

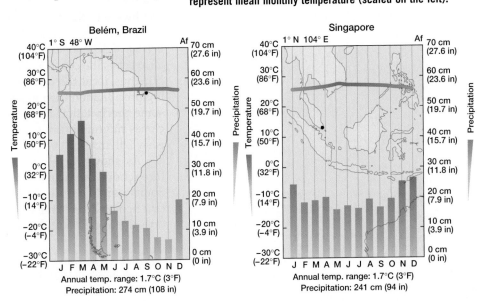

The Koeppen system is undoubtedly the most widely used for distinguishing and mapping the climates of the world, in large part because it is based on easily obtainable data and attempts to have boundaries that coincide with observable vegetation borders. But there really is nothing magical about it—it is simply one of many schemes that have been developed over the years. Like all the others, it is a human construct, created with certain goals in mind. It follows, therefore, that the climates that emerge from Koeppen's classification rules are also human constructs, rather than objective entities waiting to be discovered by Koeppen and his followers. In other words, although the mechanics of climatic classification involve the development and use of objective rules, in a larger sense climate classification is a decidedly subjective process. What constitutes a "good" classification depends very much on judgment and purpose. By no means, then, should one take Koeppen's scheme as "correct" or even "best." In fact, even its admirers readily admit to several shortcomings.

One of the most important shortcomings of the system is that it is based on vegetation boundaries that have been associated with monthly values of mean temperature and precipitation. This is problematic because these two variables alone do not directly determine the geographic limits of natural vegetation. A superior system would be based on the factors that play a more direct role in determining the geographic limits of vegetation, the most prominent of which are precipitation and potential evapotranspiration.

In tandem, the opposing effects of evapotranspiration and precipitation determine the *water balance*. Wherever evapotranspiration exceeds precipitation, the amount of moisture in the soil is reduced, as its loss is not offset by soil moisture inputs. When precipitation exceeds evapotranspiration, the soil moisture is replenished until it reaches the maximum amount of water that can be retained by the soil against the force of gravity (field capacity). Typical plots of monthly water budgets reflecting these inputs and outputs are shown in Figure 1.

Thornthwaite's classification system, based on the principle of the water balance, evolved through decades of work that culminated in its final form in 1955. The system uses four criteria for delimiting climate regions. The first criterion is a *moisture index* that compares the amount of average precipitation each month to the potential evapotranspiration. The latter value derives from a formula using mean temperatures and the monthly values of average period of daylight (a function of latitude) for each station to establish a monthly moisture index. These monthly values are then summed to produce an annual moisture index, the value of which distinguishes arid, semiarid, subhumid, humid, and perhumid climates. These divisions are based on arbitrary percentage changes

in the moisture index (20 percent changes for the humid climates, 33 percent changes for the dry climates), not from associations with plants or other nonclimatic phenomena.

The second criterion is the *thermal efficiency* of a location, or the total amount of potential evapotranspiration. The remaining two criteria are based on the seasonality of precipitation and potential evapotranspiration. When combined, the four criteria create a more physically based climate classification scheme than that of Koeppen.

So why hasn't the Thornthwaite system supplanted Koeppen's as the most popular? In part because of its greater complexity. Compared to Koeppen's system, the water-balance computations needed by Thornthwaite's method are laborious, and the resulting regions are therefore quite removed from the underlying climatic data. Thus, the pattern of climates that emerge is harder to interpret in terms of large-scale processes. Also, although the basic concept of potential evapotranspiration is widely accepted, the method developed by Thornthwaite does not follow from physical principles, but instead relies on data collected mainly in the eastern United States. The data were used to construct empirical (observation-based) equations for potential evapotranspiration. Unfortunately, Thornthwaite did not publish details regarding how the equations were developed, and there are questions as to how well they work in other areas. Thus, if

southwest India, near the eastern Bay of Bengal, and in the Philippines.[2] These areas do not extend nearly as far inland as the tropical wet climates, because their existence depends largely on the effect of speed convergence that occurs as offshore air reaches the coast. Rainfall in these climates is also enhanced by orographic uplift. Thus, localized convergence from surface heating is much less a factor in causing precipitation than it is in the tropical wet climates. During the low sun season, some precipitation may occasionally result from the passage of midlatitude cyclones migrating unusually far equatorward. Near the end of summer and into early fall, tropical cyclones and hurricanes can also bring heavy deluges.

As shown in Figure 15–3, precipitation does not occur nearly as steadily throughout the year in a monsoonal climate as it does in a tropical wet climate. Some months can experience exceedingly heavy rainfall while others are nearly dry. In many cases, the wet months in monsoonal climates yield far more rain than does the wettest month for tropical wet climates. In fact, annual precipitation totals

---

[2]Note that the designation *monsoonal climate* is not synonymous with locations subject to the reversal of the winds discussed in Chapter 8.

Thornthwaite's method were applied to the entire globe, it is not clear how meaningful the resulting regions would be.

Of course, the Koeppen and Thornthwaite systems are not the only systems that have been devised. Most of the others are designed for more specific applications than either of these two. Some, for example, have identified regions of human comfort, whereas others consider the dominance of different types of air masses. More sophisticated ones have used energy budget considerations. Regardless of the premises on which they are based, each has its own set of advantages and disadvantages.

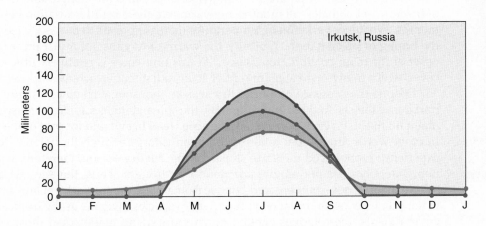

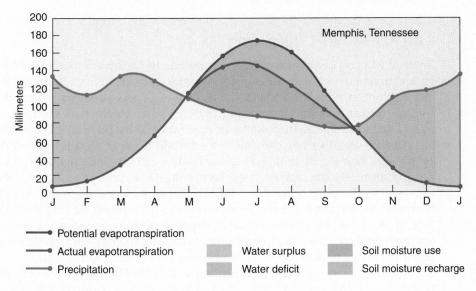

FIGURE 1

Water balance diagrams for Irkutsk, Russia, and Memphis, Tennessee.

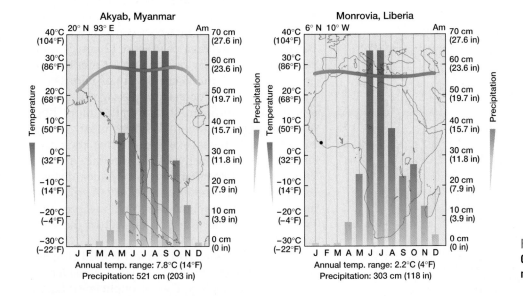

FIGURE 15–3

Climographs for Akyab, Myanmar, and Monrovia, Liberia, representative of monsoonal climates.

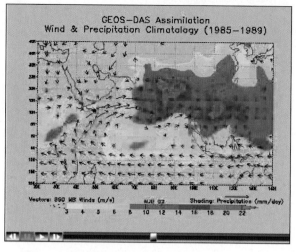

in some monsoonal regions are among the highest in the world, with monthly precipitation values during the peak rainfall periods easily exceeding 80 cm (33 in.). Seasonal totals can even surpass 10 m (400 in.)!

Monthly mean temperatures in monsoonal climates are comparable to those of tropical wet climates; all months are warm and there is little variation between months. Whatever variation in temperature there is appears to be associated with the timing of precipitation. Typically, the warmest months occur just prior to the onset of the main precipitation season. At this time there is relatively little cloud cover, which allows a greater amount of solar radiation to reach the surface.

Despite the presence of a brief dry season, monsoonal climates usually support dense forests. In these environments, the soil maintains sufficient moisture to maintain the lush vegetation even in the absence of heavy rain for part of the year. In other words, the annual total precipitation is large enough that plants do not experience pronounced moisture stress during the dry season; therefore, vegetation generally does not require adaptation to drought. Thus, the Am climate is said to have a "non-compensated" dry season, unlike the tropical wet and dry climate. While not as luxuriant or as abundant in species diversity as the tropical wet environments, these locales contain much more living matter than those of the drier, tropical wet and dry climates.

TROPICAL WET AND DRY (Aw)

Tropical wet and dry climates often occur along the poleward margins of the tropics and border dry climates on one side and tropical wet climates on the other. They are most extensive in South and Central America and southern Africa. Because they are farther from the equator, they undergo much greater seasonality in precipitation and temperature than do the tropical wet and the monsoonal climates.

As with the other two climates of the humid tropics, tropical wet and dry climates owe their existence largely to the Hadley cell. During the high sun season, the intertropical convergence zone favors the formation of afternoon thundershowers. As the position of the overhead sun shifts to the opposite hemisphere, however, the subtropical high arrives to bring descending air and the resultant lack of precipitation. These periods of dryness are more pronounced and longer lasting than those of the monsoonal climate because their distance farther from the equator puts them closer to the mean position of the subtropical high.

Localized convection by solar heating within the ITCZ is not the only process that brings precipitation to tropical wet and dry climates. Tropical depressions can bring widescale precipitation. Along coastal areas, occasional tropical storms and hurricanes can increase average accumulations. Figure 15–4 illustrates the seasonality of temperature and precipitation for typical tropical wet and dry climates. In Acapulco, for example, each of the months between May and October receive an average of at least 12 cm (5 in.) of rain. September is by far the wettest month, with an average of about 36 cm (15 in.) of rain. This is largely due to the occasional passage of tropical storms and hurricanes that can dump huge amounts of precipitation. Although most years go by without the passage of these storms, their occasional occurrence increases the monthly average. During the fall, the monthly precipitation decreases, until the dry months of February–April. On an annual basis, these climates receive less precipitation than do either the tropical wet or the monsoonal climates.

Unlike the other two tropical humid climates, the tropical wet and dry undergo considerable year-to-year variability. Thus, drought episodes can

FIGURE 15–4

Climographs for Bamako, Mali, and Acapulco, Mexico, representative of tropical wet and dry climates.

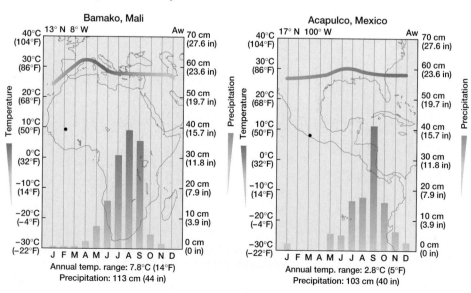

reduce even further the amount of precipitation received in the dry season—often with fatal consequences, as we have witnessed in the Sahel of Africa (see Chapter 8). Likewise, unusually wet rainy seasons can lead to severe flooding and erosion.

Within the year, monthly mean temperatures exhibit more variability than is found in the other tropical climates, but the variability is nonetheless low compared to most others. The annual temperature range is usually between about 3 and 10 °C (5 and 18 °F). Diurnal variations are likewise greater in these parts of the tropical humid environment than in the wetter regions. This is especially true during the dry season, when the absence of clouds facilitates greater daytime heating and nighttime cooling, and daily lows and highs might range between 15 and 30 °C (59 and 86 °F). During the rainy season, the combination of high humidity and cloud cover reduces the diurnal ranges to values similar to those of the tropical wet regions (about 10 °C, or 18 °F).

Tropical wet and dry climates are associated with a natural vegetation type unique among the tropical regions—the **savanna**. This vegetation consists mainly of grasses interspersed with widely separated trees or clumps of trees. While it is tempting to attribute to the lack of forest to the presence of the dry season, many ecologists doubt the causality of this relationship. Instead they believe that the vegetation complex results from numerous factors, including recurrent fire, waterlogged soils, and the development of hard layers within the soil.

Dry Climates

It may surprise many people to learn that the definition of a dry climate is not based on precipitation alone. In other words, there is no set value of annual precipitation (for example, 10 cm) that makes a region considered to be arid. Instead, aridity also depends on the potential evapotranspiration, as well as the timing of the precipitation relative to the period of peak potential evapotranspiration. For our purposes, it is sufficient to state that climates are dry when potential evaporation exceeds the annual precipitation. These climates occupy 30 percent of Earth's land surface, which is considerably more than any other climate group.

The dry regions of the world can be divided two ways: by the level of aridity (that is, true deserts versus the less arid semi-deserts) and by their latitudinal position (hot versus cooler dry areas). **Semi-deserts** are transitional zones that separate the true deserts from adjacent climates. They are also called **steppe** climates, with reference to the associated vegetation type consisting of short grasses. True deserts are so dry that only a sparse vegetation consisting entirely of xerophytic species (that is, adapted to drought conditions) can take hold.

Deserts and semi-deserts can be classified as either *subtropical* or *midlatitude*. Subtropical dry regions extend across wide expanses of land between the latitudes of about 10° to 30° in either hemisphere and result from large-scale sinking air motions during most of the year. Notable examples include the desert of southern California–Arizona–Baja California, the Sahara Desert of North Africa and the Arabian Peninsula, and most of the Australian interior.

Midlatitude deserts and semi-deserts usually occur to the east of major topographic barriers that create a strong rainshadow or over interior continental regions well removed from moisture sources. They are found over large portions of the western United States and interior Asia.

The two-tiered system of categorization yields four types of dry climates: **subtropical desert, subtropical steppe, midlatitude desert**, and **midlatitude steppe**.

SUBTROPICAL DESERTS (BWh)

As shown in Figure 15–1, the most extensive areas of desert exist in the subtropical regions, particularly within the western portions of the continents. The most important factor in the formation of the subtropical deserts is the subsidence associated with the Hadley circulation. As we saw in Chapter 8, the subtropical highs

DID YOU KNOW?

One of the driest desert climates of the world is separated from a tropical wet climate by a mere 250 km (150 mi). The coastal area of northern Peru occupies part of the Atacama Desert and receives on average less than 2.5 cm (1 in.) of precipitation annually. The desert is abruptly truncated by the rugged Andes Mountains, which separates it from the large Af climate to the east, where precipitation exceeds 285 cm (114 in.) per year.

of the Hadley circulation do not appear as continuous bands at the surface, but rather as semi-permanent cells. Although the Hawaiian and Bermuda–Azores high-pressure systems at the surface contract and weaken during the winter, subsidence nonetheless dominates within the middle troposphere. This causes stable conditions to exist in the middle troposphere, which restricts uplift and inhibits precipitation, creating the major subtropical deserts.

Some deserts occur in subtropical regions as narrow strips along the west coast of continents, adjacent to cold ocean currents. The Atacama Desert along the west coast of Chile has the lowest average precipitation on Earth and provides an excellent example of this phenomenon. As air flows out of the subtropical high pressure system in the eastern part of the South Pacific, it flows over the cold Peru current, and the lower portion of the atmosphere is cooled. This cooling can bring the air temperature down to the dew point so that the air becomes damp and foggy. It also lowers the environmental lapse rate and causes the air to be extremely stable. Though the air advected off the coast is damp, stable conditions suppress uplift so that several years can go by without any precipitation at all in some areas.

Over many subtropical deserts, the precipitation that does occur often comes in the form of localized showers from summertime convectional activity. This is not the case for all subtropical deserts, however, as illustrated in Figure 15–5. Yuma, Arizona, and Cairo, Egypt, are both located at about the same latitude, in the midst of subtropical deserts. While both receive scant precipitation over the course of the year, Cairo's precipitation occurs mainly in the winter while Yuma's is about equally divided between the late summer and winter months. As is the case over much of southern Arizona, August usually marks the peak of what is locally known as the *Arizona monsoon*. While having little similarity to the wet season of the Asian monsoon, the Arizona version does undergo a shift in the airflow that brings damp air into the area in late summer. This influx of moisture is subject to strong surface heating that can lift the air sufficiently to trigger isolated thunderstorms. While these thunderstorms can be intense and cause flash flooding, they are neither frequent nor strong enough to make the area anything other than the true desert that it is. The winter precipitation at Yuma results from the passage of midlatitude cyclones. While these systems can bring precipitation across a wide swath of the country, often in the form of heavy rainshowers or snowstorms, over the desert the moisture supply is usually too low to allow much precipitation to fall. This is because Pacific moisture is blocked by the western mountains, and Gulf of Mexico air does not usually flow this far westward.

Daytime summer temperatures in subtropical deserts can be extremely high. In fact, the hottest locations in the world are all found in these regions. During the summer, the combination of low humidities, high sun, and clear skies allows high inputs of solar radiation to be absorbed at the surface. Furthermore, the lack of soil moisture (except after recent rainshowers) causes the ground temperatures to become extremely high as little heat is expended in the evaporation of water. Thus, it is not uncommon for daytime temperatures to reach as high as 45 °C (113 °F) or higher. After the Sun sets, the clear skies and low humidities that led to rapid heating also allow the air to cool considerably. As a result, diurnal temperature ranges can be very large.

The same applies to the annual temperature ranges. At Baghdad, Iraq, for example, the mean monthly temperature in August is 35 °C (95 °F), which is 25 °C (45 °F) higher than the January mean of 10 °C (50 °F). This type of temperature range is greater than that found in any other climate of the tropics or subtropics. As we will see,

FIGURE 15–5

Climographs for Yuma, Arizona, and Cairo, Egypt, representative of subtropical deserts.

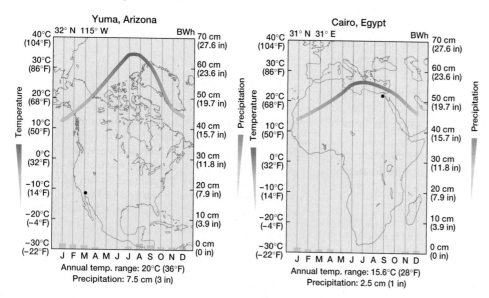

however, midlatitude deserts commonly have even greater temperature ranges.

There is a widely heard axiom in regional climatology that areas of low annual precipitation also have the greatest amount of interannual variability. This certainly applies to subtropical deserts, where many years can go by with only minimal rainfall, only to be followed by a season having numerous rainshowers, delivering several years' worth of precipitation. Thus, the annual average is a poor indicator of how much rain is likely to fall in a given year. Figure 15–6 plots annual rainfall at Borrego Springs, California, using a 45-year period of record as an example.

SUBTROPICAL STEPPE (BSh)

All the conditions that distinguish a subtropical desert also apply to subtropical steppe climates— though to a lesser degree. Like their more extreme desert counterparts, subtropical steppes are marked by aridity, high year-to-year variations in precipitation, extreme summer temperatures, and large annual and daily temperature ranges. It is therefore not surprising that subtropical steppe climates commonly border the subtropical deserts. And though they are transitional between deserts and non-arid regions in terms of their climatological characteristics, they are not necessarily narrow buffer zones. Much of the southwestern United States and northern Mexico, for example, is occupied by subtropical steppe.

Cloncurry, Australia, and Monterrey, Mexico (Figure 15–7), illustrate the greater precipitation totals, somewhat cooler conditions, and lower annual temperature ranges associated with these climates. These two examples also reveal a distinct seasonality in the precipitation regime typical of subtropical steppes located on the equatorward side of deserts. In these regions, precipitation occurs more often during the summer months than during the winter, as a result of localized convection and tropical disturbances. In contrast, steppe regions on the poleward side of subtropical deserts experience most precipitation in the winter in response to the passage of midlatitude cyclones.

MIDLATITUDE DESERTS (BWk)

Midlatitude deserts result from extreme continentality. Such regions occur deep within continental interiors or downwind of orographic barriers that cut off the supply of moisture from the ocean. The greatest expanse of midlatitude desert occurs in Asia—which is not surprising considering the immense size of that continent. The two major areas of midlatitude desert in Asia are found just east of the Caspian Sea and north of the Himalayas. Both are far to the east of the Atlantic Ocean, so much of the moisture associated with eastward-moving cyclones is depleted before reaching either area. Mountains to the south block the northward flow of moisture out of the Indian Ocean.

The second greatest expanse of midlatitude desert occurs in the western United States. The midlatitude desert extends southward and merges with the subtropical desert of the Southwest. Although

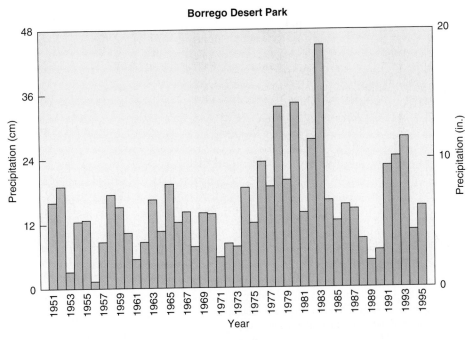

Borrego Desert Park

FIGURE 15–6

Annual precipitation at Borrego Springs, California, 1951–95. Dry climates such as this one normally have a wide year-to-year variability in precipitation.

FIGURE 15–7

Climographs for Cloncurry, Australia, and Monterrey, Mexico, representative of subtropical steppe climates.

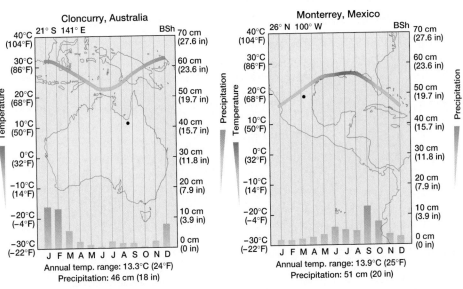

Cloncurry, Australia
Annual temp. range: 13.3°C (24°F)
Precipitation: 46 cm (18 in)

Monterrey, Mexico
Annual temp. range: 13.9°C (25°F)
Precipitation: 51 cm (20 in)

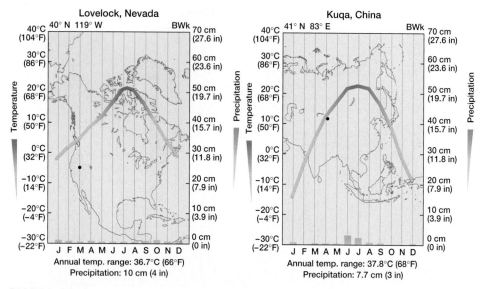

FIGURE 15-8

Climographs for Lovelock, Nevada, and Kuqa, China, representative of midlatitude desert climates.

the midlatitude desert lies poleward of the subtropical desert, it should not be assumed that it is neither as dry nor as hot in the summer as the desert to the south. In fact, Death Valley, California, one of the hottest and driest places in the world, lies within the midlatitude desert.

Midlatitude desert in the Southern Hemisphere is confined to a narrow strip in South America, east of the Andes. A quick look at Figure 15-1 reveals that the midlatitudes of the Southern Hemisphere are almost completely covered by ocean. Thus, it is rare to see any type of land climate in this region, let alone a midlatitude desert.

Figure 15-8 shows two typical climographs for midlatitude deserts. Midlatitude deserts have a greater range of temperatures, both on a daily and annual basis, than do their subtropical counterparts. While both types of deserts become extremely hot during summer days, midlatitude deserts have more rapid nighttime and winter cooling. With more precipitation and lower potential evapotranspiration, they are generally more humid than subtropical deserts. Thus, although vegetation is adapted to dry conditions, ground cover is likely to be continuous, without large patches of bare ground common in hotter (subtropical) desert regions.

MIDLATITUDE STEPPE (BSk)

The midlatitude steppe accounts for most of the arid regions of western North America. It flanks the northern portion of the midlatitude desert and merges with the subtropical steppe to the south. A large swath of steppe also extends from the Great Plains, east of the Rocky Mountains, all the way from northeast Mexico into western Canada. The midlatitude steppe region of Asia is in many places a fairly narrow strip of land surrounding the true deserts.

Midlatitude steppes have the same temperature characteristics as the midlatitude deserts. The primary difference between the two is the greater amount of precipitation in the steppes, which commonly totals about 50 cm (20 in.) annually (Figure 15-9).

FIGURE 15-9

Climographs for Denver, Colorado, and Semey, Kazakhstan, representative of midlatitude steppe climates.

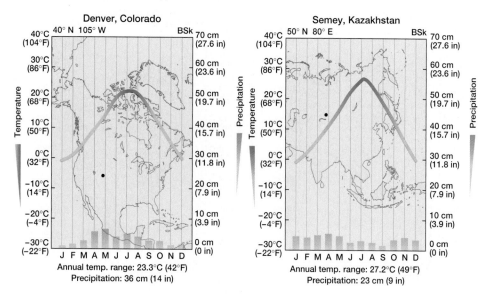

Mild Midlatitude Climates

The mild midlatitude climates are located in parts of the latitude range between 30° and 60° in either hemisphere. They occur as long, narrow strips of land along the west coasts of North and South America and southern Australia. In North and South America, the eastern border of the climate is delimited by mountains. Another west coast mild midlatitude climate—in fact the largest—surrounds the Mediterranean Sea.

Mild midlatitude climates also cover large areas of the eastern portions of the continents, especially in North and South America and Asia. While these areas extend farther inland than do their counterparts along the west coasts, these mild midlatitude climates have narrower latitudinal extents. While west coast midlatitude climates can be found as far poleward as just a few degrees shy of the Arctic Circle, most on the eastern sides of continents do not extend poleward of 40° latitude.

The individual climates within this general group do not share similar precipitation patterns. In fact, the precipitation regime can vary greatly from one region to another. To get an idea of the extent of this variability, consider the fact that along the west coast of the United States, the mild midlatitude includes San Diego, California, which receives about 25 cm (10 in.) of annual precipitation, and the Olympic rain forest in Washington State, where the precipitation locally exceeds 375 cm (150 in.) per year.

The term *mild* refers to the winter temperatures and not necessarily those of the summer. In North America, for example, this climate group is found over inland areas of California where summer temperatures routinely exceed 38 °C (100 °F), as well as in the Gulf states of Florida, Georgia, Alabama, Mississippi, and Louisiana—places hardly noted for their mild summers. Thus, in effect, *mild* refers to little or no snowcover.

This climate group is subdivided into three climates, two of which exist along the west coasts of continents and the other on the eastern sides. **Mediterranean** climates can be found along the west coasts between about 25° and 40° latitude. Within about the same range of latitude on the eastern side of continents are the **humid subtropical** climates. The **marine west coast** climates lie adjacent to and poleward of the mediterranean climates.

MEDITERRANEAN (Csa, Csb)

Mediterranean climates are the only ones that have a distinct summer dry season and a concentration of precipitation in the winter. Figure 15–10 clearly depicts this pattern for two typical sites, Athens, Greece, and Los Angeles, California. The summer aridity in mediterranean climates is attributable to the presence of semipermanent subtropical high-pressure systems offshore. For North America, the Hawaiian high-pressure system to the west "blocks" the eastward migration of summertime midlatitude cyclones and deflects them to the north. This deflection of the storms, coupled with subsidence along the eastern portion of the subtropical high, deprives coastal southern California of the uplift mechanisms necessary for precipitation. During the winter months, the Hawaiian high weakens, shrinks in size, and migrates toward the equator. This opens the way for the passage of Pacific cyclones.

Annual precipitation increases with latitude and with elevation along windward slopes in mediterranean climates. The latitudinal gradient results from the more frequent passage of midlatitude cyclones at the higher latitudes. At the same time, the mountain slopes induce orographic uplift of the predominantly westerly airflow.

While the climographs in Figure 15-10 reveal the mean precipitation patterns for the two sites, they do not reveal the extreme variability inherent with winter rainfall. Indeed, some winters can be quite dry, while others bring an onslaught of sequential storms that produce major flooding and hillside erosion. This climatic feature has been particularly prominent since the 1970s in southern California, which has experienced several record drought years along with extremely wet winters.

Winter temperatures in mediterranean climates are usually mild, especially right along the coast. Inland temperatures occasionally drop below freezing and sometimes threaten fruit-growing areas with widespread crop damage. On the other hand, precipitation results mainly from the passage of midlatitude cyclones, which

FIGURE 15–10

Climographs for Los Angeles, California, and Athens, Greece, representative of mediterranean climates.

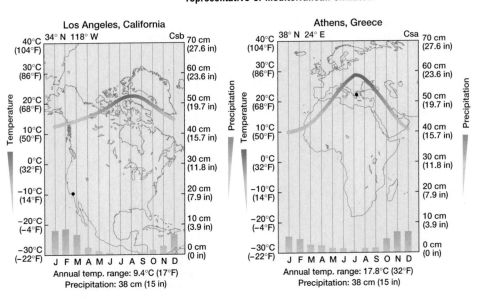

transport relatively warm, maritime polar air. Thus, precipitation in the lower elevations falls almost exclusively as rain and not snow.

Summer temperatures range from mild to hot, with daily highs typically decreasing toward the coast and higher latitudes. Central California provides a good example of how variable the summer temperatures can be in mediterranean climates. San Francisco is noted for particularly cool summers, while about 120 km (70 mi) to the northeast, Sacramento experiences hot, dry conditions.

HUMID SUBTROPICAL (Cfa, Cwa)

Humid subtropical climates occur within the lower middle latitudes of eastern North America, South America, and Asia. Typical temperature and precipitation patterns are shown in Figure 15–11.

Though they exist in the middle latitudes, these climates have a distinct tropical feel during their long summers. Located to the west of large semi-permanent anticyclones and warm ocean currents, the prevailing winds circulate hot, humid air into these climatic zones. Summer daytime temperatures are usually in the lower 30s Celsius (high 80s to low 90s Fahrenheit), and dew points in the mid-20s Celsius (mid-70s Fahrenheit) help retard nighttime cooling. Thus, hot muggy conditions remain throughout the day and night, especially along the equatorward boundaries of the climates. Fortunately, afternoon convectional thundershowers are common in these areas and bring temporary relief from the extreme heat.

Winter temperatures are typically lower than those of mediterranean climates farther to the west because of their greater continentality, and subfreezing temperatures are not uncommon. The occurrence of frost and snow decreases toward the lower latitudes, but even south Florida is not completely immune.

Humid subtropical areas receive abundant precipitation, ranging from about 75 to 250 cm (30 to 100 in.) per year. Over most areas the maximum precipitation is concentrated in the summer, but this generalization does not always hold. Over most of the southeastern United States, for example, summer is the wettest season. But the area extending from east Texas into Tennessee and Kentucky has a winter precipitation maximum. Regardless of which season receives most precipitation, summer is always the season of moisture deficit, because of greater potential evapotranspiration.

Precipitation during the summer is largely convectional in nature and tends to be scattered and brief. Winter precipitation, on the other hand, is usually triggered by the passage of midlatitude cyclones. Over the coastal areas, tropical storms and hurricanes are capable of bringing extreme rainfall from time to time during the late summer and early fall.

MARINE WEST COAST (Cfb, Cfc)

Marine west coast climates (Figure 15–12) normally occur poleward of mediterranean climates and, as the name would suggest, along the west coasts of climates. But inspection of Figure 15–1 shows that the latter is not always the case, as large areas having the climate are found along southeastern Australia and southeast Africa. While not truly located on west coasts, those areas are located east of fairly narrow strips of land that don't greatly modify the maritime nature of the predominantly westerly flow. This exposure to air passing over cold ocean currents and their location in the path of eastward migrating midlatitude cyclones gives these climates their characteristic features.

Both summers and winters are typically mild. Extremely high summer temperatures are certainly

FIGURE 15–11

Climographs for Dallas, Texas, and Guangzhou, China, representative of humid subtropical climates.

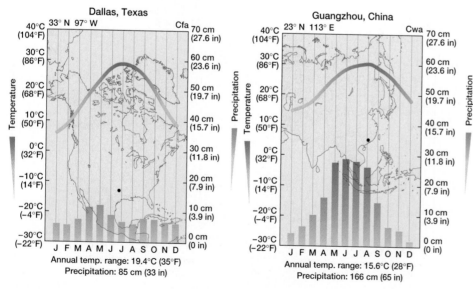

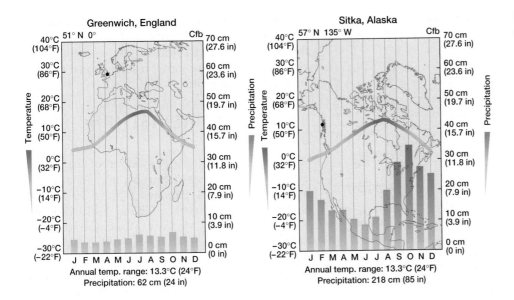

FIGURE 15—12
Climographs for Greenwich, England, and Sitka, Alaska, representative of marine west coast climates.

not unknown in these climates, but they are the exception rather than the norm. From northern California through Alaska, for example, low and mid-level cloud decks frequently keep daytime temperatures down and often bring drizzle or light rain to coastal regions. Along the lower elevations, the percentage of days experiencing precipitation can be high, but the total amount of rainfall is usually light. On the other hand, the Coast Ranges parallel the Pacific coastline and create a major barrier to the westerly flow. This creates a strong orographic effect that causes some areas to have extremely heavy precipitation amounts that rival or exceed those of the tropics. The Olympic Peninsula of Washington State is an excellent example of this phenomenon, where extremely heavy rainfalls are sufficient to support a lush, midlatitude rain forest.

Just as maritime conditions moderate temperatures during the summer, they also allow for mild winter conditions at surprisingly high latitudes. Even as far north as Sitka, Alaska (57° N), the coldest month of the year has a mean temperature above the freezing point of water. At many lower-latitude locations, low-elevation snowfall is a rarity; when it does occur, it usually melts away within a short time. In Europe, where the marine west coast climate extends farther inland than it does in North America, snow is more frequent and remains on the ground for a longer period of time.

Given the fact that both summer and winter temperatures are mild, it follows that these climates have low annual temperature ranges. This contrasts sharply with the situation for climates in the eastern portion of continents at the same range of latitudes. These climates have moderately high summer temperatures, but when combined with the extreme cold of winter, they yield extremely high temperature ranges of the severe midlatitude group of climates.

Severe Midlatitude Climates

The severe midlatitude climate group includes two climates, humid continental and subarctic, both of which are marked by very cold winters. These climates require large continental areas within the high-middle latitudes—between about 40° and 70°. Thus, they are restricted to Europe, Asia, and North America and are not found at all in the Southern Hemisphere. As expected from climates resulting from strong continentality, they exhibit large annual temperature ranges.

Both of the severe midlatitude climates receive precipitation throughout the year and have no true dry season. In many locations, there is greater precipitation

DID YOU KNOW?

Seattle, Washington, has a reputation for experiencing a lot of rain each year. But it actually receives less annual precipitation on average—93 cm (37 in.)—than does New York City, with 124 cm (50 in.). The difference is that Seattle has an average of 150 rainy days each year, but these are usually of low intensity. New York averages only 120 days with measurable precipitation annually, but that city usually experiences more intense precipitation events.

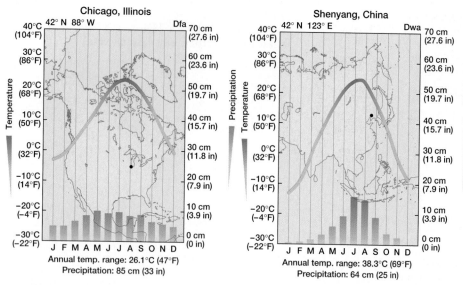

FIGURE 15-13
Climographs for Chicago, Illinois, and Shenyang, China, representative of humid continental climates.

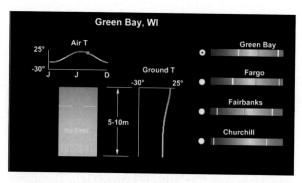

WEATHER IN MOTION
Worm Climates

in the summer than winter. Precipitation during the summer can result from local convection or by the passage of midlatitude cyclones; winter precipitation (mostly as snow) results almost entirely from cyclonic activity.

HUMID CONTINENTAL (Dfa, Dfb, Dwa, Dwb)

A huge segment of the population of the United States, Canada, eastern Europe, and Asia live in **humid continental** climates (Figure 15–13), including the inhabitants of New York, Chicago, Toronto, Montreal, Moscow, Warsaw, and Stockholm. This is the more temperate of the two severe midlatitude climates, normally found between about 40° N and 55° N in the eastern parts of continents. Summers are warm and often hot. New York City, for example, has an average high temperature in August of 29 °C (84 °F). But winter is another matter altogether, with a mean February maximum temperature of 4 °C (40 °F). It should be noted that New York City is located near the southern boundary of the climate and on the coast, so temperatures are milder than those of most other locations with a humid continental climate.

Mean annual precipitation in these climates usually ranges between 50 and 100 cm (20 to 40 in.). Over the United States and southern Canada, there is a conspicuous decrease in precipitation with increasing latitude and distance from the Atlantic shoreline, reflecting a reduced moisture content in the atmosphere.

SUBARCTIC (Dfc, Dfd, Dwc, Dwd)

Subarctic climates occupy the northernmost extent of the severe midlatitude regions, with more than half of the area of Alaska and Canada having this type of climate. In North America, the coniferous forest that dominates the region is referred to as the **boreal forest**; in Asia, it goes by the name **taiga**. Summer temperatures are somewhat lower than those of the adjacent humid continental regions, but the major difference occurs in winter, when mean monthly temperatures can be extremely low. At many locations, the monthly mean temperature can remain below the freezing level for up to seven months. Thus, winter is long and separated from the brief summer only by a short-lived autumn and spring. Figure 15–14 highlights the very large annual temperature ranges that result from the mild summers and severe winters.

Typically, precipitation is greater in the summer than winter, mostly because of the more poleward displacement of midlatitude cyclone tracks in summer. Nonetheless, annual precipitation is usually low, ranging from about 12 to 50 cm (5 to 20 in.).

Polar Climates

Polar climates exist in the highest latitudes—typically areas poleward of about 70°. Such regions occur in the Northern Hemisphere, across northern Canada, Alaska, Asia, and coastal Greenland. In the Southern Hemisphere they are almost entirely confined to the continent of Antarctica. As we mentioned earlier, the climate boundaries of the Koeppen system were set to coincide with boundaries of differing natural vegetation types. In this case, polar climates begin at the high latitude boundaries of the subarctic climates, where vast expanses of coniferous forest give way to a generally treeless landscape. The polar climate group consists of two distinct types. The most equatorward and milder of the two is the **tundra**. At the most poleward regions of the globe lie the true ice cap climates.

TUNDRA (ET)

Tundra climates are named for the associated vegetation type that consists primarily of low-growing mosses, lichens, and flowering plants, with few woody shrubs and trees. The boundary dividing the tundra from the subarctic climate occurs where the mean monthly temperature of the warmest month does not exceed 10 °C (50 °F). In the Northern Hemisphere, this occurs in the general vicinity of 60° N. This climate is almost entirely missing from the Southern Hemisphere, where there is only minimal land coverage at this latitude.

Tundra climates have severe winters in which the Sun rises only briefly each day and never gets very high above the horizon. Under these conditions, radiational cooling at the surface leads to low temperatures and strong stability—both factors that inhibit precipitation. Thus, as shown in Figure 15–15, low winter precipitation is the rule. Surprisingly, perhaps, the winters in the tundra regions are often less severe than those of the adjacent and lower-latitude subarctic climates. This is because the tundra regions of North America, Greenland, and Asia are located nearer to major water bodies and thus have a lesser degree of continentality. (Even when ice-covered, an ocean has a somewhat moderating effect on winter temperatures.)

During the summer, tundra regions have very long periods of daylight, but again the Sun never gets very high above the horizon. As a result, temperatures are normally mild in the mid-afternoon, not very much greater than those of the predawn period. Thus, despite the fact that annual temperature ranges are fairly high, daily temperature ranges are low.

One very conspicuous feature of tundra regions is the existence of **permafrost**, a perennially frozen layer below the surface. During the winter, conditions are so cold that the entire soil is completely frozen, in some places to a depth of several hundred meters. When summer arrives, there is enough warming right at the surface to melt the uppermost soil, and a relatively shallow layer of thawed soil overlies the completely impermeable frozen layer several tens of centimeters below. As a result, any rain or snowmelt that occurs at the surface is unable to permeate

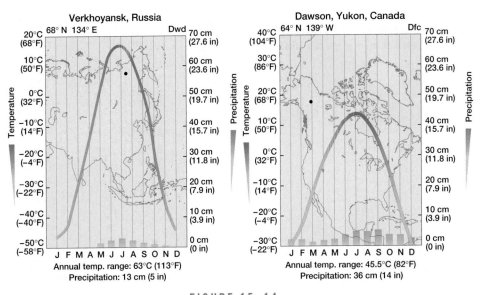

FIGURE 15–14

Climographs for Verkhoyansk, Russia, and Dawson, Yukon, Canada, representative of subarctic climates.

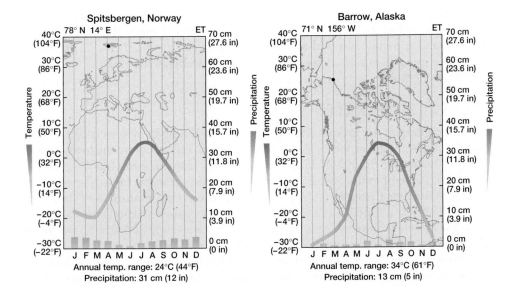

FIGURE 15–15

Climographs for Spitsbergen, Norway, and Barrow, Alaska, representative of tundra climates.

FIGURE 15–16

Climographs for Little America, Antarctica, and Eismitte, Greenland, representative of ice cap climates.

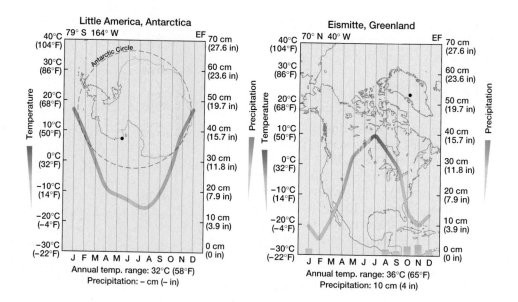

deeply into the soil, and the upper part of the surface becomes saturated. This combination of water-logged soil near the surface and a solid ice layer below precludes the establishment of any deeply rooted plants, and only low-growing vegetation can take hold.

ICE CAP (EF)

As the name rightfully suggests, **polar ice cap** areas exist where ice covers the ground throughout the entire year. They are confined to the Greenland interior and most of Antarctica. They exist where the mean temperature of the warmest month does not go above the melting temperature for ice, 0 °C (32 °F). Ice accumulations over these regions can exceed several kilometers, so in addition to being found at high latitudes they also occupy high elevations—a situation perfect for maintaining extraordinarily low winter temperatures (Figure 15–16).

The cold air overlying polar ice caps (also called *continental glaciers*) becomes extremely dense and frequently flows down the margins of the ice under the force of gravity. When funneled through narrow canyons, these katabatic winds (discussed in Chapter 8) can become extremely strong.

Most areas of ice cap receive little precipitation because of the intense cold. On the other hand, all precipitation falls as snow and has the potential to accumulate onto existing ice. Except along the margins, there is no melt and only a small amount of sublimation to remove the ice. Yet some ice does get removed. Beneath the surface, the individual ice crystals merge into an almost solid block of ice. When subjected to the constant pressure of the overlying mass, the ice gradually deforms and expands outward toward the margins of the ice sheet. At the continental margins, the ice often breaks off into large icebergs that float in the Arctic and Antarctic waters.

HIGHLAND CLIMATES (H)

Highland climates are unique among those in the Koeppen system because their distribution is not governed by geographic location, but rather by the topography. These climates are found in large mountain or plateau areas. As we saw in Chapter 1, the temperature tends to decrease with height in the troposphere. Thus, in high mountains there can be large changes in mean temperature over short distances, simply as a function of elevation. The temperature situation becomes more complicated when one considers that changes in slope angle and aspect over short distances influence the intensity of solar radiation received.

Precipitation type and intensity also vary spatially across highlands. Mountain slopes can enhance precipitation on their windward sides and simultaneously create a rainshadow downwind. Elevation difference also affects the ratio of precipitation falling as snow versus rain, with higher elevations favoring a greater amount of snow. Mount Kilimanjaro in Tanzania provides a striking example of this phenomenon. Located very near the equator at 3° S, Kilimanjaro possesses an extraordinarily wide range of climates from its base near sea level to its peak at 5895 m (19,340 ft). Near the surface, the mountain has a tropical wet and dry climate similar to the surrounding area. But the climate changes with height and the mountain eventually becomes covered with perennial ice over its higher reaches. This is an example of *vertical zonation,* the layering of climatic types with elevation in mountainous environments. Thus, although we use a single designation H, it must be understood that this category contains an extremely rich collection of climates. Obviously, the various climates within the H group are highly localized, with a spatial arrangement not related to large-scale atmospheric processes. As such, the group can be thought of as kind of a climatic "grab bag," consisting of climates that would not necessarily be expected by considering latitude and continental location alone.

Summary

People have long recognized the desirability of a classification system by which distinct climates could be delineated and mapped. This pursuit is neither as straightforward nor as easy as one might expect, because climate consists of a number of different elements. Thus, a certain degree of ingenuity is necessary to create a scheme that includes the major climatic variables and still retains enough simplicity to make the climates comprehensible. The Koeppen system has become the most widely used of all of these techniques.

Koeppen devised a system in which temperature and precipitation characteristics along vegetation boundaries were used as criteria for distinguishing one climate from another. The scheme is a hierarchical, multi-tiered system that uses combinations of capital and lowercase letters.

The first level of categorization in the Koeppen system uses a capital letter from A through E, along with a special category, H, for highland climates. All but one of the A through E climates are based on temperature characteristics. The one exception, the B climates, represent areas of aridity and are called *dry climates.* The temperature-based climates—A, C, D, and E—are referred to as tropical, mild midlatitude, severe midlatitude, and polar climates, respectively.

The second level of categorization depends in most cases on the seasonality of precipitation, although for the dry climates the second level reflects the level of aridity. The third level of classification typically refers to temperatures during particularly critical months. Having established the various climates, one can refer to a map such as that shown in Figure 15-1 to visualize their geographic distribution and understand the processes that give rise to the various types.

At the outset of this chapter we distinguished short-term changes in the atmosphere (weather) from long-term conditions (climate). This might lead us to believe that climates are permanent characteristics that never change. This is not true, however, as the climate is forever undergoing slow changes through time. In fact, looking over the history of the planet, we see that there have been many time periods in which the climates were far different from what we know today. The next chapter of this book addresses those temporal changes in climate.

Key Terms

Review Questions

1. Describe what is meant by *climate*.

2. Describe the general criteria by which the Koeppen system delineates climates.

3. The first order grouping of climates in the Koeppen system is based mainly on temperature. Which climate type departs from that rule?

4. Describe the geographical distribution of tropical climates. What features distinguish this particular group?

5. Briefly describe the fundamental differences between Af, Am, and Aw climates.

6. Of the three types of tropical climates, which occupies the smallest portion of Earth's land surface?

7. Despite their low latitudes, tropical climates are not among the hottest on Earth. Why not?

8. Where are the various dry climates located, and what geographical characteristics cause them to occur where they do?

9. Describe the four types of dry climates and explain how they differ from each other.

10. What factor other than annual precipitation is involved in a climate being defined as dry?

11. Describe the various types of mild midlatitude climates and their distribution. Why is it that two of them locate mostly along the west coast of continents, while the other tends to be on the eastern side?

12. Are the mild midlatitude climates really mild? Explain.

13. What are the two types of severe midlatitude climates, and how do they differ?

14. Why are severe midlatitude climates missing from the Southern Hemisphere?

15. Describe the three types of polar climates and their distributions.

Critical Thinking

1. Although the Koeppen system is intended to create distinct climate classes, its boundaries are based on the boundaries between vegetation types. Is this really a problem? Can you think of any alternative methods to delineate climates?

2. Do you anticipate that global warming, if it continues as expected, will substantially alter the location of the boundaries of the various Koeppen climates?

3. Western Kansas is located near the junction of B, C, and D climates. Do you suppose that a person driving around this region would notice substantial climatic differences as she crossed from one climate zone to another? What does this tell us about the applicability of large-scale classification schemes to smaller-scale analysis?

Problems and Exercises

1. Check Figure 15-1 and determine the type of climate you live in. Then log on to **http://ggweather.com/normals/index.htm** and find the climate information for the location nearest to where you live. Create a climograph for that location and compare it to the example given in the book for the type of climate you live in. How closely do they match?

2. Go to the Web page for the National Weather Service office nearest to where you live. Compare the monthly temperature and precipitation values observed there over the last year to the average for that location. Were they markedly different? How much variance do you expect to encounter between monthly observed values and climatological averages?

Useful Web Sites

http://www.fao.org/WAICENT/FAOINFO/SUSTDEV/EIdirect/climate/EIsp0002.htm
Offers a large number of climate maps, compiled by the United Nations Food and Agriculture Organization.

http://ggweather.com/normals/index.htm
A great source of information for monthly average temperature and precipitation data.

http://www.intellicast.com/Almanac/
Descriptive climate information for each month.

http://iri.columbia.edu/climate/cid/latest/
A Web-based newsletter providing information on seasonal climate and its impacts on society.

http://www.ncdc.noaa.gov/oa/climate/research/monitoring.html
Online access to many reports from the National Climate Data Center.

http://www.ncdc.noaa.gov/oa/climate/severeweather/extremes.html
Provides data on extreme weather events.

http://www.cpc.ncep.noaa.gov/
Home page of the NOAA Climate Prediction Center.

Media Enrichment

Weather in Motion

Monsoon

Seasonal changes in the mean wind (indicated by the length and direction of arrows) and precipitation (color-coded) conditions over the monsoon region are illustrated for the period from April through December. Initially, the Indian subcontinent is dry, with heavy precipitation confined to Indonesia and southeastern China. Following the northward march of solar declination, low-level winds begin ferrying moisture to India and southeast Asia, resulting in widespread areas of heavy rainfall for June though August. The wind field weakens and reverses in October, bringing a return to dry conditions.

Weather in Motion

Worm Climates

This movie shows the seasonal progression of soil temperature at four locations having D climates. The first two, Green Bay, Wisconsin, and Fargo, North Dakota, do not have permafrost. For both, we see a surface layer that freezes and thaws seasonally, underlain by unfrozen ground. (Predictably, the depth of frost is greater in the colder climate.) But Churchill, Manitoba, and Fairbanks, Alaska, are both permafrost climates, with an active layer above permanently frozen ground. Notice that the active layer is thinner at the colder climate, where summer warmth is short-lived and does not penetrate deeply. Notice also that, in each case, temperature changes are greatest near the surface and the amplitude of the seasonal oscillation decreases with depth. Look closely and you'll see soil temperatures follow behind those at the surface, with the lag increasing downward. Thus while it might be winter at the surface, some depth is experiencing its maximum temperature for the year. With sufficient climatological training a worm could move up and down, enjoying perpetual summer!

Weather Image

Mammoth Antarctic Icebergs

Two huge icebergs broke away from the Ross Ice Shelf of Antarctica in March 2000. The larger of the two (named B-15) is about the size of Connecticut (300 km by 37 km) and contains about six times as much water (3.4 trillion gallons) as the system of reservoirs serving New York City. Though enormous, B-15 is not the largest ever cited. That honor goes to a 1927 iceberg—330 km by 100 km. Icebergs of this size are estimated to appear every 50 to 100 years. Like their smaller cousins, they result from continental ice that has been pushed far over the ocean before breaking off. Similar conditions do not appear in the Arctic, so icebergs like this are exclusively a Southern Hemisphere phenomenon.

CHAPTER **16** | CLIMATE
CHANGES: PAST
AND FUTURE

15 °C (27 °F) smaller. Though by no means identical, other warm times undoubtedly had many similar traits.

All ice ages stand in stark contrast to "warm ages" such as the mid-Cretaceous. At one extreme, nearly the entire planet might have been ice-covered during the ice age of about 700 MYA, in a condition aptly called "snowball Earth." Evidence for a global snowball comes from a variety of diverse indicators of biological productivity, which are thought to show a decline to almost zero with growing ice volume, followed by a slow rebound during warmer climates that followed. Short of nearly global ice cover, it's hard to see why productivity would be so low. There is other evidence that points to snow lines near sea level in the tropics during this period, and glacial deposits are capped by particular rock deposits suggestive of rapid warming following intense cold. For many, the power of the snowball hypothesis is its ability to account for all of these along with still other indicators of a global deep-freeze. Other interpretations are possible, however, and the snowball hypothesis remains controversial insofar as the amount of ice is concerned. Regardless of how the controversy is resolved, it is clear that all ice ages have abundant year-round ice. For example, there is no dispute that during the most recent ice age, ice sheets advanced to within 40° of the equator, covering the ground to a depth of several kilometers. With so much water locked in land ice, sea level is much lower during ice ages. Temperature differences between ice ages and warm ages vary greatly by latitude, with high latitudes showing greater changes than the tropics. For example, during an ice age, polar sea surface temperatures (SSTs) might be 10 °C (18 °F) colder, but tropical SSTs only 1 °C to 5 °C (2 °F to 9 °F) colder. Beyond these generalities, little can be said with certainty about the six earliest ice ages. On the other hand, we have substantial information regarding conditions during the most recent ice age.

THE CURRENT ICE AGE

One of Earth's recent epochs, the **Pleistocene,** is often referred to as *The Ice Age.* However, as we have already seen, Earth has undergone at least several ice ages, and we are in one today. Moreover, the term *Pleistocene Ice Age* is misleading, because the roots of the last glaciation go back considerably farther than the start of the Pleistocene. In fact, the most recent ice age had its origins some 55 MYA, when the global climate began to cool following a warm episode (Figure 16–3a). The first major ice accumulation occurred in Antarctica about 34 MYA. But the cooling was very gradual, so that even as recently as 20 MYA the climate was still warm enough that areas of forest could be found on Antarctica. By about 14 MYA, east Antarctica was glaciated; by 10 MYA, the Antarctic sheet had reached its present size. Finally, by about 5 MYA a continental ice sheet covered nearly all of Greenland; thus, one would have to say the last ice age was firmly established.

Within our ice age, climate has been anything but uniform, with numerous oscillations clearly evident (Figure 16-3b). Beginning about 2.5 MYA, these oscillations began increasing in amplitude, and about 800,000 years ago (Figure 16-3c) the amplitude increased dramatically, becoming about twice as large as in the preceding million years. These oscillations in temperature and ice cover are called **glacial/interglacial cycles.** As can be seen in the diagram, they are quite irregular. For most cycles ice volume increases slowly and then terminates rapidly in a warming event. In addition, neither ice growth nor decay is uniform; "quivering change" is a better descriptor, with short-term oscillations superimposed on the longer cycles. The last three-quarters of a million years have been dominated by cycles lasting about 100,000 years, with shorter-term quivering present on a subdued scale. Taking the last two million years as a whole, there have been about 30 cycles altogether, with associated global temperature changes of perhaps 5 °C (9 °F).

Ice volume changes have been largest in the Northern Hemisphere, with the size of ice sheets growing and shrinking by a factor of 3 or so with each glacial cycle. Although the Antarctic sheet has changed far less, Antarctic air temperature

DID YOU KNOW?

Large-scale human impact on climate might have begun 8000 years ago. Following the depths of the last glacial period, the concentration of carbon dioxide and methane rose significantly, contrary to the downward trend that would have occurred if they followed cycles observed to hold for the previous 250,000 years. The expansion of irrigated (paddy) rice cultivation and later widespread land clearing are believed to have increased these greenhouse gases enough to cause global warming of about 0.8 °C (1.4 °F), long before the Industrial Age.

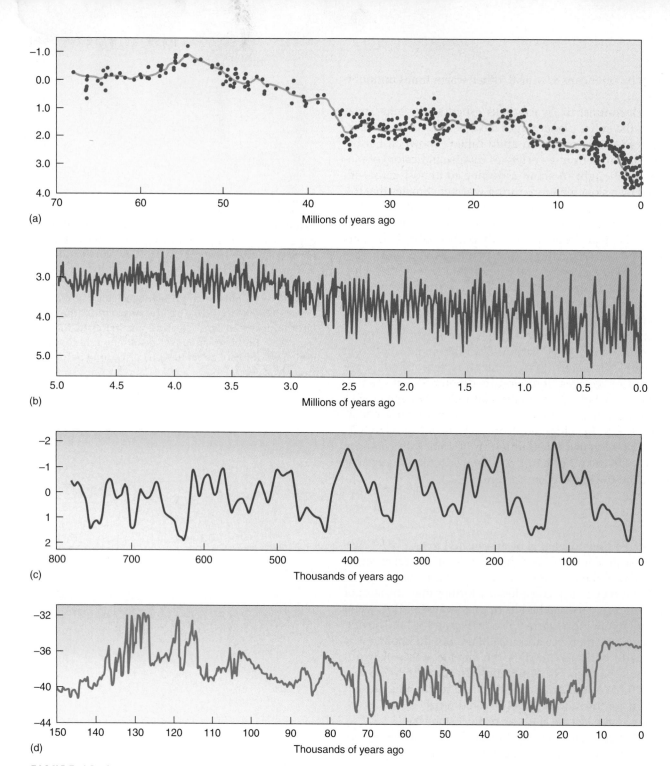

FIGURE 16-3

Indicators of climate for various parts of the last 70 million years. Curve (a) is a smoothed composite of global average ocean temperature and ice volume. Curves (b) and (c) indicate mainly global changes in ice volume. Curve (d) reflects climatic conditions over the North Atlantic but is broadly consistent with global changes, as can be seen by comparison with later portions of (c).

changes are believed to match those of north polar latitudes (about 10 °C—18 °F—cooler during a glacial). Evidence from mountain snow lines suggests that other Southern Hemisphere locations change temperature as much as their northern counterparts between glacial and interglacial times, and there is other evidence implying that the timing of major warming and cooling events have been generally in step over the last 150,000 years. When the two hemispheres are out of phase, it appears that the Southern Hemisphere leads the Northern Hemisphere by just a little more than 1000 years. But even accepting these differences in timing, we must conclude that the Southern Hemisphere, as a whole, has participated in glacial/interglacial cycles just as the rest of the planet has.

As is clear from Figure 16-3, the planet is now in a warm interglacial, rivaled only a few times in the last 2 million years. Interestingly, one of those times was the last interglacial, which reached its peak about 125,000 years ago and might have set the record for Pleistocene warmth. Global sea level was about 6 m (20 ft) higher than now, and there is evidence that midlatitude continental areas were 1°C to 3 °C (2 °F to 5 °F) warmer. In contrast, sea surface temperatures were not too different from what they are now. Between these two warm periods sits the most recent glaciation, an event that reached its maximum about 20,000 years ago.

THE LAST GLACIAL MAXIMUM

Following the last interglacial, ice volume increased, but not uniformly. There were two main pulses of glaciation, one about 115,000 years ago, and another about 75,000 years ago. It seems that most ice was added to polar caps during the first pulse and to ice caps in North America and Eurasia during the later pulse. In the depths of the last glaciation, around 20,000 years ago, many aspects of the Earth–atmosphere system were different. Most dramatically, of course, land ice covered much more area, as seen in Figure 16–4.

FIGURE 16–4
Map showing the maximum extent of ice at time of last glaciation.

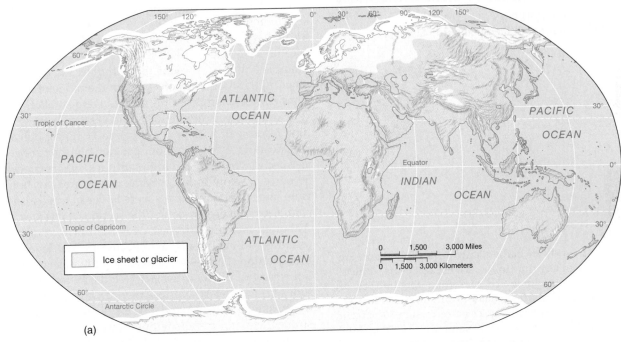

(a)

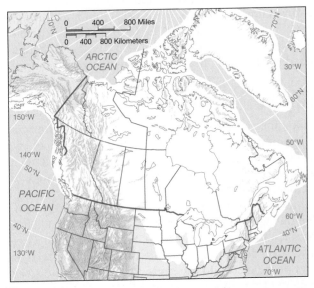

(b)

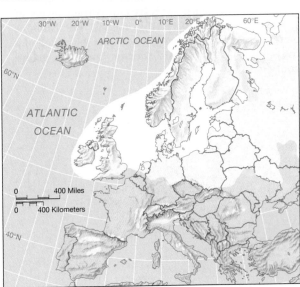

(c)

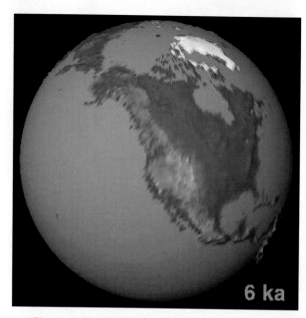

6 ka

WEATHER IN MOTION
Retreat of Continental Ice Sheets

In North America, it is certain that ice reached about as far south as present-day St. Louis, but only to the latitude of New York and Seattle on the East and West Coasts. Paradoxically, there is considerable doubt about the polar boundary of the northern sheet, with some arguing that much of the Arctic rim was unglaciated. Regardless, tremendous quantities of water were transferred from ocean to land, building sheets 3500 to 4000 m thick. Given enough time, this would be enough to depress the continental crust by more than 800 m. When the ice melted, the land surface gradually expanded upward toward its original level. (Even now, continents have yet to fully rebound from glacial depression.) The Laurentide ice sheet in North America was in some ways equivalent to a huge mountain range, running from the Rocky Mountains to the Atlantic. Sea level was about 120 m (394 ft) lower than it is now, so that a land bridge existed between Siberia and Alaska. (At equilibrium, movement of water from the ocean would cause oceanic floors to rise by about 35 m—115 ft.) There were also significant changes in sea ice, especially in the Antarctic Ocean, where winter sea ice covered about twice the area it now does. Of course, the sea ice changes had little effect on sea level, because water did not move between the land and ocean reservoirs.

On land, temperature changes varied greatly by proximity to the ice sheets and to the ocean. For example, in western North America, maritime air masses kept temperatures to within 4 °C to 5 °C (7 °F to 9 °F) of modern values. This contrasts sharply with the area that is now Tennessee and South Carolina, where temperatures were 15 °C to 20 °C (27 °F to 36 °F) colder than they are now. These two examples might represent the extremes for midlatitude changes—a number of other midlatitude locations were in the range of 5 °C to 8 °C (9 °F to 14 °F) cooler. Temperature changes in the tropics were smaller, perhaps 4 °C to 5 °C (7 °F to 9 °F). Snowline was about 1000 m (3300 ft) lower, which translates into a temperature decrease of 5 °C to 6 °C (9 °F to 11 °F) for elevations above 2000 m (6600 ft).

Most places were not only colder, but they seem also to have been drier. This is especially true for the high latitudes, where precipitation amounts were about 50 percent below today's. Some desert areas of both South America and Africa were larger, and lake levels were lower in tropical Africa and Central America. A cold desert covered much of the section of western Europe that wasn't under ice. Although increased dryness was the norm, a number of wetter areas could be found on all continents, as reflected in lake level changes and other indicators. Precipitation changes were undoubtedly determined to a large degree by circulation changes. Where prevailing winds shifted such that they blew from high latitudes (as for eastern North America), conditions were drier. What little evidence exists suggests that wind speeds were greater, as might be expected with a stronger equator-to-pole temperature gradient.

It must be emphasized that this glacial period was hardly uniform. In fact, abrupt climate changes were very common throughout this period, with polar temperatures changing by 5 °C to 8 °C (9 °F to 14 °F) over the course of just decades to centuries (Figure 16-3d). Because changes such as these are not confined to the glacial period, they are covered in more detail in a later section.

THE HOLOCENE

Like other glacial-to-interglacial transitions, warming following the last glacial maximum was fast, at least compared to times of cooling. Warming began about 15,000 years ago (prior to the start of the **Holocene**), only to be interrupted about 2000 years later when colder conditions returned. The cold event was most extreme (almost glacial) in the North Atlantic area, but it can be found in records as far away as Antarctica. Called the **Younger Dryas,** it lasted for about 1200 years. Following the Younger Dryas, starting about 11,800 years ago, came another period of abrupt warming, with temperatures in Greenland increasing by about 1 °C/decade, bringing climate into the interglacial we enjoy today. Spanning the end of the glacial to the mid-Holocene was the so-called **African Humid Period** running from about 15,000 to 5000 years before present (BP). Northern Africa was nearly completely vegetated, with rainfall far above what is seen today. Savanna landscapes were abundant, including lakes: The remains of antelope, giraffe, elephant, hippopotamus, crocodile (and humans) have been found in places that now have virtually no measurable precipitation. The transitions in and out of this wet period were very abrupt, requiring a few hundred years or less.

Two other significant events mark the Holocene. First, the early warmth was interrupted by yet another abrupt and short cooling 8200 years ago. Temperatures fell rapidly throughout the Northern Hemisphere Atlantic region and remained low for 300 to 400 years, with sea surface temperatures in the subtropical Atlantic suppressed by 7 °C to 8 °C (13 °C to 14 °F). Second, there was a long, severe drought in mid-continental North America from perhaps 4300 to 4100 years BP. Water tables fell, sand dunes became active, wildfires increased in frequency and/or intensity, and there were widespread changes in forest vegetation. Droughts lasting decades are common for this part of the world in the Holocene, but the 4200 BP drought stands unmatched for duration and depth of dryness. Many other Northern Hemisphere locations experienced persistent drought at this time, including the Middle East, northern Africa, the Mediterranean, and southern Europe. Despite limited data, it appears that much of the low- to midlatitudes of the Northern Hemisphere experienced severe drought at the same time, suggesting that this was a hemispheric event.

Global mean temperature during middle Holocene was about average for the epoch (meaning a little cooler than present). Northern Hemisphere summers were warmer, and tropical locations somewhat cooler. Since then global temperatures have been modulated by a number of events of various durations and expressions, depending on the time and place. For example, there is evidence that the period A.D. 900–1200 was warm in the North Atlantic. Called the **Medieval Warm Period,** it coincides with the Viking settlement of Greenland. Mountain glaciers in Europe advanced before and after, but not during, this time, and there is evidence for glacial retreat elsewhere as well (including the Canadian Rockies). But there is also much physical and historical information indicating that this was not a global event of any significance.

Less ambiguous is another celebrated event, the so-called **Little Ice Age.** Spanning from perhaps 1400 to 1850, this was a cold period for western Europe. During these years, alpine glaciers advanced as temperatures fell by about 0.5 °C to 1 °C (0.9 °F to 1.8 °F). Historical records indicate that this seemingly small decrease in mean temperature had a considerable effect on living conditions throughout Europe. Shortened growing seasons led to reductions in agricultural productivity, especially in northern Europe. In contrast to the Medieval Warm Period, the Little Ice Age is expressed in mountain records from around the world. Although in no sense is it a true "ice age," it does represent the largest temperature change during historical times and is considered a global event.

Of course, there have been other climate changes on smaller scales involving both precipitation and temperature. Consider, for example, Figure 16–5, which shows moisture conditions for coastal Virginia/North Carolina from A.D. 1200 to the present. Evident are numerous excursions from average conditions, some persisting for decades. It is particularly interesting to note that the driest three years of the entire 800-year period coincide with the disappearance of the Lost Colony of Roanoke sometime after August 1587. Similarly, the drought of 1606–1612 brought the driest 7-year period of the record, by happenstance coinciding with establishment of the Jamestown colony in 1607. American readers will recall that the Jamestown colony suffered tremendous mortality (60 percent died in the first year) and was nearly abandoned, in large part because of malnutrition. Standard explanations for these catastrophes mention poor planning, mismanagement, bad relations with native societies, and similar social factors. In light of the extreme droughts, one must guess that environmental factors were at least as important.

Within the Holocene, considerable attention has been devoted to understanding the last 1000 years, in large part because it sets the context for presumed global warming occurring in the last hundred or so years. There have been a number of climate

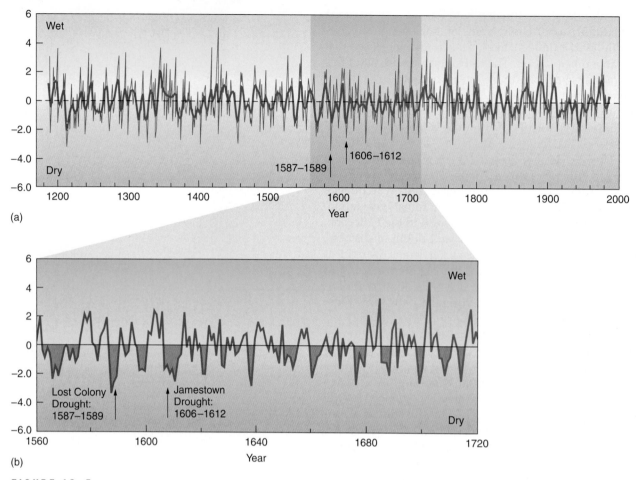

FIGURE 16–5

A longtime series representing July moisture conditions for the Tidewater region of Virginia and North Carolina.

reconstructions for this as seen in Figure 16–6. The latest (published in 2005) used what are believed to be more reliable indicators and more advanced data analysis methods. Compared to the others, it suggests ample warmth for the Medieval period followed by a more intense Little Ice Age than most other curves imply.

During the middle of the last century, a growing network of meteorological stations was established around the world. Though there are many problems in the data set, such as the movement of meteorological instruments at a given site, the data give firsthand accounts of temperature and precipitation patterns. As described in *Box 16–2, Focus on the Environment: Historic Warmth in the 1990s and Early 2000s,* there have been two episodes of general warming during the century separated by a two-decade period without a discernable trend.

There is a very important question regarding the rising temperatures of the last few decades. As we have seen, changes in global climate occur on a wide variety of time scales, each with its own range of variation. We must now ask if this recent rise in temperature is just part of the natural variability in climate, or if it marks the onset of human-induced warming from the emission of "greenhouse gases" into the atmosphere. At present, it is impossible to conclude one way or the other with any certainty. But most atmospheric scientists believe that this warming may be largely the result of increases in greenhouse gases, and others are now fully convinced.

MILLENNIAL-SCALE OSCILLATIONS

The preceding sections mostly took a sequential view, tracking climate through time. Another way of looking at climate is to focus on individual time scales of change, without regard to the state of the system (warm vs. cold, etc.). When this is done, persistent oscillations emerge for every time period examined so far, going back 500,000 years. The oscillations are called *millennial-scale* because they appear at intervals of roughly 6000, 2600, 1800, and 1450 years. During the most recent glacial period, the oscillations have been quite similar to one another; each begins with a rapid increase in temperature, taking just a few decades to centuries to develop. Air temperatures over Greenland increase by 5 °C to 8 °C (9 °F to 14 °F), and North Atlantic sea surface temperatures (SSTs) increase by about 3 °C (5 °F). Temperatures remain high for 1000 to 2000 years, after which they return quickly to former low values. Thus, rather than rounded cycles, these oscillations are more like square waves,

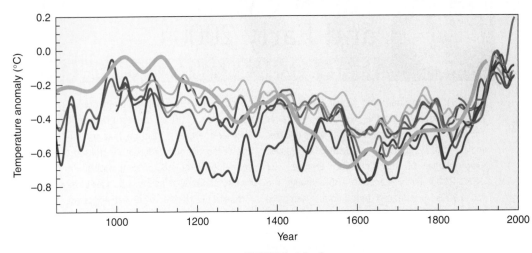

FIGURE 16–6

Temperature reconstructions for the last 1000 years, expressed as departures from the 1961–90 average. The heavy pink line is the latest, published in 2005.

with climate jumping between two states. The cycles appear in clusters, as shown schematically in Figure 16–7. Each flip-flop is a *Dansgaard-Oeschger* (D-O) cycle, thought to arise from circulation changes in the North Atlantic ocean. There is a progressive cooling from one D-O cycle to the next, ending in an especially cold period called a *Heinrich event*. Together the multiple D-O cycles form a single *Bond cycle* of about 5000 years' duration. These cycles, which are named after the researchers who discovered them, are revealed by different indicators and have somewhat different causes.

Oscillations of similar length occurred during other glacials, as well as during interglacials and transitions between glacials and interglacials. Oddly, they are largest during interglacial-to-glacial transitions, with SST changes in the range of 4 °C to 4.5 °C (7 °F to 8 °F). During the interglacials, the oscillations are just as persistent, but much smaller, with North Atlantic SST warming and cooling by about 0.5 °C to 1 °C (1 °F to 2 °F). They appear throughout the Holocene and are expressed most recently as the Little Ice Age cooling.

Millennial-scale oscillations are significant because they suggest that the Earth–atmosphere system has a tendency to flip back and forth between warm and cold states, independent of whatever the long-term climate is doing. Thus, rather than a regular progressive change that might follow changes in some boundary condition, it's as if the system suddenly reorganizes itself and moves quickly to another mode. No one knows why this happens, although a widely held view is that it arises from processes internal to the system, most likely involving feedbacks between atmosphere and ocean. The alternative would call for something external to turn on and off at more-or-less regular intervals, with effects that are amplified during transitional and glacial times. Regardless of the cause, the size and speed of change is a matter of some concern. If the planet really does have this bipolar behavior, a switch from the present warm state to the other could have profound effects on many social systems.

Before concluding this section, we must stress again that climatic changes are not restricted to mean values; there can also be changes in the frequency of rare events. Thus, for example, computer models suggest that many areas may experience an increase in the incidence of heavy precipitation and drought in association with an increase in temperature. A recent study has shown that just such

FIGURE 16–7

Idealized millennial-scale climate cycles. Four Dansgaard-Oeschger cycles of increasing intensity terminate in a Heinrich event. The entire sequence forms a single Bond cycle. The time axis covers an arbitrary period of about 7000 years, with the Heinrich event appearing as the last of five oscillations.

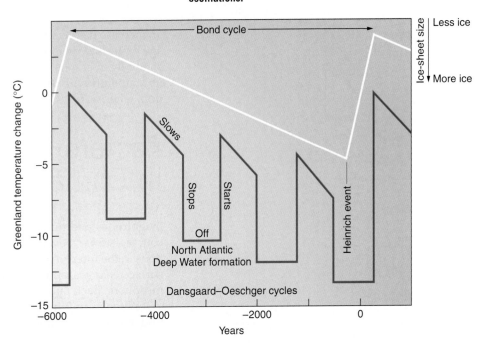

There is no question that global temperatures have increased during the twentieth century, as shown in Figure 1. The values on the vertical axes indicate the difference in temperature for a given year relative to the "climatic average" for 1880–2005. Every year from 1880 through the mid-1920s experienced below-normal temperatures, but temperatures began to increase at about 1900. Following the late 1930s, most years had above-normal temperatures, with marked warming beginning in 1975. Over the entire period global surface temperatures increased at a rate of 0.6 °C (1.1 °F) per century, but during the last thirty years the rate of increase was much greater, at 2.0 °C (3.6 °F) per century.

The last decade of the twentieth century and the first 5 years of the twenty-first century were remarkably warm, with 12 of the 13 warmest years on record all during that period. The most recent, 2005, was either the warmest year or a close second to 1998. The extreme warmth of 2005 is all the more impressive considering that the 1998 warmth got a boost from a strong El Niño. The overall warmth of the last 15 years would have been even greater were it not for the eruption of Mt. Pinatubo in the Philippines in 1991. Stratospheric aerosols from the eruption reduced the amount of solar radiation reaching the surface of the Northern Hemisphere during

1992 and 1993, making them the 2 coolest years of the 1990s (though still warmer than the long-term average). Notice that although land and ocean temperatures both increased over the entire period of record, temperatures over land rose by about twice as much as those over ocean.

Human health threats due to rising temperatures may be more serious than temperature data alone would suggest. A recently published study looked at trends in both temperature and apparent temperature (based on the combination of temperature and humidity, as described in Chapter 3) in the United States between 1949 and 1995. The study found that the incidence of extreme apparent temperatures has increased more dramatically than the incidence of extreme temperatures alone. To make matters worse, the increase in nighttime apparent temperature has been greater than the increase in daytime apparent temperatures, meaning that the high apparent temperatures, instead of dropping at sunset, persisted overnight. This is important because experts believe that the most dangerous health threat occurs not from very high daytime temperatures, but from the persistence of high apparent temperatures for several days without intervening cool nights. The results of the study are even more noteworthy when one takes into account that data from the record warm years of 1997 through 1999 were not included.

The big question with regard to warming is whether it is largely the result of human activities or natural variation. While the answer is not known for certain, each new warm year lends further support to the general circulation models' predictions of a warmer atmosphere in response to increasing concentrations of greenhouse gases. In 2001 the Intergovernmental Panel on Climate Change (IPCC) issued its third assessment report, which summarized the current knowledge about global warming. This prestigious panel of scientists, under the auspices of the World Meteorological Organization (WMO) and the United Nations Environment Programme, stated in its report that "the warming over the past 100 years is very unlikely to be due to internal variability alone . . . ." It further stated that "most of the observed warming over the last 50 years is likely to have been due to the increase in greenhouse gas concentrations." In June 2002, the U.S. Environmental Protection Agency (EPA) further concluded that warming over the last several decades is "likely due mostly to human activities . . . ." The report added that even if greenhouse emissions were to be reduced in upcoming years, the effects of their increase in the atmosphere over the last few decades would not soon be reversed.

an increase has occurred over North America during the last century. The percentage of annual precipitation accounted for by heavy rainfall events has risen substantially over most of the United States in conjunction with the observed increase in temperature. Extreme events now yield about 12 percent of annual precipitation, whereas in 1910 they accounted for about 9 percent.

Factors Involved in Climatic Change

As we have seen, Earth's climate has undergone significant changes of varying magnitudes and time scales over the course of its existence. The big question, of course, is *why?* Several possible causes are easy to identify. These include variations in the intensity of radiation emitted by the Sun, changes in Earth's orbit, land surface changes, and differences in the gaseous and aerosol composition of the atmosphere. Each operates on a different time scale, as we will now see.

While examining the list of factors effecting climate change, it is important to understand that many of them do not operate independently. This means, first of

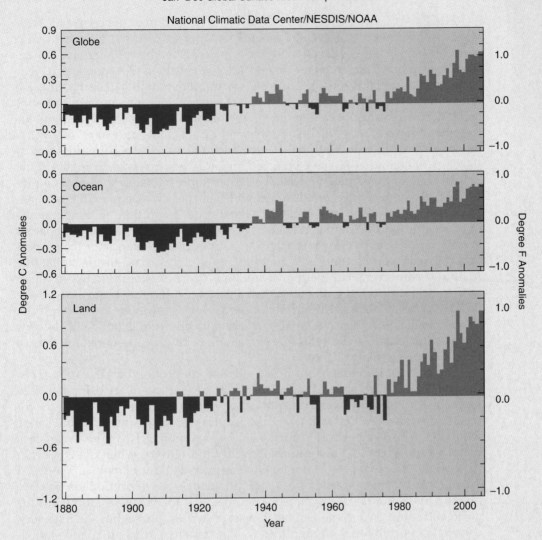

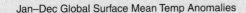

Jan–Dec Global Surface Mean Temp Anomalies

National Climatic Data Center/NESDIS/NOAA

FIGURE 1

Average global air temperatures from 1880 to 2005 for the entire Earth surface (a), over the oceans (b), and over land (c).

all, that they operate simultaneously. Thus, for example, while one agent might be leading to warming, another might counteract or enhance that warming. Second, it means that agents of change might interact with one another, so that the effects are not merely additive. For example, the effect of tropospheric aerosols produced by humans might differ depending on whether or not they are overlaid by a plume of stratospheric volcanic aerosols. The issues of simultaneous and interacting factors will become especially evident in the discussion on changes in Earth's orbital characteristics.

VARIATIONS IN SOLAR OUTPUT

As we implied earlier, Earth's climate is quite sensitive to the Sun's output. Considering that the amount of energy emitted by the Sun is not truly constant, this mechanism of climate change has considerable theoretical appeal. For example, some changes in the solar output, on the order of 0.1 to 0.2 percent, appear to be related to the occurrence of sunspots. As mentioned in Chapter 2, sunspots are relatively cold regions of the photosphere, about the size of Earth's diameter. The abundance of sunspots rises and falls on several time scales, including the very striking 10.7-year cycle. (As is customary, we will refer to this as the 11-year cycle.)

Satellite measurements show that when measured over the course of a few weeks, solar radiation decreases as sunspots increase. This is consistent with sunspots being cold—when more of the Sun is covered by cold regions, less radiation is emitted. But at longer time scales, such as those of a complete 11-year cycle, increasing sunspots are correlated with more radiation. Clearly, at these longer time scales, there must be solar changes that compensate for the increased area of sunspots. (The most likely explanation is an increase in surrounding bright areas.) The link between sunspot activity and solar output has led to considerable speculation that such phenomena could account for some of the climatic changes that have occurred on Earth. For example, droughts in the Great Plains of the United States have shown some tendency to recur at an interval that roughly corresponds to a double sunspot cycle, and earlier research has noted similar periodicities in Nile River flows. In addition, air temperatures over eastern North America rise and fall by about 0.2 °C (0.4 °F) in apparent synchronicity with the 11-year cycle. Some supporting evidence for the connection between climate and solar activity is also given by the fact that the **Maunder Minimum,** the period of minimal sunspot activity between about 1645 to 1715, coincided with one of the coldest periods of the Little Ice Age (see *Box 2–2, Physical Principles: The Sun,* on page 40). However, there have been other episodes in which variations in sunspot activity did not coincide with changes in climate, and alternative explanations have been offered for the apparent 22-year climate changes.

The equivocal evidence for a Sun–climate connection became stronger in the late 1980s, when several scientists observed that the relationship between tropospheric conditions and sunspot activity was much stronger when the direction of stratospheric winds over the tropics was taken into account. These winds tend to reverse their direction in approximately two-year cycles in a pattern known as the **quasi-biennial oscillation (QBO).** When the QBO is in its west-to-east mode, for example, there appears to be a relationship between the number of sunspots and winter conditions over northern Canada. Surface pressure rises and falls with sunspot number, and the mean storm track shifts north and south. When the QBO is in its east-to-west phase, however, no such connection is evident. Although these associations are intriguing and statistically very strong, no causal mechanisms have been proven to explain these relationships.

Somewhat stronger evidence for an Earth–Sun connection comes from consideration of millennial-scale variations in solar output and climate. Data for the last 12,000 years shows a definite 1500-year cycle in solar radiation that matches well with debris deposited on the floor of the North Atlantic by icebergs. This has led some researchers to conclude that solar forcing underlies at least the Holocene portion of the North Atlantic's millennial oscillation.

On an entirely different time scale, it is believed that the rate of solar output from the Sun has increased by about one-third since the formation of the solar system. This brings up an interesting paradox because, as we saw earlier in this chapter, Earth is believed to have been warmer than present throughout most of its early history. This counterintuitive association between a less radiant Sun and a warmer Earth is called the "early faint Sun paradox." Two very different explanations of the paradox are currently in vogue. One calls for a massive CO_2 greenhouse effect, with an early atmosphere having CO_2 partial pressures up to ten times the current total surface pressure. Organic molecules are much more difficult to generate in such an atmosphere; thus, this idea makes the appearance of life hard to fathom. In addition, there is geologic evidence that says such levels were never reached. Thus, an alternative view is popular as well, one calling for elevated ammonia levels creating an early greenhouse. Disputed for many years on the grounds that ammonia would be broken up by ultraviolet radiation, this was considered an implausible explanation. However, in 1997 it was shown that high-altitude shielding by other gases may have allowed ammonia to accumulate at lower levels. Regardless of which (if either) hypothesis is correct, it's interesting that the Sun–climate connection runs from the extreme of wondering whether there is any effect at all (sunspots) to wondering why the effect was formerly so weak (early faint Sun).

CHANGES IN EARTH'S ORBIT

In Chapter 2 we saw that the seasons occurred primarily because of the tilt of Earth's axis relative to the Sun. If we imagine a plane on which Earth makes its revolution around the Sun, we can see that the axis of rotation is oriented 23.5° from the perpendicular to the plane (that is, it has an obliquity of 23.5°). The orientation of the axis is constant through the course of the year, so no matter where Earth is relative to the Sun, its axis points toward the North Star, Polaris. For the six months following the March equinox, the Northern Hemisphere is inclined toward the Sun, and during the rest of the year the Southern Hemisphere has a greater exposure to the Sun. This is the primary factor in causing the seasons. Obviously, if the axis of rotation were greater than 23.5°, this effect would be stronger and lead to a greater seasonality. Likewise, the effect would disappear if the axis was exactly perpendicular to the plane of the orbit.

We also saw that Earth's orbit is elliptical, rather than circular, so that on about January 4 Earth is about 3 percent closer to the Sun than on July 4 (see Figure 2–10). This change in Earth–Sun distance causes the planet as a whole to receive about 7 percent more solar radiation at the top of the atmosphere in early January (perihelion) than during early July (aphelion). It should be readily apparent that a greater eccentricity would result in greater differences in incoming radiation available at the top of the atmosphere during the course of a year.

As far as the Northern Hemisphere is concerned, the Earth–Sun distance is largest during the summer and smallest during the winter, causing winters to be somewhat warmer and summers to be cooler than they otherwise would be. Thus, not only is the amount of eccentricity important with regard to seasonality, but so is the timing of the minimum and maximum Earth–Sun distances with respect to the equinoxes and solstices (as will soon be explained).

In sum, there are three astronomical factors that influence the timing and intensity of the seasons: eccentricity in the orbit, the tilt of Earth's axis off the perpendicular to the plane of the orbit, and the timing of aphelion and perihelion relative to the timing of the equinoxes. As it happens, these three factors all change slowly over time on a variety of time scales.

ECCENTRICITY

The **eccentricity** of Earth's orbit changes cyclically on several time scales, with a cycle of about 100,000 years being especially prominent. Though the Earth–Sun distance at aphelion is currently about 3 percent greater than at perihelion, the relative distance has varied between about 1 and 11 percent over the last 600,000 years. Over about the last 15,000 years, there has been a steady decrease in eccentricity, which will continue for about another 35,000 years.

OBLIQUITY

The tilt of the Earth's axis, **obliquity,** also varies cyclically, but with a dominant period of about 41,000 years, during which it varies between 22.1° and 24.5° off the perpendicular. While the range of the axis tilt may seem small, it is capable of producing substantial differences in summer and winter insolation. In particular, high-latitude regions can undergo changes in available solar radiation at the top of the atmosphere about 15 percent due to variations in obliquity. The most recent peak in obliquity occurred roughly 10,000 years ago. Thus, we are about midway in the half cycle from maximum to minimum obliquity.

PRECESSION

Though the summer solstice for the Northern Hemisphere currently occurs near the time of aphelion, this changes through time because the axis wobbles on a 27,000-year cycle. In other words, the axis of rotation gyrates so that in about

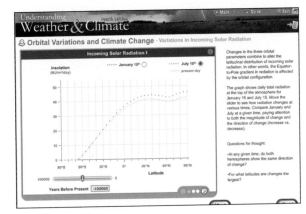

TUTORIAL
Orbital Variations

FIGURE 16–8
Precession.

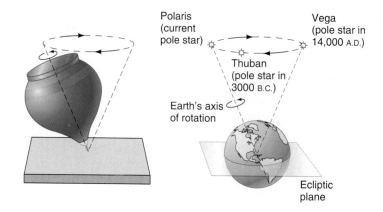

13,500 years it will point to a different star, Vega, instead of Polaris (Figure 16–8). This change in orientation of the Earth axis, called **precession,** directly alters the timing and intensity of the seasons. Combined with changes in the orientation of the elliptical orbit, the result is a 23,000-year cycle in radiation. If orientation of the axis toward Vega were to exist today along with the current timing of aphelion and perihelion, the winter solstice for the Northern Hemisphere would nearly coincide with aphelion. The resultant increase in seasonality would cause warmer summers and cooler winters in the Northern Hemisphere. At the same time, the Southern Hemisphere would experience less seasonality because its summer solstice would occur near aphelion.

It is important to note that the importance of precession on influencing radiation receipts depends on the magnitude of the eccentricity of the orbit. Low values of eccentricity (nearly circular orbits) mitigate the importance of precession; greater eccentricity amplifies it. Given the current trend toward decreasing eccentricities, we can expect that this effect will be relatively small over the next 50,000 or so years.

These three cycles are collectively called the **Milankovitch cycles,** in honor of the early twentieth-century astronomer who expounded on their potential influence on Earth's climate. Most scientists believe that the Milankovitch cycles have played an important role in the expansion and retreat of glaciers during the Pleistocene because of the way they work together to influence seasonality. Large glaciers, such as those currently occupying most of Greenland and Antarctica, are most likely to expand when seasonality is low. With less seasonality, warmer winter temperatures foster a greater amount of snowfall over much of the ice sheets due to a greater availability of water vapor. Cooler summers also promote glaciation because the rate of melt along the margins of the ice sheets is slowed.

Considerable observational evidence for the role of Milankovitch cycles is seen in the climate record of the Pleistocene. Using radiation in the northern middle latitudes as a surrogate for orbital forcing, there is good agreement between the timing of ice advances and retreats. In addition, when the variability in the climate record is broken down according to various time scales, the three main Milankovitch periods emerge as containing most of the climate variation. There is also some understanding of the pathways by which orbital forcing influences climate. For example, computer modeling suggests that increased summer radiation during the early Holocene led to enhanced monsoon circulations, especially in North Africa. These would have brought increased moisture to the continent, as is observed in the climate record. For these and other reasons, changing orbital parameters are widely accepted as driving glacial/interglacial cycles. To the extent that episodes of glacial advance and retreat are affected by Milankovitch cycles, the scenario is for us to return to another episode of glacial advance sometime during the next few millennia. But the Milankovitch theory is not without its problems, chief of which is one that concerns the eccentricity cycle. The 100,000-year cycle has the smallest effect on radiation reaching the planet but emerges as the strongest in the climate record. If this is the true cause, processes yet to be fully explained must amplify the relatively small radiation cycle. A related question concerns the observed amplitude increase in the 100,000-year climate cycle. If eccentricity cycles are an explanation, why did they suddenly become more important about 800,000 years ago? Finally, there is the question of precession, which is out of phase between the Northern and Southern Hemispheres. Precession effects that would cause cooler conditions in one hemisphere should produce the opposite response in the other hemisphere. Yet, as we have seen, climate changes have been similar in both hemispheres, not opposed to one another.

CHANGES IN LAND CONFIGURATION AND SURFACE CHARACTERISTICS

Many climatologists believe that climatic changes occurring over the longest time spans were at least partly in response to changes in the size and location of Earth's continents. Support for an older view, that continental drift was the primary forcing variable, has eroded in the face of quantitative estimates of change obtained from computer simulations. The breakup of **Pangaea** (the early supercontinent) and the slow movement of the resultant continents undoubtedly caused major climatic changes, even if not as large as observed in the geologic record. Such would have to be the case, of course, because all the factors that affect temperature and other climate variables (such as latitude and continentality) were themselves greatly affected by the movement of the continents. Though dramatic, the climatic changes resulting from continental displacement would be extremely slow. Like changes in position, episodes of continental mountain building have almost certainly produced significant climate change. For example, computer modeling suggests that the presence of large mountain regions (the Rockies, Himalayas, and Andes) would amplify Rossby waves during the winter season and promote enhanced monsoon circulations in the summer, both of which are consistent with observational data.

At shorter time scales, modification of Earth's surface, especially by human activity, can greatly influence the disposition of solar radiation. One such activity is deforestation, in which large tracts of land are cleared of trees. The loss of vegetation reduces evapotranspiration from the surface. This in turn leads to higher temperatures near the surface, as the amount of energy channeled into the latent heat of evaporation is reduced and also decreases precipitation. In addition, decomposition of cleared vegetation directly increases atmospheric CO_2, an important greenhouse gas. Compounding this is the loss of vegetative surfaces formerly providing photosynthesis, a process that removes CO_2 from the atmosphere. Even if reforestation occurs, the conversion of mature to immature trees still reduces the rate of CO_2 removal from the atmosphere.

Alteration of arid and semi-arid land surfaces by the overgrazing of cattle may also lead to changes in the regional climate. Soil compaction associated with vegetation can increase the amount of runoff, thus making less water available for evaporation into the atmosphere. Also, one hypothesis suggests that the removal of vegetation leads to an increase in the surface albedo. As the albedo increases, according to this line of reasoning, the reduction in the energy absorbed by the ground causes cooling of the surface and a reduction in the environmental lapse rate. Because a lowered lapse rate would make the air more stable and less susceptible to convectional precipitation, overgrazing could enhance the vulnerability of these regions to drought. On the other hand, the reduction in vegetation due to overgrazing could have the opposite effect, in which a reduction in evaporation leads to a general warming of the surface and an increase in the environmental lapse rate. Of course, the instability associated with the increased lapse rate will have little effect if there is not enough water available to yield precipitation. Yet another effect arises from the change in surface roughness, which alters the momentum transfer between atmosphere and ground.

CHANGES IN ATMOSPHERIC TURBIDITY

Atmospheric **turbidity** refers to the amount of suspended solid and liquid material (aerosols) contained in the air. Some aerosols are released into the atmosphere through natural processes, such as large-scale volcanic eruptions; others are released by human activity, such as smokestack emissions. Aerosols can enter the atmosphere directly as in the examples just named. They can also enter the atmosphere indirectly as a result of chemical processes in which certain gases—usually sulfates, but also nitrates and hydrocarbons—react in the presence of sunlight to form solid and liquid aerosols. Some aerosols are no more than tiny clumps of molecules, whereas others are millimeters across. This is a huge range in size, roughly comparable to the difference between tennis balls and planets. Regardless of their source, composition, or size, however, prolonged variations in aerosol contents can have important ramifications for climate.

Existing both in the stratosphere and the troposphere, aerosols directly affect the transmission and absorption of both solar and infrared radiation. By absorbing incoming sunlight they can cause a heating of the atmosphere around the aerosols, but they can also increase the amount of backscattering and thereby reduce the amount of radiation reaching the surface. The relative effects of aerosol absorption vs. backscatter are hard to assess and depend on numerous factors, including the albedo of the aerosols and the underlying surface.

To a lesser extent, aerosols can increase the absorption of outgoing longwave radiation that would otherwise escape to space and thereby increase the longwave radiation radiated from the atmosphere to the surface. In this way, they can have the direct effect of increasing nighttime temperatures.

Aerosols can also affect climate indirectly, through their ability to serve as cloud condensation nuclei. A cloud with a greater number of condensation nuclei might contain as much liquid water as one with few nuclei, but it will have more droplets of mostly smaller sizes. Clouds containing a large number of small droplets are less likely to yield precipitation and, therefore, are likely to persist for longer periods. In addition, the total droplet surface area of such clouds is greater, and this greater surface area increases the reflectivity of the cloud (assuming the same amount of liquid water). Thus, tropospheric aerosols might give rise to more extensive, longer-lived, brighter clouds.

Support for the indirect effect of aerosols comes from satellite images of ship trails, which are tracks of low-level clouds produced by ocean-going vessels (Figure 16–9). In the relatively clean ocean atmosphere, aerosols enhance condensation, creating long cloud trails that stand out against the surrounding clear sky. With their high albedo, these clouds reduce the amount of solar radiation reaching the surface. Additional support for indirect aerosol effects comes from a recent study of pollution tracks on land, which appear on satellite images as plumes of brighter clouds downwind of cities. By analyzing data from a number of sensors, it was discovered that aerosols substantially reduce precipitation in downwind areas by suppressing both ice formation and droplet coalescence. (Recall that these are important precipitation growth mechanisms.)

Moving beyond these generalities, let us now look in more detail at the climatological effects of tropospheric and stratospheric aerosols as agents of climatic change.

TROPOSPHERIC AEROSOLS

Natural sources of tropospheric aerosols include the spraying of salt particles by ocean waves and bubbles, soot and gases from fires, the erosion of soil by wind, the dispersal of spores and pollen, the emission of sulfides by marine plankton, and other biospheric processes. (In some cases, the original emission is gaseous but is converted to liquid or solid.) Of course, volcanic eruptions can suddenly discharge a huge amount of material into the troposphere, but it settles and precipitates out soon after the eruption and thus exerts no long-term effects on climate (unless many volcanoes erupt sequentially).

FIGURE 16−9
Ship tracks appear as white streaks embedded in a low-level cloud deck of speckled light gray clouds. The image shows an area just offshore of the northwestern United States (colored green), with small amounts of cloud-free ocean (blue).

Although we have no direct long-term records of tropospheric aerosols, human activities have unquestionably increased their concentrations. This is especially true over industrialized land areas. Unlike some gases, however, which have very long residence times (Chapter 1), individual aerosols do not have long life spans in the troposphere. Aerosols produced over urban or industrial centers settle out before they can be widely dispersed across the globe, thereby leading to concentrations that vary widely both temporally and spatially. As a result, their climatic effects tend to be isolated to areas near the pollution sources.

Until recently it was not known whether the direct effects of tropospheric aerosol increases would lead to an overall warming or cooling near the surface. Numerical models now tell us that increased aerosol contents have the net effect of reducing surface temperatures globally. In fact, the cooling due to anthropogenic aerosols appears to have been about the same order of magnitude as the warming that would theoretically accompany observed increases in greenhouse gases over recent decades. Thus, if the addition of greenhouse gases into the atmosphere has helped promote warming during the last century, its effect may have been masked by the cooling effects of aerosols. But we cannot expect the aerosol effect to offset the greenhouse gas effect indefinitely. If there were to be a stabilization in the release of anthropogenic emissions, atmospheric CO_2 contents could take up to 200 years to reach a new equilibrium level, while aerosol levels would respond almost immediately. Thus, the greenhouse effect would continue to increase while the aerosol effect would stabilize and cease to offset the opposing effect. Despite the possibly beneficial effects of aerosols on climate change, it should be remembered that aerosols exert their own negative effects on humans and the environment (Chapter 14).

STRATOSPHERIC AEROSOLS

Unlike tropospheric aerosols, those in the stratosphere result primarily from natural processes; human activities are not as important. The stratosphere maintains a background level of aerosols introduced by the upward diffusion of sulfur gases from the troposphere, which then undergo gas-to-particle conversion. Though background levels remain fairly constant through time, significant increases can occur in the months following volcanic eruptions.

Stratospheric aerosols can remain in the stratosphere longer than their tropospheric counterparts for two reasons: First, they tend to be smaller and therefore have lower terminal velocities. More importantly, the most effective agent of aerosol removal, precipitation, is not important in the stratosphere.

As previously stated, aerosols affect the energy balance by absorbing and backscattering solar radiation and by absorbing and radiating longwave radiation. The reduction in solar radiation reaching the surface would favor a lowering of tropospheric air temperatures, while the absorption of outgoing longwave radiation and the increase in downward emission would promote surface warming. So which actually dominates in the case of stratospheric aerosols? It turns out that the relative importance of shortwave and longwave radiation changes depending on the size of the aerosols. If the stratospheric aerosols are small, the reduction in solar radiation reaching the surface exceeds the gain in longwave radiation. In fact, the aerosols *are* small and promote lower temperatures for up to several years after volcanic eruptions.

Several recent volcanic eruptions have provided scientists with useful observational data on the effects of stratospheric aerosols. Mount St. Helens in Washington underwent a major eruption on May 18, 1980. Despite the massive ejection of solid material (much of one side of the mountain was blown away), it caused little if any major impact on hemispheric weather because it released relatively small amounts of sulfuric gases, which can ultimately transform to aerosols. Though the daytime sunlight was blocked out downwind of the eruption for a few days, no effects were noted much beyond that time frame.

The April 4, 1982, eruption of El Chichón in Mexico was far more violent than that of Mount St. Helens and, more significantly, it released a particularly large amount of sulfuric gases. The result was an increase in atmospheric albedo, which decreased global temperatures by about 0.2 °C (0.4 °F) for several months. The Mt. Pinatubo (Philippines) eruption of June 12, 1991, is believed to have released about twice as much sulfuric gas as did El Chichón, along with the predictable effect of an even greater reduction in global temperatures. The eruption increased the atmospheric albedo both directly (by enhanced aerosol backscatter) and indirectly (by an increase in the reflectivity of clouds). The cooling effect was greatest by about August 1992, when the global mean tropospheric temperature decreased by about 0.73 °C (1.6 °F) from that of June 1991 (despite the fact that August is climatologically warmer than June in the Northern Hemisphere). Figure 16–10 shows the change in aerosol content over the globe prior to (a) and during four periods (b), (c), and (d) following the eruption. Aerosol contents increased 100-fold during the peak concentration.

CHANGES IN RADIATION-ABSORBING GASES

The issue of potential climatic warming due to increases in carbon dioxide and other greenhouse gases has been the subject of intense scientific scrutiny and political controversy. On both human and geologic time scales, there is little doubt that large changes in CO_2 have occurred. For example, ice cores show that CO_2 rises and falls in concert with glacial/interglacial cycles. (The CO_2 changes lag behind the ice volume changes, and thus they are believed to amplify the glacial cycles, but they are not thought to be the fundamental cause of those cycles.) On human time scales the burning of fossil fuels and the clearing of forested areas has led to a steady increase in the carbon dioxide content of the atmosphere (see Figure 1–5).

Carbon dioxide is an effective absorber of longwave radiation. Thus, increases in its content through time has the potential to exert a warming effect on Earth's climate: As CO_2 builds up, there can be an increase in the absorption of outgoing radiation and the establishment of a higher equilibrium temperature in the lower troposphere. On the other hand, the stratosphere loses energy mainly by emission of longwave radiation. Because increasing CO_2 leads to greater emission, the stratosphere (and upper troposphere) are expected to cool as CO_2 levels grow. Given the complexity of the atmosphere, it should not be surprising that there are a host of processes and conditions that are not known well enough for us to establish with certainty the exact outcomes of the buildup. What follows is an overview of some of the important components of the greenhouse gas question.

April 10, 1991 to May 13, 1991

June 15, 1991 to July 25, 1991

August 23, 1991 to September 30, 1991

December 5, 1993 to January 16, 1994

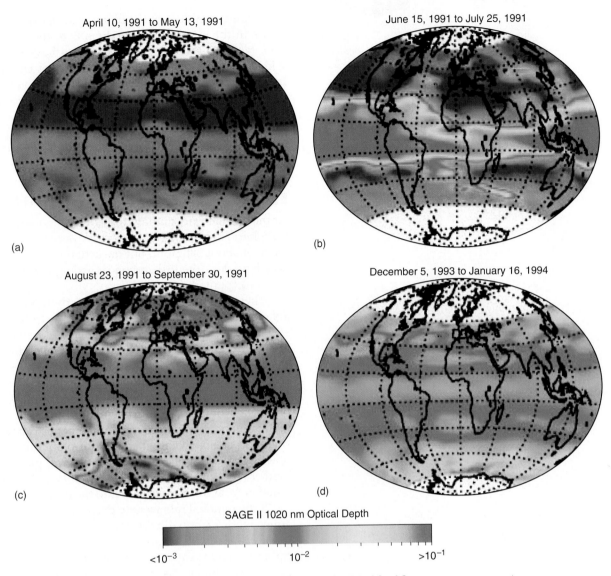

(a)

(b)

(c)

(d)

SAGE II 1020 nm Optical Depth

$<10^{-3}$ 10^{-2} $>10^{-1}$

FIGURE 16-10

The average aerosol content over the globe prior to and after the 1991 Mt. Pinatubo eruption. Redder areas indicate greater aerosol contents.

ANTHROPOGENIC CONTRIBUTIONS OF CO_2

Although much of the media discussion about the current increase in greenhouse gases centers on carbon dioxide, CO_2 is only one of several anthropogenic greenhouse gases that absorb outgoing longwave radiation. Methane (CH_4), nitrous oxide (N_2O), and chlorofluorocarbons (CFCs) are also effective absorbers whose contents have increased over historical times. In fact, only little more than half of the increase in net radiation accounted for by these gases is due to carbon dioxide. The other gases combine to produce nearly as great an effect as carbon dioxide.

As we will see, the potential for these gases to promote a global warming is only partly due to their absorption of longwave radiation. They also can promote warming through the indirect effect of increasing water vapor contents in the atmosphere. That is, if their absorption of radiation increases the atmospheric temperature, net evaporation should increase and thereby raise the water vapor content of the air. This would lead to even greater absorption, because water vapor is even more effective at absorbing longwave radiation than the anthropogenic trace gases. Though water vapor is an important contributor to any potential increase in greenhouse warming, it will be excluded from the rest of this discussion because its increased level in the atmosphere is an indirect response to human activities.

Since the middle of the nineteenth century, there has been an exponential increase in the input of carbon dioxide to the atmosphere by fossil fuel consumption. Not surprisingly, most emissions have come from the developed nations, with 90 percent originating in the Northern Hemisphere. Between 1980 and 1989, fossil fuel combustion released about 5.4 billion tons (gigatons, Gt) of carbon (plus or minus 0.5 Gt) into the atmosphere each year. This is supplemented by another 2 Gt per year resulting from deforestation. Deforestation adds CO_2 to the atmosphere through direct burning of the logged trees, decomposition of biomass, and other processes. Moreover, the removal of trees reduces the ability to remove subsequent inputs of carbon dioxide by photosynthesis.

EXCHANGE OF CO_2 BETWEEN THE ATMOSPHERE AND OCEAN

At the current rate, humans are adding enough carbon dioxide to increase the atmospheric CO_2 content by 3.5 parts per million (ppm) per year over the current value of about 380 ppm. But the actual rate of increase is only about half as large. So where does the lost carbon dioxide go? About half enters the ocean, with the rest unaccounted for.

There is a constant exchange of carbon dioxide between the ocean and atmosphere as gas molecules cross back and forth across the ocean surface. The rate of exchange depends on several variables, including wind speed, water temperature, and the difference between the CO_2 content of the atmosphere and surface layer of the ocean. Overall, the atmospheric CO_2 pressure is higher than the ocean values, so there is a net transfer into the ocean, making the oceans a net sink for the greenhouse gas. Obviously, carbon cannot accumulate indefinitely in the upper ocean, or else it would cease to receive atmospheric carbon. Photosynthesis by phytoplankton and other marine plants removes carbon, whereupon it enters the marine food chain. Eventual settling of plant and animal remains transfers carbon downward, acting like a kind of biological pump. One of the important issues in predicting future atmospheric CO_2 levels is whether the net rate of carbon dioxide removal from the atmosphere into the ocean will change in the future.

If the removal rate were to decrease, there would be an accelerated increase in atmospheric carbon dioxide levels. One estimate is that increasing ocean temperatures will lead to a future reduction in the ocean's ability to take in atmospheric carbon dioxide, causing a 5 percent increase in the rate of atmospheric CO_2 growth. But this is clearly a very complicated question, depending as it does on both purely physical processes and on the vigor of the biological pump. One can imagine that climate changes might perturb marine ecosystems in ways that overwhelm the effects of changing solubility.

As mentioned earlier, carbon dioxide is not the only greenhouse gas released to the atmosphere by human activities. Unlike carbon dioxide, several of these come from sources quite different from fossil fuel burning. The important point to keep in mind here is that collectively they are about as important as CO_2 with regard to potential climatic change (despite lesser mention in press reports). Yet, even so, changes in the concentration of these gases are just part of the puzzle of climatic change. And our ability to predict future changes is further complicated by the interconnections of different components of the atmospheric systems. These interconnections are known as **feedbacks.**

Feedback Mechanisms

Consider a simple system that consists of only two variables, and assume that changes in either one affect the state of the other. If there is a change in the first variable, it will produce a change in the second, which will in turn affect the first variable.

Some feedbacks tend to inhibit further changes in the system, while others support further changes in the system. Those that inhibit further change, called negative feedbacks, are self-regulating. In other words, when the second variable responds to initial change in the first variable, its response will suppress further change in the first. Feedbacks that amplify change in the initial variable are called **positive feedbacks.** When a positive feedback situation exists, the response of the second variable causes the initial change to grow (you can also think of this as a "snowball effect").

Of course, the Earth system involves many feedbacks, some positive, some negative. The situation is further confounded by the fact that the various feedbacks operate at vastly different rates: some are instantaneous, whereas others operate over thousands of years. In some cases relationships seem relatively straightforward, but for others it is difficult to determine if a single connection between two variables is positive or negative. With these points in mind, we will examine some of the important feedback mechanisms related to climatic change.

ICE-ALBEDO FEEDBACK

Much of Earth's surface is occupied by large continental ice sheets and floating sea ice (Figures 16–11 and 16–12). If the atmosphere were to cool along the margins of these ice masses, there could be an expansion of the ice-covered area. Similarly, any warming would likely lead to melting and a retreat of the ice margin. Because ice has a higher albedo than most other natural surfaces, an expansion of the ice would lead to a reduction in the amount of insolation absorbed by the surface, which in turn would lead to further cooling (a positive feedback). Likewise, a retreat of the ice associated with a warming climate would increase the amount of surface no longer covered by the reflective ice and, therefore, lead to enhanced warming. Thus, there is a positive feedback mechanism that favors continued warming or cooling once such a temperature trend is initiated. It is largely for this reason that middle and high latitudes show larger changes than low latitudes. Where there is little prospect for ice (the tropics), this feedback does not operate. Of course, ice advances and retreats do not continue unchecked, because other feedback mechanisms prevent the ice from totally covering the globe or from completely disappearing.

EVAPORATION OF WATER VAPOR

Another positive feedback mechanism involved in climatic change concerns the water vapor content of the atmosphere. As we pointed out in Chapter 5, more water can exist as a vapor in warm air than in cold air. If the climate were to warm, there would likely be an increase in evaporation from the surface and an increase in the atmospheric water vapor content. Because water vapor is an extremely effective absorber of radiation emitted by the surface, higher water vapor contents should increase the atmospheric absorption of thermal energy and thus amplify the initial warming. In fact, computer models tell us that water vapor feedback contributes about half of the warming associated with increasing carbon dioxide, methane, nitrous oxide, and chlorofluorocarbons.

OCEAN–ATMOSPHERE INTERACTIONS

One obvious relationship between the oceans and the atmosphere relates to the effect that warming temperatures has on sea levels. Increases in global temperatures can raise the mean sea level, primarily by the expansion of warmer waters. Increases in atmospheric temperature can also contribute to rising sea levels by causing glaciers to melt and release water back to the oceans. A recent study estimates that increasing temperatures associated with

FIGURE 16–11
Sea ice.

FIGURE 16–12
The distribution of Northern Hemisphere sea ice, as obtained by satellite based radar imagery.

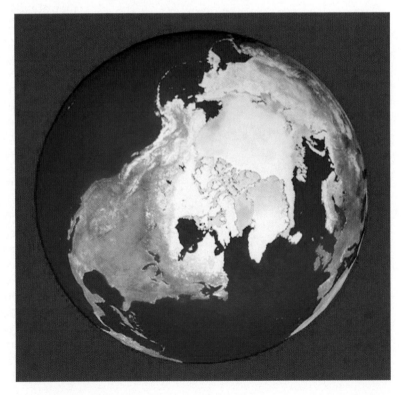

a projected doubling of carbon dioxide within the next half century could cause the average sea level to increase by 19 cm (7 in.), but the rise in sea level would not be uniform throughout the oceans, and the rise could be as much as 35 cm (14 in.) off the coast of Europe.

On longer time scales, 500 to 1000 years, melting of the West Antarctic ice sheet becomes a possibility, which would raise sea levels by 5 m (16 ft) or more. Unlike any other ice sheet, the West Antarctic sheet is a concern because its base is below sea level. It is therefore susceptible to melting by ocean waters eroding its periphery. Until recently, this idea was not widely accepted, in large part because there was no evidence that such a thing had ever happened before. In mid-1998, however, information gleaned from sediments beneath the ice sheet strongly suggested that part or all of the sheet collapsed sometime in the last 2 million years, most likely in the interglacial period 400,000 years ago. Sea level changes, whether large or small, feed back on climate patterns through a number of mechanisms, including effects on surface water currents and the movement of deep water below the surface, the proportion of land and ocean, and marine and terrestrial productivity, to name a few.

A very important feedback, only recently recognized, concerns global-scale ocean circulations. As we discussed in Chapter 8, there is a large flow of water northward in Atlantic surface waters that delivers tremendous heat to the overlying atmosphere. After cooling and sinking in the North Atlantic, water flows southward at great depth to the rim of Antarctica, where it joins deep water forming at the edge of the ice cap. Other branches of deep water flow northward into the Indian and Pacific Oceans, eventually rising and returning to the southern ocean at the surface. This huge, belt-like circulation system is partly driven by temperature differences, but also by salinity variations; hence, it is called **thermohaline circulation.**

There is now convincing evidence to suggest that this pattern is not steady but instead makes rapid jumps from one mode of operation to another. Rates and locations of sinking water change abruptly, which in turn affect other aspects of the pattern. The new configuration lasts for some time before changing abruptly again. Numerous reorganizations of such ocean circulation have been found in a marine record running from about 60,000 years ago until about 10,000 years ago. Because they appear at nearly the same times as large swings in Greenland air temperature, they are associated with the millennial-scale climate oscillations described earlier. In this way, changes in ocean circulation are implicated in abrupt climate jumps, at least during the last glacial period. For example, Heinrich events are believed to be triggered by pulses of fresh water released during melting episodes of the Larentide ice sheet. A layer of less dense fresh water at the surface shuts off downwelling, which in turn interrupts the flow of warm surface waters northward across the Atlantic. By implication, the same processes are responsible for millennial-scale changes in other periods, though that has yet to be demonstrated.

Feedback from atmosphere to ocean arises as changing climate alters patterns of runoff, precipitation, and evaporation, which in turn influence ocean temperature and salinity. In addition, changing winds and other atmospheric processes affect the distribution of sea ice, which has a huge impact on the ocean heat balance. For example, modeling suggests that removal of sea ice can account for the magnitude of the D-O warmings in Greenland. If these changes are large enough, ocean circulation shifts to a new mode, which in turn influences global climate. If true, this feedback mechanism provokes a concern that relatively modest global warming might cause another rapid reorganization perhaps 100 years from now, when world population will demand about three times today's food supply. If the resulting climate change turns out to be as large as the one 8200 years ago (which occurred when the climate was even warmer than now), the results could be disastrous.

It should be noted here that the huge mass and high specific heat of the oceans, along with their vertical and horizontal mixing, create a strong thermal inertia. This means that warming of the ocean in response to increasing atmospheric temperatures would be very slow. Thus, oceanic temperature changes would be retarded in response to global warming, and there could be a time lag of several decades before any increase in atmospheric temperature would be observed in the oceans. Once the warming is realized, it would modify atmospheric pressure and wind distributions.

Oceans are implicated as affecting climate in other ways as well. For example, there is a form of water-methane ice stored in ocean sediments at cold temperatures and high pressure. If seafloor temperature rises above a threshold, say because of changes in deep-water circulation, these methane hydrates decompose into water and gaseous methane. A sudden release of methane to the atmosphere by this process matches spikes in atmospheric methane observed with rapid temperature change of D-O cycles. Similarly, but with vastly larger amounts involved, it has been suggested that a huge release of oceanic methane led to a strong warming episode that occurred 55 million years ago. Paradoxically, very recent analysis suggests that so much carbon was added to the atmosphere (about 80 percent of all fossil fuel reserves) that methane hydrates could not be the sole source. In other words, the most recent evidence confirms the role of carbon in that warming but requires a supply larger than seemingly possible from methane hydrates.

Another oceanic process is thought to operate in the Antarctic ocean. During glacial times the deep ocean was much more stable than now, with a salty, dense layer found in bottom waters around Antarctica. Excessively briney conditions are attributed to salt left behind as water ice froze at the margin of the ice sheet. Regardless of how it formed, a dense layer like that would suppresses vertical mixing, and it is therefore certain that the glacial ocean was very stable. Furthermore, it likely would have remained especially so at times when the thermohaline circulations was shut down. But gradually, over thousands of years, geothermal heat from the Earth below would have warmed the salty bottom layer. It is thought that a temperature change of 2 °C (3.6 °F) would have been enough to destabilize the ocean, leading to a pulse of warm water and salt delivered to the upper ocean and surface. The salt pulse would restart the thermohaline circulation, which would lead to warming in the Greenland and North Atlantic area. After the pulse the ocean stabilizes, which leads to another cycle of slow deep-water warming and eventual discharge. The calculated timing for this—roughly every 10,000 years—fits with Bond cycles, as do deep ocean warmings known to have occurred before a number of Heinrich events. This hypothesis is attractive for its ability to explain the sudden warmings so characteristic of glacial times and for its explanation of why the warm periods following Heinrich events are longer and warmer than others.

ATMOSPHERE–BIOTA INTERACTIONS

Changes in climate are linked with land–vegetation patterns. The influence of climate on vegetation is easy to comprehend—banana trees do not grow in Greenland and there is no tundra vegetation in the Amazon Basin. Less obvious, perhaps, is the fact that vegetation likewise influences the climate in a number of ways. By transpiring moisture to the air, the presence of plants affects the moisture content of the atmosphere and the likelihood of precipitation. Plant assemblages also affect the surface albedo and the transfer of heat at the surface. It is easy to see why this is important in arid and semi-arid locations, where the abundance and type of vegetation is very sensitive to precipitation. Less obviously, vegetation–albedo feedback effects are also important at high latitudes, because forest cover greatly reduces the high albedo that would otherwise occur during winter in a snow climate. For example, it is believed that poleward migration of forests following the last glacial maximum had a significant amplifying effect on Holocene warming.

One of the most important feedback mechanisms involved in the possible warming of the atmosphere involves the influence of CO_2 concentration on photosynthetic rates. It has been known for some time that many plant species undergo accelerated growth and enhanced photosynthesis in an environment rich in carbon

Plant Migrations and Global Change

As global temperatures have warmed and cooled, plant communities have responded by migrating poleward during periods of glacial retreat and equatorward as glaciers advanced. For example, large tracts of species such as spruce that once inhabited much of what is now the northern and northeastern United States have shifted northward into Canadian regions formerly covered by glacial ice.

The rate at which plant communities can migrate in response to changing climates is important to their survival. If plant communities are too slow to adapt to changing conditions, the vegetation type faces possible extinction. Because of the possibility of future, rapid increases in global temperatures, scientists are concerned about the potential for major changes in Earth's plant communities. But recently obtained evidence suggests that plants migrate more rapidly in response to climate changes than previously believed. Thus, the threat of extinctions may be less menacing than once feared.

On the other hand, it is clear that some species in a plant community might be better able to expand their boundaries than others. Such changes in the rate of migration for individual species would lead to a change in the plant community, in which the overall composition of the plants in the new environment is different from what it previously was. Moreover, the expansion of suburbs and agriculture into previously undisturbed areas has caused some plant communities to become fragmented. In other words, instead of having extensive areas of a particular plant community, those vegetation associations now occur in isolated patches. The fragmentation of these communities hinders their ability to migrate in response to climate change and could make some future migrations impossible.

An interesting example of rapid vegetation change due to increasing temperatures has been observed in the grasslands of northeastern Colorado. A trend toward increasing nighttime temperatures has caused the average date of the last killing frost to occur earlier in the spring. This has put the normally dominant grass species—the blue grama, which historically has accounted for 90 percent of the ground cover—at a competitive disadvantage compared to various weeds. The weeds that are taking over the grasslands landscape are far more susceptible to drought. This is highly significant to ranchers who rely on the grass cover to supply most of their livestock's food needs.

dioxide, and some species also do better under warmer conditions. This could create a negative feedback (with potentially positive results as far as people are concerned) in which increasing CO_2 contents and temperatures allow plants to suppress further increases in the greenhouse gas.

Scientists believe that this "fertilization" process may already be appearing, especially in the latitudes between about 45° and 70° N. Satellite observations indicate that the region's supply of green vegetation is increasing and that its growing season is beginning about a week earlier and ending a week later than it did prior to the 1980s. These effects would be consistent with those expected from increased warmth and CO_2 levels observed in the area. At the same time, the seasonal oscillations in Northern Hemisphere carbon dioxide contents described in Chapter 1 have undergone changes in the intensity and timing of their cycles. The spring decrease in carbon dioxide associated with the leafing of deciduous plants is also occurring about a week earlier, and the difference between springtime maxima and late-summer CO_2 minima is increasing. Changes in the seasonal cycle of CO_2 are most conspicuous near the Arctic, where the magnitude of the oscillations has increased by 40 percent over the past few decades.

On the other hand, a lack of nutrients can become a limiting factor in plants' ability to respond to an enriched CO_2 atmosphere. This has recently been demonstrated in the Alaskan Arctic, where tundra vegetation was artificially subjected to twice the carbon dioxide level of the normal atmosphere. At first, the vegetation responded to the fertilization with increased growth rates. But by the third year, elevated CO_2 ceased to have any effect.

In a separate experiment, it appeared that high CO_2 contents might even *harm* tropical rainforest plants by causing a loss of nutrients in the soil. And another recent study has shown that trees in tropical rain forests have been growing, maturing, and dying more rapidly than in the past. This leads to a more open environment that favors replacement of trees with vines requiring greater amounts of sunlight. Ironically, the vines are less effective at photosynthesis than the trees they replace, and they don't even consume enough CO_2 to offset that released by the decaying trees.

Increases in atmospheric CO_2 could lead to other unwelcome results. It is possible, for example, that certain weeds would benefit more from higher carbon dioxide concentrations than would agricultural plants. Insects and other pests might also enjoy a warmer environment associated with increased CO_2, becoming more troublesome to agriculturalists than they are today. (For more information on climate change impacts on vegetation, see *Box 16–3 Focus on the Environment: Plant Migrations and Global Change.*)

General Circulation Models

General circulation models (GCMs) are mathematical representations of the atmosphere run on supercomputers. The models couple known laws of physics with prescribed initial and boundary conditions of the atmosphere to compute the evolution of the atmosphere through time. By averaging the output over a number of years, an estimate of climatic response can be obtained. These models help us understand the types of equilibrium and nonequilibrium conditions the atmosphere could assume in response to designated changes (such as a doubling of the carbon dioxide content).

At present, there are several such models in operation at various research institutions and universities worldwide. Although the models are all based on the same laws of physics, they each have different ways of dealing with processes that cannot be explicitly represented based on physical laws. For example, they all must make certain assumptions about what sets of conditions would lead to the formation of clouds and the height at which those clouds would exist. Variations in some of these underlying assumptions result in different regional outputs by the various models. Another important difference concerns the oceans. Some use fixed ocean temperatures, some allow SSTs to change but have no ocean currents, some have only fixed ocean currents, and some allow the oceans to respond fully to changes in the atmosphere. The appendix to Chapter 13 deals with many of the issues involved in weather forecast models that are similar to GCMs.

In 2001 the Intergovernmental Panel on Climate Change (IPCC) used the results of recent GCM runs to predict the range of possible outcomes from elevated atmospheric CO_2 concentrations. Their models indicated a "worst-case" outcome resulting in an increase in global average temperature of as much as a 5.8 °C (10.4 °F) in the year 2100, relative to the 1990 value. In its previous report issued in 1995, the panel predicted a worst-case increase of 3.5 °C (6.3 °F). The difference in the two forecasts arose because the new estimate incorporated the reduction in sulfuric particulates in the atmosphere due to environmental regulations. These particulates, though responsible for acid rain and other environmental hazards, have suppressed global warming by reducing the amount of insolation reaching the surface. The reduction of these aerosols may reduce the degree to which they suppress further warming.

Figure 16–13 illustrates the climatological changes in temperature for December through February projected by three GCMs in response to a presumed doubling of

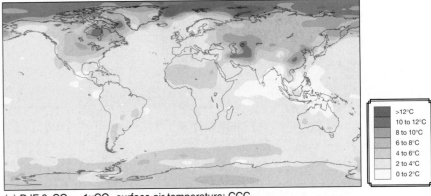

(a) DJF $2xCO_2 - 1xCO_2$ surface air temperature: CCC

	>12°C
	10 to 12°C
	8 to 10°C
	6 to 8°C
	4 to 6°C
	2 to 4°C
	0 to 2°C

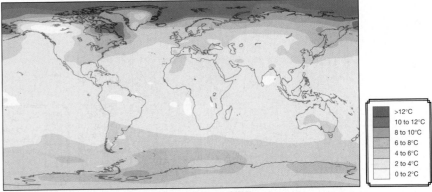

(b) DJF $2xCO_2 - 1xCO_2$ surface air temperature: GFHI

	>12°C
	10 to 12°C
	8 to 10°C
	6 to 8°C
	4 to 6°C
	2 to 4°C
	0 to 2°C

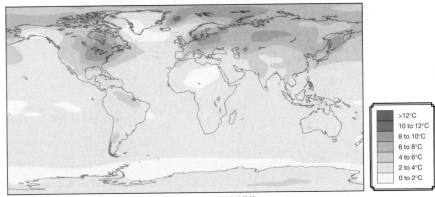

(c) DJF $2xCO_2 - 1xCO_2$ surface air temperature: UKHI

	>12°C
	10 to 12°C
	8 to 10°C
	6 to 8°C
	4 to 6°C
	2 to 4°C
	0 to 2°C

FIGURE 16–13

Temperature changes projected by three GCMs in response to a doubling of atmospheric carbon dioxide.

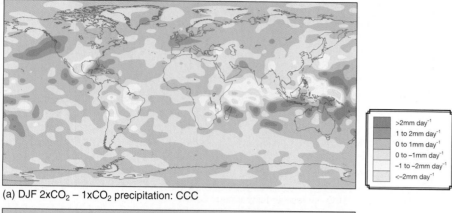

(a) DJF 2xCO_2 – 1xCO_2 precipitation: CCC

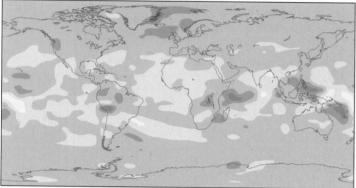

(b) DJF 2xCO_2 – 1xCO_2 precipitation: GFHI

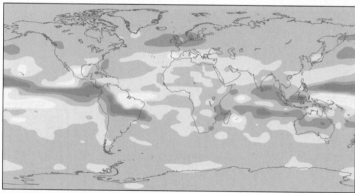

(c) DJF 2xCO_2 – 1xCO_2 precipitation: UKHI

FIGURE 16–14

Precipitation changes projected by three GCMs in response to a doubling of atmospheric carbon dioxide.

carbon dioxide. The models all predict substantial warming, particularly in the high latitudes. Although the various models are in reasonably close agreement with regard to worldwide changes in temperature, there is considerable variation in their regional patterns. This is clearly evident over North America, where the location of greatest temperature increase is highly variable from model to model. Another clear difference can be seen in western Asia, where the GFHI model indicates little increase in temperature, in distinct contrast to the output of the other two models.

There is even less consensus among the models for the global distribution of precipitation in response to a doubling of CO_2 (Figure 16–14). All three models show large areas of enhanced precipitation over the tropics, but in the midlatitudes there is considerable variability. Again, using North America as an example, there is good agreement between the GFHI and UKHI predictions, but the two differ substantially from the CCC model output.

Methods for Determining Past Climates

Though there are huge gaps in our knowledge of past climates, scientists have learned much about the climatic history of Earth. It is really quite remarkable that we are able to describe past climates as well as we can. Consider the fact that although Earth formed about 4.5 billion years ago, people are believed to have inhabited the planet for only about 2 million years. While 2 million years is an extremely long period of time relative to our own experiences, it is actually very brief in terms of all geologic history. In fact, people have walked the face of Earth for only about one-half of a thousandth of the planet's existence.

You can also view the relative time spans another way: Imagine that the planet formed at midnight on January 1 of some vast cosmic year, and it is now midnight one year later. In this scenario people would have first appeared on the scene about $3\frac{1}{2}$ hours ago, at 8:27 P.M., December 31. The first thermometer would have appeared about two seconds ago. Thus, throughout the vast majority of Earth's history, nobody was around to observe and record the climate. The lack of firsthand information is further worsened by the fact that, throughout most of human history, no direct records of climate were preserved.

We continue to gain insights into past climates based on information left in the geological and biological records. The evidence has been obtained in different places, using different techniques, and brought together by scientists in various disciplines. Taken all together, the information gives us some idea of what **paleoclimates,** the climates of the past, must have been like. Not surprisingly, the further back we go in time, the less detailed is the information provided by the various indicators.

We now briefly describe some methods for studying past climates, beginning with techniques useful for uncovering change on the longest time scales.

OCEANIC DEPOSITS

Scientists have been drilling into the ocean floor since the 1970s. They extract deep cores of material that has been deposited over very long time periods, with more recent material constantly burying older material previously laid down. Included in the deposited material are the bones and shells of plankton and other animal life, made largely of calcium carbonate ($CaCO_3$). The information contained in the oxygen in the calcium carbonate is most important for determining past climates.

Most oxygen atoms have an atomic weight of 16 (^{16}O), but a small percentage of oxygen atoms contain two additional neutrons per atom, making their atomic weight 18 (^{18}O). Both *isotopes* exist in ocean water, and both get incorporated into the shells and bones of marine animals. If the ratio of ^{18}O to ^{16}O in the water is relatively high, it will also be high in the sea life living in that water. Because ^{16}O is lighter than ^{18}O, water containing ^{16}O evaporates more readily than does water containing ^{18}O. Thus, when glaciers are expanding, the oceans (and the oxygen-containing calcium carbonate in shells and bones) will have relatively high $^{18}O/^{16}O$ ratios, as more of the ^{16}O water is removed from the ocean and deposited as snow onto the growing ice sheets. When the organisms die, they sink to the ocean bottom where their calcium carbonate is deposited. The ocean bottom thereby maintains a record of climate through the varying ratios of $^{18}O/^{16}O$ in its layers. Scientists extract cores of the ocean-bottom material, note the isotope ratios, and infer past changes in global ice volume. Information obtained from sea cores was instrumental in overturning a long-held, but mistaken, idea that there were four glaciation episodes during the Pleistocene.

Other ocean indicators of past climates comes in the form of material removed from land surfaces by glaciers. Icebergs shed by continental glaciers carry rocks, pebbles, and other debris equatorward, and this material is deposited on the ocean floor when the iceberg melts. Called **ice-rafted debris,** this material was a key in discovering Heinrich events.

ICE CORES

Scientists have also determined $^{18}O/^{16}O$ ratios for deep **ice cores** obtained from the Greenland and Antarctic ice sheets and from alpine glaciers at lower latitudes. When snow is deposited onto existing glaciers, the oxygen in the H_2O may be either ^{18}O or ^{16}O. Snow that falls under relatively warm conditions contains a higher ratio of the heavier isotope. On the ice caps, scientists from Europe and the United States have been drilling through the ice and extracting cores nearly 3 km (1.8 mi) deep to infer past temperature patterns.

Oxygen isotope studies have shown that at the end of the Pleistocene, the temperature over the Greenland ice sheet warmed about 9 °C (16 °F) over several decades. It was not known, however, whether the Arctic warming preceded or followed climatic changes in the tropics. A 1999 study based on nitrogen and argon isotope ratios ($^{15}N/^{14}N$ and $^{40}Ar/^{36}Ar$)[3] in the Greenland ice sheets strongly suggests that Arctic warming probably preceded tropical warming or increased precipitation (or both) by two to eight decades. If true, this finding is consistent with the notion that changes in thermohaline circulation can be an important mechanism for triggering rapid climatic shifts.

In addition to the temperature data obtained from isotope ratios, ice cores provide information on the past chemistry of the atmosphere and on the incidence of past volcanic eruptions. As new snow falls onto a glacier, bubbles of the ambient air become permanently trapped in the ice. The concentration of carbon dioxide and other trace gases in these bubbles yields a long-term record of their varying levels in the air. Among the more interesting results from chemical analyses at the cores is the strong correlation between past temperatures and concentrations of carbon dioxide and methane. Past periods of high temperature coincide with high

---

[3]A description of how these ratios are used is beyond the scope of this discussion.

concentrations, whereas glacial periods coincide with reduced concentrations of these greenhouse gases. Generally the changes in gas concentration lag behind the temperature changes by 800 to 1000 years, which implies that changes in composition are not the root cause of the temperature changes. Instead it indicates a positive feedback process, in which changes in the greenhouse gases amplify the climate change. Thus, for example, warming triggers release of the gases to the atmosphere, and the elevated greenhouse concentration produces more warming.

Ice sheets also provide valuable information about major volcanic eruptions. When such eruptions occur, some of the dust ejected into the atmosphere settles on the tops of glaciers. Researchers can determine when heavy volcanic activity occurred by noting the depth of the dust layers within the cores. The chemical composition of aerosols deposited in glacial ice also provides information on past events, with high acidities implying increased volcanic activity. This information may prove to be valuable in determining the importance of volcanic activity as a causal factor for climatic change.

REMNANT LANDFORMS

All of Earth's landforms are the end result of processes that build up and wear down features at the surface. The largest features, such as mountains and valleys, are produced by *tectonic* forces, those that produce a deformation of Earth's crusts. Once formed, all such features become subject to erosional processes that remove material at the surface and transport it to other locations. When the forces transporting the material are no longer capable of moving the material, *deposition* occurs. There are several mechanisms for eroding and depositing material, including the movement of water, the slow-moving ice sheets expanding across the surface, wave action along coastlines, wind, and floating icebergs carrying land debris. Each of these mechanisms leaves certain telltale characteristics, and trained field scientists can use the evidence to infer climatic conditions at the time of erosion or deposition.

FEATURES ASSOCIATED WITH ICE AND WATER

Obviously, glaciers can exist only under cold climatic conditions. Fortunately for the earth scientists, erosional and depositional features caused by the movement of glaciers often leave very distinct signatures on the landforms that last long after the glaciers have fully retreated. Some of the features associated with glaciation are related to erosion. For example, as alpine glaciers expand down preexisting valleys, they widen out the lower reaches of the valleys. This can transform typical V-shaped valleys (looking up or down the valleys), associated with those cut by running water, to U-shaped valleys. Glaciers also often leave scratch marks on solid rock walls and valley floors, or polish exposed rocks to a very smooth finish.

Unlike running water, flowing ice sheets are capable of moving very large-sized sediment as effectively as small particles of sand, silt, and clay. When the ice sheet melts away, its deposits (called *till*) can contain a very wide assortment of sediment sizes, from microscopically small clays to very large boulders. Thus, poorly sorted deposits often exist along the past margin of an ice sheet. Where glaciers terminate in an ocean, land materials are rafted away and eventually deposited on the seafloor as icebergs melt.

Streams and rivers also provide useful evidence for reconstructing past climates. Depositional features can be particularly useful in this regard, because the size of material that can be transported by running water depends on the speed of flow. Rapidly moving streams can carry large rocks. If flow of water in a stream begins to slow down, the maximum size of sediment it can transport decreases, and the largest material it contains will be deposited. This allows us to infer that layers made up of very large sediment must have been deposited by large streamflows. Thus, by examining the layering, or *stratigraphy,* of stream banks or road cuts, one can get some idea of the sequencing of high and low precipitation episodes.

Waves along coastlines also leave distinct features that can be left behind after sea level rises or lowers. Thus, by examining the elevation of terraces or submerged coastal platforms, we can infer how much glacial ice has accumulated or melted.

CORAL REEFS

Coral reefs are hard ridges extending from the ocean floor to just below the water surface, along the shallow margins of warm, tropical oceans. They form by the growth of small marine organisms (coral) having hard shells, composed largely of calcium. Colonies of coral live together atop the hard material left behind by past coral. When these coral die, their shells remain at the top of the reef and provide the foundation upon which subsequent coral grow.

Coral reefs have been useful in several ways for the inference of past conditions. Because they exist along shallow waters, relic coral reefs can provide information on the location of past sea levels. Moreover, because the chemical composition of growing coral is affected by the water temperature, analysis of the changing chemistry of coral reefs with depth provides useful information on past conditions. They have been particularly useful in providing information on past El Niño events.

PAST VEGETATION

Climate is by no means the only important factor affecting the distribution of vegetation, but it does exert a strong influence on the distribution of vegetation communities. When a vegetation community occupies a region, some of its pollen and spores can be deposited and preserved indefinitely in lake beds or bogs. This preserved pollen can be extracted and identified by *palynologists*, who might then subject it to a technique called **radiocarbon dating.** Radiocarbon dating provides a good estimate of the age of material younger than about 50,000 years and allows a determination of the distribution of vegetation species that existed at various times during the past. This information can then be displayed in pollen diagrams (Figure 16–15) that depict the sequence of past vegetation assemblages. Because many types of vegetation stands are identified with particular climate types, these diagrams provide useful information for deciphering the climatic history of an area.

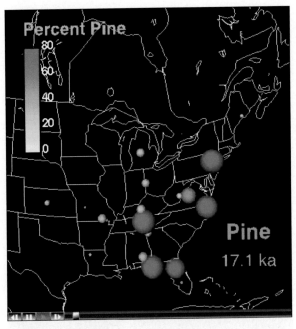

WEATHER IN MOTION
20,000 Years of Pine Pollen

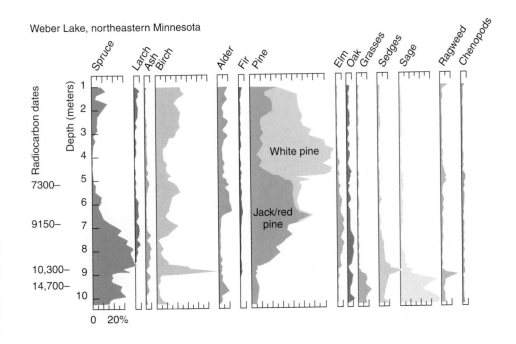

FIGURE 16–15
Pollen diagrams provide information on past vegetation at a site, which is useful for determining past climates.

FIGURE 16—16
Tree rings.

FIGURE 16—16
Tree rings.

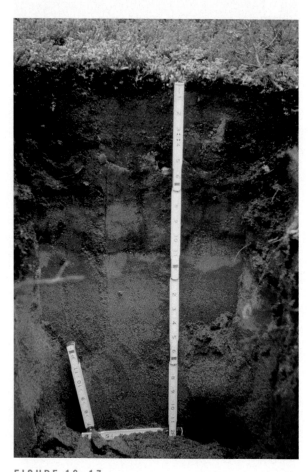

FIGURE 16—17
Buried soils along the Tundra-Forest ecotone in Canada record 7000 years of climate change.

Much information about past climates extending back for several thousand years can also be obtained from **tree rings.** Each year many trees increase the width of their trunks by the growth of concentric rings, each distinct from the previous rings (Figure 16–16). The width of each ring depends on how favorable temperature and/or moisture conditions were during a given year for the particular tree species. Under climatic stress conditions resulting from a lack of moisture or excessive warmth, the growth of these rings will be retarded. When conditions are favorable for growth, the rings will be relatively thick.

Some tree species are sensitive to temperature variations, whereas other species are affected primarily by changing moisture conditions. In either case, the extraction of cores from the trunks of very old trees (the oldest ones date back more than a millennium) yields a continuous record of annual tree growth, which correlates with precipitation and/or temperature. The correlation depends on the species and on the environmental setting. At high elevations, for example, low temperatures might slow annual growth, while for the same species at lower elevation, warm temperatures might create slower growth through increased moisture stress. Although obviously a complicating factor, differential climatic response can be used to advantage by permitting researchers to isolate various types of climatic change from one another. Indeed, such records have been obtained from old stands of trees around the world and provide climatologists with a wealth of information about past conditions.

RELIC SOILS

Just as particular vegetation communities tend to locate in areas of particular climates, certain soil types develop in response to the local climate and vegetation. A number of variables, such as the color and texture of soils, are used to identify particular soil types. If a local climate changes, there will eventually be a change in the type of vegetation and its associated soil. Through time, new vegetation–climate patterns may develop and existing soils can be buried by new soil material. By digging pits and observing the buried soils (called **paleosols**), a sequence of past conditions can be inferred (Figure 16–17).

Summary

The climate of Earth has undergone numerous changes that have occurred over differing time scales, and there is every reason to expect that the climate will continue to change long into the future. Climatic changes have occurred with differing magnitudes over varying time frames. In general, the longer period oscillations have had more extreme changes than those over shorter time intervals.

At the longest time scales, Earth has been mostly warm, with little permanent ice. But at least seven times the planet has experienced an ice age, the most recent of which continues to the present. Within an ice age, glaciers grow and shrink in what are called glacial/interglacial cycles. Over the last 5 million years, these cycles have varied considerably, both in amplitude and length. Present times correspond to an interglacial, which followed the last glacial maximum of 20,000 years ago. Within this period (the Holocene), the most notable changes were a large abrupt cooling 8200 years ago and another much smaller one from 1400 to 1850 (the Little Ice Age). Over the historical period, temperature records suggest global warming by a few tenths of a degree Celsius. Throughout both glacial and interglacial times, there have been rapid changes in climate, as Earth jumps back and forth between two more-or-less stable states (millennial-scale oscillations).

We know of several processes that can cause the climate to change. For example, changes in solar radiation receipts over hundreds of thousands of years are caused by variations in Earth's orbit around the Sun—the Milankovitch cycles. Radiation receipts also vary in response to changes in radiation emitted by the Sun. Changes in the land surface of the planet may have been responsible for shifts in climate over both very long and relatively short time scales. Atmospheric turbidity increases from volcanic eruptions are believed by some scientists to have affected climate, although this viewpoint is highly controversial. The absorption of solar and thermal radiation by aerosols and radiation-absorbing gases may be important determinants in recent and future climatic conditions. All of these factors that influence climatic conditions do not act separately, but as part of complex feedback mechanisms connecting many parts of the Earth system. The effects of these feedbacks can be examined by the use of General Circulation Models (GCMs).

Most of what we know about past climates has been inferred through indirect evidence of past conditions. Such evidence includes remnant landforms, botanical information, the presence of old soils, oceanic deposits, and cores obtained from the Greenland and Antarctic ice sheets. Hypotheses regarding future climatic conditions and the impacts of increasing carbon dioxide and other gases can be analyzed with the aid of general circulation models.

Key Terms

intransitivity   page 476
ice ages   page 478
Pleistocene   page 479
glacial/interglacial cycles   page 479
Holocene   page 482
Younger Dryas   page 482
African Humid Period   page 482
Medieval Warm Period   page 483

Little Ice Age   page 483
millenial-scale oscillations   page 485
Maunder Minimum   page 488
quasi-biennial oscillation   page 488
eccentricity   page 489
obliquity   page 489
precession   page 490

Milankovitch cycles   page 490
Pangaea   page 491
turbidity   page 492
feedbacks   page 496
negative feedbacks   page 497
positive feedbacks   page 497
thermohaline circulation   page 498

general circulation models   page 501
paleoclimates   page 502
ice-rafted debris   page 503
ice cores   page 503
coral reefs   page 505
radiocarbon dating   page 505
tree rings   page 506
paleosols   page 506

Review Questions

1. Explain how it is possible that the global climate can be both cooling and warming at the same time.

2. How does the present climate compare to past climates over the course of geologic history? Is it correct to say that the ice age is over?

3. Describe the frequency at which glacial/interglacial cycles have occurred during the Pleistocene.

4. What time frames constitute the Pleistocene and Holocene epochs?

5. What magnitude of mean temperature differences coincided with the various glacial/interglacial episodes of the Pleistocene?

6. Which regions of North America experienced the greatest temperature differences from current values during the last glacial episode of the Pleistocene?

7. List the factors that can lead to climatic change. At what time scales do each of these occur?

8. Describe the factors that can lead to variations in the amount of solar radiation available at the top of Earth's atmosphere.

9. What evidence is there that variations in sunspot activity do or do not lead to climate changes on Earth?

10. How do changes in eccentricity and obliquity and precession interact to influence Earth's climate? What time scales apply to each?

11. What types of occurrences on Earth can affect atmospheric turbidity? How do turbidity differences affect global temperatures?

12. Describe positive and negative feedbacks and provide examples of each.

13. Explain how cores taken from ocean deposits and ice sheets can be used to infer past climate conditions.

14. Describe two types of remnant landforms that can provide information on past climates in a region.

15. What is radiocarbon dating, and how does it help in determining past climates?

16. How are tree rings used as indicators of past climates?

17. Describe how pollen samples obtained from old soils provide information on past climates.

Critical Thinking

1. The opening to this chapter discussed the concept of intransitivity. What would be the implications if Earth's climate were indeed found to be intransitive? Do you think that the question of whether the climate is intransitive can ever be answered?

2. The vast majority of atmospheric scientists now believe that at least some of the warming of recent decades is due to the activities of humans, but some are not convinced that the effect of this warming will be problematic for society. How would you argue for and against this viewpoint?

3. In addition to changes in average temperature and precipitation, it is believed that the frequency of unusual events (such as droughts and floods) might be more common as a result of global warming. What regions of North America are more vulnerable to economic losses from these events than from increased temperatures?

4. As recently as 20,000 years ago a continental glacier extended as far south as the central United States. Describe what impacts the ice might have had on the average position and magnitude of the polar jet stream.

Problems and Exercises

1. View the Web site at **http://www.pewclimate.org/** and check to see if it includes any new press releases. If so, what are the major findings contained in them? Are these findings actually new or do they expand on known information?

2. Open the Weather in Motion module, "Retreat of Continental Ice Sheets," on this book's CD-ROM. Move the cursor on the map so that you can observe the retreat of the Northern Hemisphere glaciers over the last 21,000 years. Did the ice retreat uniformly across North America? Where in the Northern Hemisphere did the ice first begin to retreat?

3. Check your favorite newspaper or news magazine to see if there are any articles describing new findings about climate change or political issues related to the topic. How would you rank the importance of climate change relative to other major political issues?

Quantitative Problems

The atmosphere is an extremely complex system with numerous feedbacks. Atmospheric scientists use sophisticated models to better understand the interrelationships between its different components. The Quantitative Problems section for this chapter on the book's Web site at **http://www.prenhall.com/aguado** allows you to use a very simplified model to see how some of atmospheric variables may respond to changes in other variables.

Useful Web Sites

http://www.ipcc.ch/
Contains a vast amount of interest from the latest and previous reports of the Intergovernmental Panel on Climate Change. An extremely valuable resource for anybody interested in any aspect of climate change.

http://www.pewclimate.org/
Web page for the Pew Center for Global Climate Change, an advocacy group to promote action on issues related to the topic.

http://www.usgcrp.gov/
Comprehensive site of the U.S. Global Change Research Program. Offers numerous links, including summary reports of major research initiatives.

http://www.gcrio.org/ipcc/qa/cover.shtml
A useful primer on climate change from the U.S Global Change Research Information Office.

http://www.ncdc.noaa.gov/paleo/pollen/viewer/webviewer.html
Animations of North American vegetation change over the last 20,000 years. Based on pollen analysis, many plant taxa are represented.

http://www.exploratorium.edu/climate/index.html
An interesting and well-designed site with pages dedicated to various aspects of global climate change.

http://www.unep.ch/iuc/submenu/infokit/factcont.htm
Published by the United Nations Environment Programme, this page links to 30 climate information pages discussing various aspects of climate change and its impacts.

http://www.climatechange.gc.ca/english/index. shtml
The Climate Change Web site produced by the Canadian government.

http://www.iisd.org/climate/climate_canada/
Web-based newsletter providing information on climate change as it affects Canada.

Media Enrichment

Tutorial
Orbital Variations
Solar radiation received by the planet is affected by cyclical changes in three primary elements of Earth orbit: eccentricity, obliquity, and precession. There is little doubt that such changes have been the main driver of climate change over the last few million years. This tutorial uses animations to depict orbital changes and to show how they combine to influence the seasonal and latitudinal distribution of incoming solar radiation.

Weather in Motion
Retreat of Continental Ice Sheets
This movie illustrates the transition from the last glacial maximum to the present, showing changes in sea level and contintental ice over the last 21,000 years. Drag the mouse left/right to rotate Earth and up/down to move forward and backward in time. Note that at the outset nearly all of Canada and considerable portions of the north-eastern United States were under ice. Asia and North America were joined by a land bridge, thanks to lower sea level. By 11,000 years ago, the western portion of North America is ice-free, and total ice extent is not too different from today at 7000 years BP. By then, rising sea level accompanying deglaciation had opened the Bering Strait.

Weather in Motion
20,000 Years of Pine Pollen
The relative abundance of pollen from various plant species provides a useful indicator of past climatic conditions. This movie, in effect, shows the change in the distribution of pine forest over eastern North America during the most recent deglaciation. During the glacial maximum, pine forests were mainly found over the southeastern United States. With the onset of warming, they slowly came to occupy more northerly climes, achieving greatest abundance about 7000 years BP in a continuous swath from the eastern seaboard to the Great Plains. Since then, pines have given way to other species, though they are still found in much of the region.

PART SEVEN

Special Topics
and Appendices

A rainbow, one of many optical phenomena resulting from the systematic bending of light rays.

I magine yourself driving along a straight, two-lane highway on a hot, sunny, summer afternoon. The car ahead of you is moving just a little too slowly, so you take a look into the oncoming traffic lane and you decide that it is safe to pass. But as you cross over into the lane of oncoming traffic, another car appears, seemingly out of nowhere. Though you are surprised by the sudden appearance of the other vehicle, you have enough time to get back into your own lane—perhaps a bit unnerved but otherwise no worse off. But you can't help but wonder why you didn't see that car before you started to pass. Were your eyes playing tricks on you? Or perhaps the visibility was not as great as you thought it was. The answer could very well be that the atmosphere altered the path of the visible radiation reflected off the oncoming car so that the rays were deflected away from your eyes, and that it was not until you got sufficiently close to the vehicle that the light was able to meet with your eyes. Sometimes the atmosphere can indeed make objects appear to be in a different position from where they really are or even alter their appearance entirely. This final chapter describes the processes by which the atmosphere affects the path of visible radiation passing through it and the resultant images we see. We refer to these topics collectively as **atmospheric optics.**

Clear Air Effects

In Chapter 3 we saw that the atmosphere scatters and absorbs incoming solar radiation. Both processes have an important effect on the energy obtained by the surface and the atmosphere. In addition to scattering and absorbing, the atmosphere also *refracts* solar radiation, where **refraction** is defined as the bending of rays as they pass through the atmosphere.

Refraction occurs whenever radiation travels through a medium whose density varies or whenever it passes from one medium to another having a different density. Refraction in air occurs because radiation speed varies with density—the denser the atmosphere, the slower the radiation. To visualize how differential speeds cause the radiation to bend, imagine two people in a canoe, with one of them paddling on the right side of the canoe and the other on the left. If the person on the right paddles more vigorously, the canoe steers to the left. The same thing happens to a wave of electromagnetic energy passing through the atmosphere. Under most circumstances the density of the atmosphere decreases with height above the surface. This causes the radiation to refract slightly, forming an arc with the concave side oriented downward (Figure 17–1). The rate at which density changes with height varies considerably from place to place and from day to day because of differences in the temperature profile above the surface (the relationship between temperature and density was discussed in Chapter 4). Thus, the amount and direction of refraction vary with atmospheric conditions. Let's look now at several notable results of refraction.

REFRACTION AND THE SETTING OR RISING SUN

Refraction of incoming solar radiation is greatest when the Sun is low over the horizon, because the low solar angle causes the rays to pass through a greater amount of atmosphere (as described in Chapter 3). At sunset, refraction is sufficient to cause direct rays to be visible even after the Sun

FIGURE 17–1

Refraction due to air density differences causes light to be refracted. In this case, the light from the top of the tall building is bent downward, so its path is concave downward. The light reaches the viewer's eye at an angle slightly greater than what it would be without refraction, making the top of the building appear higher up than it really is.

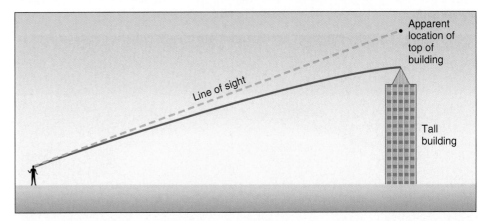

Apparent location of top of building

Line of sight

Tall building

◀ **A mirage in southern New Mexico.**

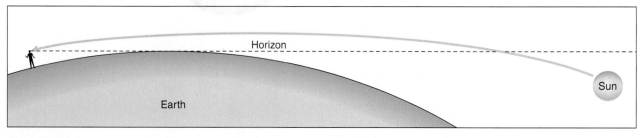

FIGURE 17–2
When the Sun is slightly below the horizon, it can still be seen at Earth's surface because of refraction. The various wavelengths are refracted differentially so that the bottom of the Sun appears redder than the top.

has dropped below the horizon (this is also true just prior to sunrise). In Figure 17–2 the Sun is positioned below the horizon. Without an atmosphere this would bring nightfall, but refraction causes the Sun to appear to be just above the horizon. When the Sun is positioned slightly farther below the horizon, its direct rays cannot be seen at the surface, but diffuse radiation can illuminate the sky to create **twilight** conditions. A further lowering of the Sun puts it far enough below the horizon to bring total nighttime. The period of twilight varies; it is longest during the high Sun season and increases with latitude.

In addition to slightly shifting the apparent position of the Sun near the horizon, refraction can also affect its apparent shape and color. You may have noticed that near dawn or sunset the Sun seems to have horizontal bands of different colors, with redder coloration near the bottom. This occurs because longer wavelength colors (such as reds and oranges) undergo less refraction than do shorter wavelength colors (such as blue and green). The shorter wavelengths, undergoing a greater amount of refraction, concentrate near the top of the apparent sun, while the longer wavelengths locate near the bottom. Under some atmospheric conditions the Sun appears momentarily to be capped by a bright green spot, known as the **green flash** (Figure 17–3).

MIRAGES

We are all familiar with movie and cartoon images of a parched, emaciated man crawling through a hot desert who perceives an oasis on the horizon, only to be disappointed by the reality of it being a mere **mirage.** Such mirages are caused not by excessive optimism, but rather by the refraction of visible light when the temperature decreases rapidly with increasing height. And contrary to what some people think, the false oasis is not the only type of mirage; that term applies to any apparent upward or downward displacement of an object due to refraction.

To understand how mirages form, let's first consider a midday situation in which the air temperature near the surface decreases rapidly with increasing height, as shown in Figure 17–4a. Near the surface, vertical changes in air density with height are controlled primarily by vertical temperature gradients. With a very high temperature near the surface, density is low. The steep temperature gradient in this example results in increasing density with height, so the path of rays moving through the atmosphere are curved upward, as indicated by the yellow lines on Figure 17–4b.

In Figure 17–4b the viewer at the left (let's call her Lauren) perceives distant objects to be slightly lower than they actually are. This is because the light rays reflected off distant objects approach Lauren's eyes at an angle slightly below horizontal. Thus, a person standing at position A (we'll call him William) appears slightly shorter than he really is. Despite the minor distortion in the perceived height of William, Lauren has no trouble seeing him in his entirety.

FIGURE 17–3
A green flash.

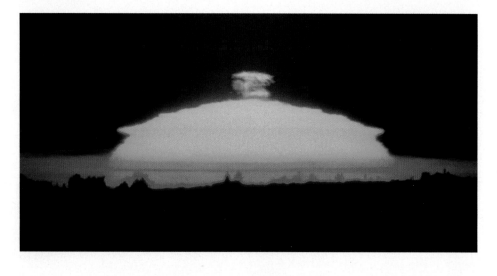

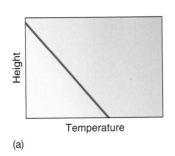

(a)

Temperature

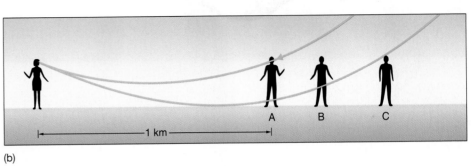

(b)

FIGURE 17-4

A steady, steep drop in temperature with height (a) can cause a refraction pattern in which rays are bent concave upward (b). Though there is some distortion of the perceived image, a person standing at position A can be viewed in his or her entirety by the person on the left. As the person on the right moves to position B, the visible light reflected off the lower legs does not reach the viewer on the left because it is absorbed at the ground. At position C, the person's image disappears entirely to the viewer on the left. Note the extreme vertical exaggeration of the diagram.

Now look at how one type of mirage develops as William moves over to position B. The lower portion of his body appears to have disappeared because the light reflected off his legs is bent all the way to the ground, where it is absorbed before it can reach Lauren. As William walks farther away, more and more of his body disappears from the bottom upward until he completely vanishes from sight at position C.

A different type of mirage appears from still more intense heating near the surface, such as that over an asphalt road on a hot afternoon. The heated air in the shallow layer just above the surface has an extremely steep temperature profile, while the air immediately above the shallow layer is somewhat cooler and has a less steep vertical temperature profile (Figure 17–5a). The steeper temperature gradient of the lower layer causes it to refract air more strongly than does the air above it. When this happens, a two-image **inferior mirage** can be seen, in which the viewer perceives not only a true image of an object, but also an inverted image directly below. This can be seen in Figure 17–5b, where light is reflected off the top of the tree in all directions. Some of the light is directed toward the viewer's eyes after undergoing only a small amount of refraction. This produces the regular image of the tree. But some of the reflected light is also directed toward the ground, where the steep temperature gradient causes very strong refraction. This light is refracted upward and reaches the viewer's eye from below, making the top of the tree appear below the ground and upside-down as if the viewer were looking at the tree's reflection on the calm surface of a pond. Light reflected off all parts of the tree undergo this type of refraction to produce a mirror image in the "pond."

So what actually happens when one sees a mirage that resembles a puddle of water? In that case a viewer perceives an inferior mirage of the sky, which looks very similar to a water surface. This chapter's opening photo of the car on the hot desert road shows an inverted image of the car apparently immersed in a puddle of water.

Note that the amount of refraction need not be the same for light leaving the top and bottom of the object. If light reflected by the bottom is refracted more strongly, the image will appear vertically stretched, or taller than reality. But if refraction increases vertically, the object will appear compressed. Obviously, the

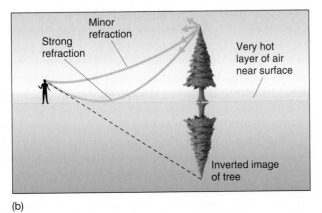

(a)

Height

Temperature

(b)

Strong refraction

Minor refraction

Very hot layer of air near surface

Inverted image of tree

FIGURE 17-5

If there is a very steep decrease in temperature immediately near the surface along with a lesser temperature gradient just above (a), a two-image inferior mirage can occur (b). Diffuse, visible radiation off the top of the tree goes in many directions. Some of the light passes directly to the viewer with minimal refraction (the upper arc from the treetop to the viewer). Some of the light approaches the surface, where intense refraction (the lower arc) gives a second image of the tree. This creates an inverted image beneath the normal image.

FIGURE 17–6
A superior mirage.

question of whether an object is stretched or compressed is independent of displacement, which is always downward for an inferior mirage.

A **superior mirage** forms when images are displaced upward. Light rays are bent concave downward as a result of decreasing density with increasing height. This is the normal situation described earlier that caused the Sun to appear higher above the horizon than it really is. But for a mirage to be noticeable, the normal density gradient must be enhanced by a temperature profile in which warm (less dense) air lies above cold air. As with inferior mirages, there may be stretching or compression of the image, depending on how refraction varies with altitude. Here, however, there is compression when refraction decreases with altitude and stretching when refraction increases vertically. In extreme situations small objects (boats, houses, etc.) can be lifted and stretched, giving the appearance of floating cities or mountains, as in Figure 17–6.

Cloud and Precipitation Optics

Refraction is not limited to clear air, nor is it the only process that can create interesting optical effects. In the remainder of this chapter, we will see how refraction and two other processes produce some very familiar (and not so familiar) optical effects.

RAINBOWS

One of the most striking features to appear in the atmosphere is the familiar **rainbow** (Figure 17–7). Rainbows are sweeping arcs of light that exhibit changes in color from the inner part of the ring to the outer part. Rainbows only appear when rain is falling some distance away, and with a clear sky above and behind the viewer that allows sunlight to reach the surface unobstructed. You might have observed that rainbows are always located in exactly the opposite direction of the Sun. In other words, if the Sun is to your back in the southwest, your shadow will point toward the center of the rainbow to the northeast.

The brightest and most common rainbows are **primary rainbows**. These rainbows are always the same size, so that at that horizon the angular distance from one end to the other extends about 85 degrees of angle (for visualization purposes, think of this as an angle that extends almost all the way from due north to due east). In a primary rainbow, the shortest wavelengths of visible light (violet and

FIGURE 17–7
Rainbows. Note that the brighter primary rainbow is surrounded by a dimmer secondary rainbow.

blue) appear at the innermost portion of the ring, and the longer wavelengths (orange and red) frame the outermost portion. A primary rainbow is often surrounded by a less distinct **secondary rainbow** that covers about 100 degrees of arc at the horizon and has the reverse color scheme of the primary rainbow (that is, the reds appear on the inner portion of the ring and the blues on the outer portion). Of course, if the precipitation shaft is not large enough or is too far in the distance, only a partial rainbow will appear.

The big question, of course, is how do these form? The answer lies in the way in which sunlight is refracted (bent) and reflected as it enters and penetrates a raindrop. When light passes through a medium of varying density, it is subject to refraction. The same phenomenon occurs as light penetrates a boundary separating substances of dissimilar density, such as air surrounding a raindrop. Let's first look at how this creates a primary rainbow. As sunlight enters a raindrop, it undergoes some refraction, with longer wavelengths being refracted less than shorter wavelengths. The refracted light penetrates the raindrop, with the majority exiting at the opposite side from which it entered. However, a small portion of the light hitting the back of the raindrop is reflected back from the interior of the surface, penetrates the droplet once again, and is refracted a second time as it exits the front side of the droplet at a position somewhat lower than where it entered the droplet. This process is shown for two hypothetical raindrops in Figure 17–8a. Because each wavelength is refracted differentially, only one particular wavelength of light exiting a raindrop is directed toward a particular viewer at a particular location. Thus the upper raindrop directs red light toward the viewer, while the lower raindrop directs violet light to that person. Because the lower raindrop appears at a lower angle above the horizon, its violet light forms the lower (inner) side of the ring, and the red light from the upper raindrop appears at the upper (outer) portion. The red light is refracted 42.3° and the violet light is refracted 40.6°. As a result, the ring is only 1.7° wide.

Secondary rainbows are formed in much the same manner as are primary rainbows, except that

FIGURE 17–8
Sunlight from behind the viewer undergoes reflection and refraction (a) to produce a primary rainbow. The amount of total refraction is different for each wavelength, causing the familiar color separation of a rainbow. A viewer at ground level observes two concentric arcs creating a primary and a secondary rainbow (b).

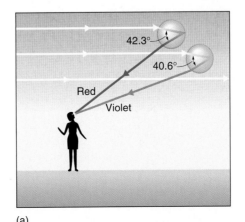

(a)

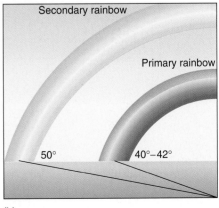

(b)

FIGURE 17–9
A secondary rainbow requires two reflections within raindrops.

two reflections occur at the back of the raindrop, as shown in Figure 17–9. This results in a reverse color scheme than that of the primary rainbow, with longer wavelengths of light situated on the inner portion of the band and with the shorter wavelengths on the outer portion. Sunlight exits the same side of the raindrop as it enters but is directed downward at a 50° angle, thus putting the top of the rainbow at 50° above the horizon (Figure 17–8b).

HALOS, SUNDOGS, AND SUN PILLARS

Cirrostratus clouds produce circular bands of light that surround the Sun or Moon, called **halos** (shown in Figure 6–19), with radii of 22° or 46° (46° halos are less common and not as bright as 22° halos). Unlike rainbows, whose appearance requires that the Sun be directly behind the viewer, halos occur when ice crystals are between the viewer and the Sun or Moon. Figure 17–10a illustrates the refraction within ice crystals that produce a 22° halo. Sunlight (or moonlight) passes through the sides of column-shaped and platelike ice crystals, in which each of the six edges form a 60° angle. The ice crystal acts as a prism that refracts the sunlight 22°. Crystals that are 22° away from the sight path of the Sun or Moon and have the necessary orientation will refract light toward the viewer. Because ice crystals are so numerous and randomly aligned within the cloud, a sufficient number will direct the light toward the observer to make the halo bright enough to be visible from the ground. Figure 17–10b shows how column-

FIGURE 17–10
Column-shaped and platelike crystals refract light to produce a 22° halo (a). Refraction where ice crystals have 90° angles produces a 46° halo (b).

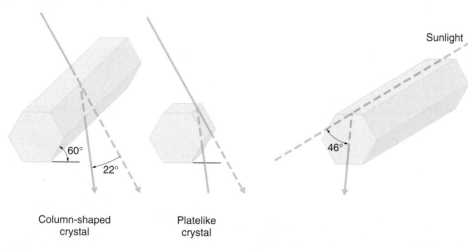

60°

22°

Column-shaped
crystal

Platelike
crystal

Sunlight

46°

(a)

(b)

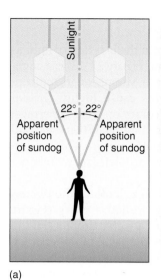

(a) (b)

FIGURE 17–11
Nearly horizontally oriented ice crystals refract light from a setting or rising Sun to produce sundogs (a), so named because of the way they accompany the Sun (b).

shaped ice crystals refract light at 46° when the crystal is oriented lengthwise toward the incoming light. The 46° halo can only result from column-shaped crystals, not by platelike crystals.

Platelike ice crystals larger than about 30 μm across tend to align themselves horizontally. If the Sun is slightly above the horizon and behind these crystals, bright spots appear 22° to the right and left of the Sun (Figure 17–11). These **sundogs** (or *parhelia*) often appear as whitish spots in the sky, but sometimes they exhibit color differentiation, with redder colors located on the side of the sundog nearest the sun and the blues and violets located on the outer side. Platelike crystals between a low Sun and an observer can also *reflect* (as opposed to refracting) sunlight off their tops and bottoms to produce **sun pillars** (Figure 17–12). The many ice crystals are aligned almost, but not exactly, horizontally, with each reflecting a portion of the incoming light differently to produce the apparent columns stretching upward and downward from the Sun.

FIGURE 17–12
A sun pillar.

FIGURE 17–13
A corona.

CORONAS AND GLORIES

Coronas and glories are optical phenomena resulting from the bending of light as it passes around water droplets (**diffraction**). The **corona** (Figure 17–13) is a circular illumination of the sky immediately surrounding the Moon—or in rarer instances, the Sun. Clouds having uniform droplet sizes cause highly circular coronas that concentrate shorter wavelength (bluish) colors on their innermost portions and longer (redder) wavelengths on their outer margins. When the cloud contains a wide assortment of droplet sizes, the illumination appears white and irregularly shaped. The size of the corona is also related to droplet size, with larger droplets producing smaller coronas.

If you are ever in an aircraft flying above a cloud deck, look for the plane's shadow on the clouds. You may see a series of rings called a **glory** (Figure 17–14). Glories occur when sunlight entering the edge of a water droplet (Figure 17–15) is first refracted, then reflected off the inside of the back of the droplet, and refracted again as it exits the droplet. In this regard, the process that produces a glory is very similar to that which creates a primary rainbow. However, a glory requires that the combination of these processes redirect the incoming light a full 180°, so that the viewer sees the returned light from the top of the cloud with the Sun behind her. In order to accomplish the additional bending, diffraction must occur along the edge of the droplet as the light makes its return back toward the Sun.

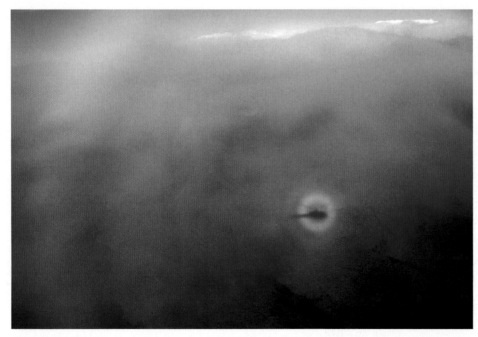

FIGURE. 17–14
A glory.

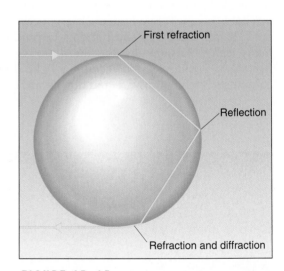

FIGURE 17–15
Glories require diffraction along the edge of a cloud droplet as the sunlight exits the droplet. The bending from refraction returns the sunlight almost 180° from the direction at which it entered the droplet.

Summary

As rays of light travel through the atmosphere, they undergo a certain amount of bending, or refraction, as a result of air density differences away from the surface. Refraction causes some interesting sunrise and sunset effects. First, the bending of solar radiation allows the Sun to be "visible" even though it is truly below the horizon. Also, the fact that different wavelengths of sunlight are refracted differentially causes a setting or rising Sun to display a sequence of horizontal bands with different colors. At its most extreme, refraction can cause only the green light from the Sun that is below the horizon to reach a viewer, resulting in a green flash.

Cloud droplets and ice crystals, along with precipitation, can produce their own unique optics. Rainbows form by the combination of refraction and reflection in raindrops that cause multicolored bands to be seen when the viewer stands between the Sun and the drops. Refraction within ice crystals causes halos and sundogs, while reflection on ice-crystal surfaces can produce sun pillars that seem to emanate from the setting or rising Sun. Diffraction by water droplets creates a corona around the Sun or Moon, while a combination of refraction, reflection, and diffraction produces glories.

Key Terms

atmospheric optics   page 513

refraction   page 513

twilight   page 514

green flash   page 514

mirage   page 514

inferior mirage   page 515

superior mirage   page 516

rainbow   page 516

primary rainbow   page 516

secondary rainbow   page 517

halo   page 518

sundog   page 519

sun pillar   page 519

diffraction   page 520

corona   page 520

glory   page 520

Review Questions

1. What is refraction and why is it related to variations in atmospheric density?

2. Describe the way refraction alters the apparent position of the setting or rising Sun.

3. Do longer or shorter wavelengths of light undergo greater refraction when passing through the atmosphere? How does the differential refraction cause an apparent banding of the Sun near the horizon?

4. Which type of vertical temperature gradients promote the appearance of superior and inferior mirages?

5. How do some mirages create the appearance of standing water on hot days?

6. Explain why the Sun must be behind you when you see a rainbow.

7. Describe the difference in the way primary and secondary rainbows form.

8. How does the color pattern of a secondary rainbow differ from that of a primary rainbow?

9. In addition to refraction, what process must occur within raindrops to produce a rainbow?

10. Why is it that some halos appear at 22° angles and others at 46° around the Sun or Moon?

11. How are sundogs formed? Describe the color patterns associated with them.

12. Describe the formation of sun pillars. Does refraction play a role in their formation?

13. Explain how coronas are formed around the Sun or Moon. What factor or factors determine their size?

14. What are glories, and how are they formed? Are they the result of refraction alone or is another process also involved?

Critical Thinking

1. Consider the way the apparent position of the Sun sweeps across the sky over the course of the day (see Chapter 2). How will the period of twilight vary between summer and winter where you live? Will twilight conditions generally last longer in the tropics or in the high latitudes? After answering this question, go to Problems and Exercises question 1, and check to see if your answer was correct.

2. Can falling ice crystals produce rainbows? Explain why or why not.

3. Can altostratus clouds produce halos? Explain why or why not.

4. Which of the optical phenomena described in this chapter are most likely to occur where you live? Are they equally likely to appear at all times of the year?

5. In Chapter 3 we discussed Rayleigh, Mie, and nonselective scattering. What similarities and dissimilarities exist between those scattering processes and the optical effects caused by refraction, reflection, and diffraction that were discussed in this chapter?

6. Explain why superior mirages do not occur over land on hot, sunny days.

Problems and Exercises

1. Refer to the Web site **http://aa.usno.navy.mil/faq/docs/RST_ defs.html#top**, and look up the definitions of civil, nautical, and astronomical twilight. Then use the available tables to determine the length of day where you live for March 21, June 21, September 21, and December 21. Does the length of day show significant differences using each of the three definitions of twilight? How do these differences vary through the year?

2. On hot, sunny days, look for the presence of mirages. Are they equally apparent in all directions? If not, why do you think that might be the case? Also, check to see how long they remain visible. Do they still persist at sunset?

Useful Web Sites

http://aa.usno.navy.mil/faq/docs/RST_defs.html#top
Provides tables of sunrise and sunset for any location, incorporating the effects of twilight. Also defines three types of twilight.

http://www.weather-photography.com/gallery.php?cat=optics
Offers photos depicting many different types of optical phenomena.

http://ww2010.atmos.uiuc.edu/(Gh)/guides/mtr/opt/home.rxml
An interesting site from the University of Illinois with much information on various aspects of atmospheric optics.

http://www.sundog.clara.co.uk/atoptics/phenom.htm
Extensive information on the effects of ice crystals, water droplets, and other topics.

http://virtual.finland.fi/finfo/english/mirage.html#refr
A Finnish site with explanations and photos of different types of mirages.

http://www.polarimage.fi/
Contains numerous photographic examples of the phenomena discussed in this chapter, as well as other interesting images.

Appendix A

Units of Measurement and Conversions

SI Units

I. Basic

Length	meter (m)
Mass	kilogram (kg)
Time	second (s)
Electrical Current	ampere (A)
Temperature	kelvin (K)

II. Derived

Force	newton $(N = kg\,m/s^2)$
Pressure	pascal $(Pa = N/m^2)$
Energy	joule $(J = N\,m)$
Power	watt $(W = J/s)$
Electrical Potential Difference	volt $(V = J/C)$
Electrical Charge	coulomb (C)

III. Some Useful Conversions

Length

1 centimeter = 0.39 inches

1 meter = 3.281 feet
= 39.37 inches

1 kilometer = 0.62 miles

1 inch = 2.54 centimeters

1 foot = 30.48 centimeters
= 0.305 meters

1 mile = 1.61 kilometers

Mass/Weight

1 gram = 0.035 ounces

1 kilogram = 2.2 pounds

1 ounce = 28.35 grams

1 pound = 0.454 kilograms

Speed

1 meter/second = 2.24 miles/hour
= 3.60 km/hour

1 mile/hour = 0.45 meters/second
= 1.61 km/hour

Temperature

Celsius Temperature = $(°F-32)/1.8$
= $K-273.15$

Fahrenheit Temperature = $1.8°C + 32$

Kelvin Temperature = $°C + 273.15$

Energy

1 joule = 0.239 calories

1 calorie = 4.186 joules

Appendix B

The Standard Atmosphere

Altitude (km)	Temperature (°C)	Pressure (mb)	p/p_0*	Density (kg/m^3)	ρ/ρ_0*
30.00	−46.60	11.97	0.01	0.02	0.02
25.00	−51.60	25.49	0.03	0.04	0.03
20.00	−56.50	55.29	0.05	0.09	0.07
19.00	−56.50	64.67	0.06	0.10	0.08
18.00	−56.50	75.65	0.07	0.12	0.09
17.00	−56.50	88.49	0.09	0.14	0.12
16.00	−56.60	103.52	0.10	0.17	0.14
15.00	−56.50	121.11	0.12	0.20	0.16
14.00	−56.50	141.70	0.14	0.23	0.19
13.00	−56.50	165.79	0.16	0.27	0.22
12.00	−56.50	193.99	0.19	0.31	0.25
11.00	−56.40	226.99	0.22	0.37	0.30
10.00	−49.90	264.99	0.26	0.41	0.34
9.50	−46.70	285.84	0.28	0.44	0.36
9.00	−43.40	308.00	0.30	0.47	0.38
8.50	−40.20	331.54	0.33	0.50	0.40
8.00	−36.90	356.51	0.35	0.53	0.43
7.50	−33.70	382.99	0.38	0.56	0.45
7.00	−30.50	411.05	0.41	0.59	0.48
6.50	−27.20	440.75	0.43	0.62	0.50
6.00	−23.90	472.17	0.47	0.66	0.54
5.50	−20.70	505.39	0.50	0.70	0.57
5.00	−17.50	540.48	0.53	0.74	0.60
4.50	−14.20	577.52	0.57	0.78	0.63
4.00	−11.00	616.60	0.61	0.82	0.67
3.50	−7.70	657.80	0.65	0.86	0.70
3.00	−4.50	701.21	0.69	0.91	0.74
2.50	−1.20	746.91	0.74	0.96	0.78
2.00	2.00	795.01	0.78	1.01	0.82
1.50	5.30	845.59	0.83	1.06	0.86
1.00	8.50	898.76	0.89	1.11	0.91
0.50	11.80	954.61	0.94	1.17	0.95
0.00	15.00	1013.25	1.00	1.23	1.00

*p/p_0 of air pressure to sea level value; ρ/ρ_0 = ratio of air density to sea level value.

Appendix C

Weather Map Symbols

Explanation of Codes

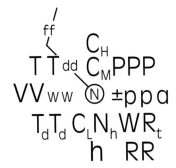

Symbol station model

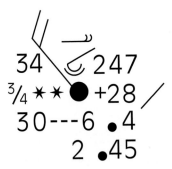

Sample report

N	Total cloud cover		h	Height in feet of the base of the lowest clouds
dd	Wind direction		C_M	Middle clouds
ff	Wind speed		C_H	High clouds
VV	Visibility in miles		$T_d T_d$	Dewpoint temperature in °F
ww	Present weather		a	Pressure tendency
W	Past weather		pp	Pressure change in mb in preceding 3 hr (+28 = +2.8)
PPP	Barometric pressure reduced to sea level (add an initial 9 or 10 and place a decimal point to the left of last number)		RR	Amount of precipitation in last 6 hr
TT	Current air temperature in °F		R_t	Time precipitation began or ended (0 = none; 1 = <1 hr ago; 2 = 1–2 hr ago; 3 = 2–3 hr ago; 4 = 3–4 hr ago; 5 = 4–5 hr ago; 6 = 5–6 hr ago; 7 = 6–12 hr ago; 8 = >12 hr ago; 9 = unknown)
N_h	Fraction of sky covered by low or middle clouds			
C_L	Low clouds or clouds with vertical development			

Air Pressure Tendency

Rising, then falling; same as or higher than 3 hr ago

Rising, then steady; or rising, then rising more slowly

Rising steadily, or unsteadily

Falling or steady, then rising; or rising, then rising more rapidly

Steady; same as 3 hr ago

Falling, then rising; same as or lower than 3 hr ago

Falling, then steady; or falling, then falling more slowly

Falling steadily, or unsteadily

Steady or rising, then falling; or falling, then falling more rapidly

Cloud Abbreviations

St	stratus
Fra	fractus
Sc	stratocumulus
Ns	nimbostratus
As	altostratus
Ac	altocumulus
Ci	cirrus
Cs	cirrostratus
Cc	cirrocumulus
Cu	cumulus
Cb	cumulonimbus

Cloud Types

Cu of fair weather, little vertical development and seemingly flattened

Cu of considerable development, generally towering, with or without other Cu or Sc, bases all at same level

Cb with tops lacking clear-cut outlines, but distinctly not cirriform or anvil-shaped; with or without Cu, Sc, or St

Sc formed by spreading out of Cu; Cu often present also

Sc not formed by spreading out of Cu

St or StFra, but no StFra of bad weather

StFra and/or CuFra of bad weather (scud)

Cu and Sc (not formed by spreading out of Cu) with bases at different levels

Cb having a clearly fibrous (cirriform) top, often anvil-shaped, with or without Cu, Sc, St, or scud

Thin As (most of cloud layer semitransparent)

Thick As, greater part sufficiently dense to hide sun (or moon), or Ns

Thin Ac, mostly semitransparent; cloud elements not changing much and at a single level

Thin Ac in patches; cloud elements continually changing and/or occurring at more than one level

Thin Ac in bands or in a layer gradually spreading over sky and usually thickening as a whole

Ac formed by the spreading out of Cu or Cb

Double-layered Ac, or a thick layer of Ac, not increasing; or Ac with As and/or Ns

Ac in the form of Cu-shaped tufts or Ac with turrets

Ac of a chaotic sky, usually at different levels; patches of dense Ci usually present

Filaments of Ci, or "mares' tails," scattered and not increasing

Dense Ci in patches or twisted sheaves, usually not increasing, sometimes like remains of Cb; or towers or tufts

Dense Ci, often anvil-shaped, derived from or associated with Cb

Ci, often hook-shaped, gradually spreading over the sky and usually thickening as a whole

Ci and Cs, often in converging bands, or Cs alone; generally overspreading and growing denser; the continuous layer not reaching 45° altitude

Ci and Cs, often in converging bands, or Cs alone; generally overspreading and growing denser; the continuous layer exceeding 45° altitude

Veil of Cs covering the entire sky

Cs not increasing and not covering entire sky

Cc alone or Cc with some Ci or Cs, but the Cc being the main cirriform cloud

Height of Base of Lowest Cloud

Code	Feet	Meters
0	0–149	0–49
1	150–299	50–99
2	300–599	100–199
3	600–999	200–299
4	1000–1999	300–599
5	2000–3499	600–999
6	3500–4999	1000–1499
7	5000–6499	1500–1999
8	6500–7999	2000–2499
9	8000 or above or no clouds	2500 or above or no clouds

Cloud Cover

○ No clouds

◔ 1/8

◔ Scattered

◑ 3/8

◑ 4/8

◒ 5/8

◕ Broken

◕ 7/8

● Overcast

⊗ Sky obscured

Wind Speed

	Miles per hour	Kilometers per hour
◎	Calm	Calm
	1–2	1–3
	3–8	4–13
	9–14	14–19
	15–20	20–32
	21–25	33–40
	26–31	41–50
	32–37	51–60
	38–43	61–69
	44–49	70–79
	50–54	80–87
	55–60	88–96
	61–66	97–106
	67–71	107–114
	72–77	115–124
	78–83	125–134
	84–89	135–143
	119–123	192–198

Fronts

Fronts are shown on surface weather maps by the symbols below. (Arrows—not shown on maps—indicate direction of motion of front.)

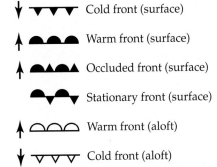

Cold front (surface)

Warm front (surface)

Occluded front (surface)

Stationary front (surface)

Warm front (aloft)

Cold front (aloft)

Weather Conditions

○ Cloud development NOT observed or NOT observable during past hour

○ Clouds generally dissolving or becoming less developed during past hour

○ State of sky on the whole unchanged during past hour

○ Clouds generally forming or developing during past hour

〰 Visibility reduced by smoke

═ Light fog (mist)

═ Patches of shallow fog at station, NOT deeper than 6 feet on land

═ More or less continuous shallow fog at station, NOT deeper than 6 feet on land

< Lightning visible, no thunder heard

• Precipitation within sight, but NOT reaching the ground

' Drizzle (NOT freezing) or snow grains (NOT falling as showers) during past hour, but NOT at time of observation

• Rain (NOT freezing and NOT falling as showers) during past hour, but NOT at time of observation

✳ Snow (NOT falling as showers) during past hour, but NOT at time of observation

• Rain and snow or ice pellets (NOT falling as showers) during past hour, but NOT at time of observation

∿ Freezing drizzle or freezing rain (NOT falling as showers) during past hour, but NOT at time of observation

S Slight or moderate dust storm or sandstorm, has decreased during past hour

S Slight or moderate dust storm or sandstorm, no appreciable change during past hour

S Slight or moderate dust storm or sandstorm has begun or increased during past hour

S Severe dust storm or sandstorm, has decreased during past hour

S Severe dust storm or sandstorm, no appreciable change during past hour

(═) Fog or ice fog at distance at time of observation, but NOT at station during past hour

═ Fog or ice fog in patches

═ Fog or ice fog, sky discernible, has become thinner during past hour

═ Fog or ice fog, sky NOT discernible, has become thinner during past hour

═ Fog or ice fog, sky discernible, no appreciable change during past hour

' Intermittent drizzle (NOT freezing), slight at time of observation

'' Continuous drizzle (NOT freezing), slight at time of observation

' Intermittent drizzle (NOT freezing), moderate at time of observation

'' Continuous drizzle (NOT freezing), moderate at time of observation

' Intermittent drizzle (NOT freezing), heavy at time of observation

• Intermittent rain (NOT freezing), slight at time of observation

•• Continuous rain (NOT freezing), slight at time of observation

• Intermittent rain (NOT freezing), moderate at time of observation

•• Continuous rain (NOT freezing), moderate at time of observation

• Intermittent rain (NOT freezing), heavy at time of observation

✳ Intermittent fall of snowflakes, slight at time of observation

✳✳ Continuous fall of snowflakes, slight at time of observation

✳ Intermittent fall of snowflakes, moderate at time of observation

✳✳ Continuous fall of snowflakes, moderate at time of observation

✳ Intermittent fall of snowflakes, heavy at time of observation

▽ Slight rain shower(s)

▽ Moderate or heavy rain shower(s)

▽ Violent rain shower(s)

▽ Slight shower(s) of rain and snow mixed

▽ Moderate or heavy shower(s) of rain and snow mixed

 Moderate or heavy shower(s) of hail, with or without rain, or rain and snow mixed, not associated with thunder

⎤ Slight rain at time of observation; thunderstorm during past hour, but NOT at time of observation

⎤ Moderate or heavy rain at time of observation; thunderstorm during past hour, but NOT at time of observation

⎤ Slight snow, or rain and snow mixed, or hail at time of observation; thunderstorm during past hour, but NOT at time of observation

⎤ Moderate or heavy snow, or rain and snow mixed, or hail at time of observation; thunderstorm during past hour, but NOT at time of observation

Weather Conditions

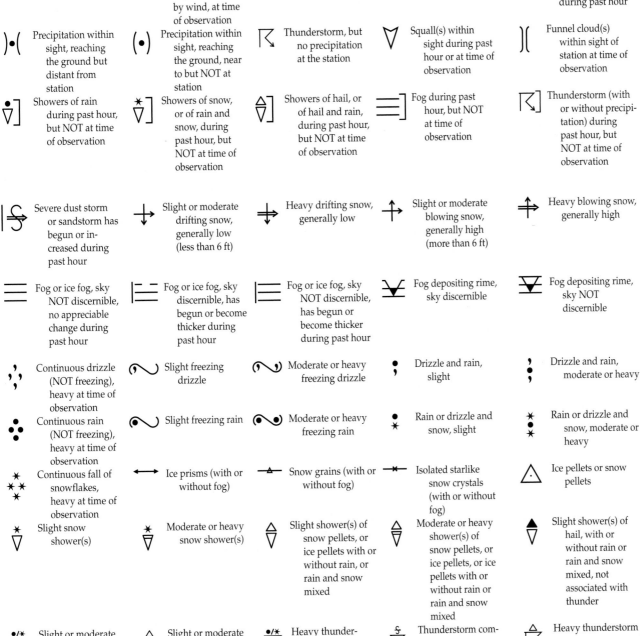

Haze	Widespread dust in suspension in the air, NOT raised by wind, at time of observation	Dust or sand raised by wind at time of observation	Well-developed dust whirl(s) within past hour	Dust storm or sandstorm within sight of or at station during past hour
Precipitation within sight, reaching the ground but distant from station	Precipitation within sight, reaching the ground, near to but NOT at station	Thunderstorm, but no precipitation at the station	Squall(s) within sight during past hour or at time of observation	Funnel cloud(s) within sight of station at time of observation
Showers of rain during past hour, but NOT at time of observation	Showers of snow, or of rain and snow, during past hour, but NOT at time of observation	Showers of hail, or of hail and rain, during past hour, but NOT at time of observation	Fog during past hour, but NOT at time of observation	Thunderstorm (with or without precipitation) during past hour, but NOT at time of observation
Severe dust storm or sandstorm has begun or increased during past hour	Slight or moderate drifting snow, generally low (less than 6 ft)	Heavy drifting snow, generally low	Slight or moderate blowing snow, generally high (more than 6 ft)	Heavy blowing snow, generally high
Fog or ice fog, sky NOT discernible, no appreciable change during past hour	Fog or ice fog, sky discernible, has begun or become thicker during past hour	Fog or ice fog, sky NOT discernible, has begun or become thicker during past hour	Fog depositing rime, sky discernible	Fog depositing rime, sky NOT discernible
Continuous drizzle (NOT freezing), heavy at time of observation	Slight freezing drizzle	Moderate or heavy freezing drizzle	Drizzle and rain, slight	Drizzle and rain, moderate or heavy
Continuous rain (NOT freezing), heavy at time of observation	Slight freezing rain	Moderate or heavy freezing rain	Rain or drizzle and snow, slight	Rain or drizzle and snow, moderate or heavy
Continuous fall of snowflakes, heavy at time of observation	Ice prisms (with or without fog)	Snow grains (with or without fog)	Isolated starlike snow crystals (with or without fog)	Ice pellets or snow pellets
Slight snow shower(s)	Moderate or heavy snow shower(s)	Slight shower(s) of snow pellets, or ice pellets with or without rain, or rain and snow mixed	Moderate or heavy shower(s) of snow pellets, or ice pellets, or ice pellets with or without rain or rain and snow mixed	Slight shower(s) of hail, with or without rain or rain and snow mixed, not associated with thunder
Slight or moderate thunderstorm without hail, but with rain, and/or snow at time of observation	Slight or moderate thunderstorm, with hail at time of observation	Heavy thunderstorm, without hail, but with rain and/or snow at time of observation	Thunderstorm combined with dust storm or sandstorm at time of observation	Heavy thunderstorm with hail at time of observation

Glossary

A

Absolute Humidity   Mass of water vapor per unit volume of air, usually expressed in grams per cubic meter (g/m^3).

Absolute Vorticity   The sum of vorticity relative to the surface and vorticity arising from Earth's rotation.

Absorption   A process in which radiation is captured by a molecule. Unlike reflection, absorption represents an energy transfer to the absorbing molecule.

Acceleration   A change in velocity; a change in speed or direction, or both.

Acceleration of Gravity   Acting alone, gravity would accelerate all objects by the same rate, about 9.8 m/sec/sec. *See* gravity.

Adiabatic   Term for processes in which no heat is added or removed. For example, a rising air parcel cools adiabatically as it expands.

Advanced Weather Interactive Processing System   (AWIPS) A system for the display and manipulation of weather information at Weather Service Offices.

Advection   Horizontal transport of some atmospheric property (heat, moisture, etc.).

Aerosols   Small, suspended particles in the atmosphere.

Aggregation   The process in which ice crystals join together to form snowflakes. If these snowflakes melt, they fall as rain.

Air Mass   A large body of air having little horizontal variation in temperature and moisture.

Air Mass Thunderstorms   Relatively small, short-lived thunderstorms that do not produce very strong winds, large hail, or tornadoes.

Albedo   The fraction of solar radiation arriving at a surface that is reflected.

Aleutian Low   A semipermanent cell found in the North Pacific in winter.

Altocumulus   A mid-level layered cloud with some rolls or patches of vertical development.

Altostratus   A mid-level layered cloud.

Aneroid Barometer   A device used to measure air pressure. The barometer's elastic chamber expands and contracts in response to the surrounding pressure.

Antarctic Circle   A line of latitude 66.5° S, the southern limit of which experiences 24 hours of daylight or darkness.

Aphelion   Earth's position when it is farthest from the Sun (July 4).

Arctic Circle   A line of latitude 66.5° N, the northern limit of which experiences 24 hours of daylight or darkness.

Arctic Oscillation   An oscillation in the temperature distribution of Arctic sea surface temperatures in the North Atlantic. Sometimes called the North Atlantic Oscillation.

Atlantic Multidecadal Oscillation   A periodic reversal of North Atlantic sea surface temperature anomalies on the order of several decades. Associated with the frequency of Atlantic hurricanes.

Atmosphere   The gases, droplets, and particles surrounding Earth's surface.

Atmospheric Window   The range of wavelengths (about 8 to 12 μm) that are not readily absorbed by the gases of the atmosphere.

Aurora Borealis or Aurora Australis   An illumination of the sky found in the high northern (borealis) or southern (australis) latitudes, which is produced as charged particles arriving from the Sun react with the upper atmosphere.

AWIPS   Acronym for Advanced Weather Interactive Processing System, the computer system used by NWS forecasters for the display of weather maps, satellite and radar imagery, and other types of data.

B

Banner Clouds   A cloud formed near the top of a topographic barrier by orographic uplift.

Barometer   An instrument for measuring air pressure.

Beam Spreading   The process whereby a beam of radiation is distributed over a larger horizontal area as the angle of incidence departs from vertical. Reduces the intensity of radiation absorption by the surface.

Bergeron Process   The primary mechanism for precipitation formation outside the tropics; this process involves the coexistence of ice crystals and supercooled water droplets.

Bermuda Azores High Low   A semipermanent cell found in the Atlantic in summer.

Blackbody   An object or substance that is perfectly efficient at absorbing and radiating radiation. Blackbodies do not exist in nature, but represent an ideal.

C

Calorie   Amount of heat required to raise the temperature of 1 gram of water 1 °C (about 4.2 J).

Carbon Dioxide   An important variable gas in the atmosphere, made up of one atom of carbon bound to two atoms of oxygen. An important greenhouse gas.

Celsius Scale   The temperature scale that designates 0° as the freezing point and 100° as the sea level boiling point of water.

Charge Separation   The separation of positive and negative ions into different parts of a cloud. A necessary precursor for lightning.

Chromosphere   The layer of the Sun immediately surrounding the photosphere.

Cirrocumulus   A high cloud composed of ice that is generally layered but with some rolls or pockets of vertical development.

Cirrostratus   A high, layered cloud consisting of ice crystals.

Cirrus   A high cloud made up entirely of ice crystals.

Climate   The statistical properties of the atmosphere, including measures of average conditions, variability, etc.

Climatology   The study of long-term atmospheric conditions.

Cloud   An area of the atmosphere containing sufficient concentration of water droplets and/or ice crystals to be visible.

Cloud-to-Cloud Lightning   Lightning that flows from one part of a cloud to another or from one cloud to another. Distinct from cloud-to-ground lightning.

Cloud Seeding   An attempt to stimulate precipitation by introducing certain materials into existing clouds.

Cold Cloud   A cloud with a temperature below 0 °C from top to bottom.

Cold Front   Transition zone between cold and warm air masses; forms when a cold air mass advances on a warm air mass.

Collector Drop   A relatively large falling raindrop that collides and coalesces with smaller, slower-moving droplets beneath it.

Condensation   Change from vapor to liquid phase. Condensation releases the energy required for evaporation. *See* latent heat of vaporization.

Condensation Nuclei   Small, airborne particles that enhance condensation. Without condensation nuclei, condensation would occur only at very high relative humidity (at about 200% or more), while condensation nuclei allow condensation to occur at or slightly below 100% relative humidity.

Conduction   Heat transfer from molecule to molecule, without significant movement of the molecules.

Convection   Heat transfer by fluid flow (movement of a gas or liquid).

Convective Outlooks Information on the probability of different types of severe storms issued by the Storm Prediction Center.

Convergence Horizontal motions of air, the result of which is a net inflow of air (with more air imported than exported). Causes rising or sinking motions.

Conveyor Belt Model The modern description of air flow through mid-latitude cyclones.

Cool Cloud A cloud in which the lower reaches have temperatures above 0 °C and in which the temperatures in the upper portions are below 0 °C.

Cooling Degree-Day An index of the amount of seasonal air conditioning required for a location. It is obtained by summing the number of degrees between an arbitrary temperature value and the number of degrees that the daily mean temperature was above that value.

Core The interior of the Sun, where nuclear fusion produces energy that is ultimately radiated to Earth.

Coriolis Force An imaginary deflective force arising from Earth's rotation that is necessary to account for motions measured relative to the surface.

Corona A circular illumination of the sky immediately surrounding the Moon, or in some instances the Sun, caused by diffraction.

Cumulonimbus A cumulus cloud with very deep vertical development extending into the lower stratosphere, distinguished by an anvil at its top consisting of ice crystals.

Cumulus Any cloud having substantial vertical development.

Cutoff Low An upper-level area of low pressure that takes on a circular flow distinct from the general flow around it.

Cyclogenesis The beginning of cyclone formation.

Cyclone A region of low pressure relative to the surrounding area.

D

Dalton's Law This law of physical science states that the pressure of a combination of gases is equal to the sum of the partial pressures of each of the gases.

Dart Leader A zone of ionized air that serves as a conduit for a lightning stroke subsequent to the initial one in a lightning flash.

Density The mass of a substance per unit volume, expressed as kilograms per cubic meter (kg/m^3) in the International System of Units (SI).

Deposition Change from the vapor phase to the solid phase (frost is an example). Deposition releases the energy of vaporization and fusion. *See* latent heat.

Derechos Powerful, large-scale winds that flow in a straight line.

Dew Point The temperature to which the air must be cooled to become saturated.

Dew Point Temperature (also called *dew point*) The temperature at which saturation will occur, given sufficient cooling.

Diabatic Processes that involve the addition or removal of heat. For example, air in contact with a cold surface loses heat diabatically by conduction.

Diffluence A type of horizontal divergence that occurs when streamlines spread apart in the downstream direction.

Diffuse Solar Radiation Sunlight that is scattered downward to the surface. *See* scattering.

Direct Solar Radiation Sunlight that passes through the atmosphere without absorption or scattering.

Divergence Horizontal motions of air, the result of which is a net outflow of air (with more air exported than imported). Causes rising or sinking motions.

Doppler Radar A type of radar that can measure horizontal motions as well as the internal characteristics of clouds.

Double Eye Wall A feature on a hurricane in which there are two eye walls—an inner one and an outer one. The appearance of a double eye wall is often a precursor to hurricane intensification.

Dry Adiabatic Lapse Rate (DALR) Temperature decrease experienced by a rising unsaturated parcel (about 1 °C/100 m). Sinking parcels warm at the same rate. The DALR is a constant.

Dry Line A boundary between humid air and denser dry air. A favored location for thunderstorm development.

Dynamic Low A low-pressure system created by divergence in the middle or upper troposphere.

E

East Greenland Drift An ocean current that flows southward in the North Atlantic.

Ecliptic Plane The imaginary surface swept by Earth's orbit around the Sun.

El Niño A recurrent event in the tropical eastern Pacific in which sea surface temperatures are significantly above normal. The inverse event (cold sea surface temperatures) is called a *La Niña*.

Electromagnetic Radiation Energy emitted by virtue of an object's temperature. Radiation is unique in that it does not require a transfer medium and can travel through a vacuum. The energy transfer is accomplished by oscillations in an electric field and a magnetic field.

Emissivity The property of a substance or object that expresses, as a fraction or percentage, how efficient it is at emitting radiation.

Entrainment The incorporation of surrounding, unsaturated air into a cloud.

Environmental Lapse Rate (ELR) The rate of vertical temperature decrease in the air column. The value is highly variable, depending on local conditions. For the troposphere, the global average is about 0.65 °C/100 m.

ENSO An acronym for the El Niño Southern Oscillation phenomenon. Involves the interaction of Tropical Pacific Sea Surface Temperatures and atmospheric pressures.

Equation of State The equation relating air pressure to temperature and density.

Equinoxes The two days of the year on which Earth's axis is not tilted toward or away from the Sun. On the equinoxes every latitude receives 12 hours of sunlight, and the Sun is overhead at the equator. The equinoxes occur March 21–22 and September 22–23.

Escape Velocity The rate of movement required of an air molecule to escape Earth's gravity.

European Center for Medium-Range Weather Forecasting Weather forecasting agency for the European Union.

Evaporation The change in phase of liquid water to water vapor.

Evapotranspiration The combined processes of evaporation and transpiration; the delivery of water to the atmosphere by vegetation and by direct evaporation from wet surfaces.

Eye The center of a hurricane, marked by generally clear skies and light winds.

Eye Wall The portion of a hurricane immediately adjacent to the eye; usually the region of highest wind speed and most intense precipitation.

F

Fahrenheit Scale A temperature scale that assigns values of 32° to the freezing point of water and 212° to the sea level boiling point of water.

Fetch Distance traveled by wind over a uniform surface, such as a water body.

First Law of Thermodynamics Most generally, the law that states that energy is a conserved property. In a meteorological context it states that heat added to a gas results in some combination of expansion of the gas and an increase in its internal energy.

Flares Intensely hot eruptions on the solar surface.

Foehn Wind A synoptic scale wind that flows downslope and warms by compression.

Fog Air that is adjacent to the surface and contains suspended water droplets, usually formed by diabatic cooling.

Force The product of mass and acceleration, as expressed in Newton's Law ($F = ma$).

Freezing Rain A form of precipitation in which rain droplets freeze as they fall below an inversion and pass into air having a temperature below 0 °C.

Friction Force that acts to slow wind but does not change its direction. Friction develops between the atmosphere and surface and between layers of air moving at different velocities.

Front A transition zone between two dissimilar air masses (that is, air masses with differing temperature, moisture, or density).

Frost A coating of ice crystals on a surface when the air adjacent to the surface becomes saturated at temperatures below 0 °C.

Frost Point The temperature at which saturation occurs, provided that that temperature is less than 0 °C.

Frozen Dew A coating of ice on a surface that occurs when a layer of dew freezes as temperatures drop below 0 °C.

Fujita Scale The scale for categorizing tornado intensity.

Funnel Cloud A column of rapidly rotating air similar to a tornado, except that the column has not extended to the ground.

G

Gamma Rays Electromagnetic radiation at wavelengths far shorter than those of visible light (from about 0.0000001 μm to 0.000001 μm). Gamma rays, which make up only a tiny proportion of the Sun's energy, are absorbed hundreds of kilometers above the surface.

General Circulation A term that refers to planetary-scale winds and pressure, features that appear in the time-averaged state.

Geopotential Height Loosely defined, the altitude at which atmospheric pressure takes on a particular value, as in "500 mb height." Because pressure reflects the mass of overlying atmosphere, geopotential heights reflect the potential energy atmosphere above that height.

Geostrophic Flow An idealized condition in which the upper-level air flows at constant speed and direction, parallel to straight isobars. There is no acceleration in geostrophic flow, and frictional forces are negligible.

Glory A series of rings formed around the shadow of an aircraft on the top of a cloud, formed by refraction, reflection, and diffraction.

Gradient A change in some quantity (temperature, moisture, pressure) over space. For example, the temperature gradient is the rate of change of temperature per unit distance, and might be expressed in degrees Celsius per kilometer (°C/km).

Gradient Wind Wind flowing parallel to curved isobars. Frictional forces are negligible. With gradient flow there is a constant adjustment between the pressure gradient force and Coriolis force, causing the wind to change speed and direction as it flows along the isobars.

Graupel Ice crystals that have grown by riming to produce a spongy, somewhat translucent particle.

Gravity The force that attracts objects to Earth's surface. Although the acceleration of gravity is constant, the force of gravity varies from object to object. The force of gravity per unit volume of air is directly proportional to density.

Graybodies Bodies or substances that are not 100 percent efficient at absorbing or radiating energy. In reality, all bodies are graybodies.

Green Flash The brief appearance of green light near the top of the Sun sometimes observed at sunrise or sunset.

Greenwich Mean Time Also called *universal time* (*UT*). An international reference for time-keeping used for weather observations, satellite imaging, etc. It corresponds to local time at 0° longitude, a meridian passing through Greenwich, England.

Growing Degree-Day An index for estimating when crops will have undergone enough growth to send them to market. Growing degree-days are calculated by subtracting a base temperature for a particular crop from the daily mean temperature and summing the differences.

H

Hadley Cell A somewhat idealized, large-scale wind and pressure pattern found in tropical latitudes of both hemispheres. Air rises above the equator, flows poleward to about 25° latitude, subsides, and flows back to the equator at low levels.

Hail Precipitation in the form of ice crystals, almost always associated with thunderstorms. Hail falls rapidly to the surface and thus does not melt during its descent.

Halo A circular band of light surrounding the Sun or Moon, caused by ice crystal refraction.

Hawaiian High A semipermanent cell found in the Pacific, most notably during the summer months.

Heat The kinetic energy of atoms or molecules that make up a substance.

Heat Index A measure of apparent temperature used for warm conditions, incorporating temperature and humidity.

Heating Degree-Day An index of the amount of seasonal heating required for a location. It is obtained by summing the number of degrees between an arbitrary temperature value and the number of degrees that the daily mean temperature was below that value.

Heterogeneous Nucleation The condensation of liquid droplets or the deposition of ice crystals onto condensation or ice nuclei.

Heterosphere The high atmosphere (above 80 km or so), where gases are not well mixed, but rather are stratified according to molecular weight. Vertical motions are too weak to overcome gravitational settling, so heavier gases are found beneath lighter gases.

High Another term for *anticyclone*.

Homogeneous Nucleation The condensation of water droplets or deposition of ice crystals without condensation or ice nuclei.

Homosphere The lowest 80 km (50 mi) of the atmosphere, in which the relative abundance of the permanent gases is constant.

Horse Latitudes Areas associated with the oceanic subtropical highs, generally characterized by clear skies and light winds.

Hot Tower An area of intense convection within a hurricane. Often associated with hurricane intensification.

Humidity An expression of the amount of water vapor in the air.

Hurricane An intense tropical cyclone (warm-core low), with sustained winds of at least 120 km/hr.

Hydrologic Cycle The perennial movement of water in its three phases between the atmosphere, Earth's surface, and groundwater.

Hydrostatic Equilibrium When the vertical pressure gradient force is balanced by the force of gravity. Because the forces balance, there is no acceleration upward or downward.

Hygrometer An instrument that measures humidity.

Hygroscopic Nuclei Airborne particles having an affinity for water, serving as condensation nuclei.

Hygrothermograph An instrument that records humidity and temperature.

I

Ice Nuclei Particles onto which ice crystals can form when the air becomes saturated. In the absence of freezing nuclei, water droplets freeze only at very low temperatures (near −40 °C). Ice nuclei allow ice to form at relatively "high" temperatures (around −10 °C).

Icelandic Low A semipermanent cell found in the North Atlantic in winter.

Ideal Gas Law (also known as the *equation of state*) Important law describing the relationship between pressure, temperature, and density.

Infrared Radiation Electromagnetic radiation at wavelengths longer than visible radiation, from about 0.7 μm to 1000 μm.

Insolation Incident, or incoming, solar radiation.

Interglacial A warmer segment of time between episodes of glaciation.

Inversion *See* temperature inversion.

Ionosphere Region in the upper atmosphere from about 80 to 500 km (50 to 300 mi) where charged particles (ions) are relatively abundant.

Ions Electrically charged atom or group of atoms.

Isobar A line on a weather map connecting points of equal pressure. Moving along an isobar, there is no change in pressure. The pressure gradient force acts perpendicular to isobars.

Isotherm   A line on a weather map connecting points of equal temperature. Moving along an isotherm, there is no change in temperature. Temperature gradients are perpendicular to isotherms.

J

Joule   Basic unit of energy in the International System of Units (SI). A joule (J) is the energy needed to accelerate 1 kg at a rate of 1 m/sec/sec across a distance of 1 m. A joule is equivalent to about 0.25 calories. *See* calorie.

K

Katabatic Winds   Air flow down a slope under the influence of gravity.
Kelvin Scale   An absolute temperature scale, where a value of 0 K implies an absence of thermal energy. The Kelvin scales assigns 100 units between the melting and sea level boiling points of water.
Kilopascal   A unit of pressure equal to 1000 pascals or 0.1 millibars.
Kinetic Energy   Energy of motion.

L

La Niña   The opposite pattern to an El Niño, in which below-normal sea surface temperatures exist in the tropical eastern Pacific.
Land Breeze   A wind that blows from the land toward the water along the coastal zone during the night and early morning.
Latent Heat   (1) Energy present in water vapor, used in converting water from liquid to gas. Latent heat is released upon condensation. (2) Energy associated with the change of phase of a substance. *See* latent heat of fusion *and* latent heat of vaporization.
Latent Heat Flux   Heat transfer that occurs whenever water vapor moves from one place to another. Energy used to evaporate water travels with the water vapor and is released upon condensation.
Latent Heat of Fusion   Also called *latent heat of melting*. Energy released when a substance freezes, consumed when a substance melts. For water, the latent heat of fusion is about 334,000 J/kg.
Latent Heat of Vaporization   Also called *latent heat of condensation*. Energy consumed when a substance evaporates, released when a substance condenses. For water, the latent heat of vaporization is about 2,500,000 J/kg.
Leader   A column of ionized air that approaches the surface and precedes cloud-ground lightning.
Lenticular Cloud   A lens-shaped cloud that usually forms downwind of topographic barriers.
Level of Free Convection   The level to which conditionally unstable air must be lifted so that it can continue to rise due to its own buoyancy.
Lifting Condensation Level   Altitude to which an air parcel would need to be lifted for condensation to occur.
Limb-Darkening   The phenomenon in which the edge of the Sun appears darker than its center.
Little Ice Age   A period in Earth's history from about 1400 to 1850 characterized by low temperatures.
Long-Range Forecast   A weather prediction extending beyond seven days.
Long Waves   *See* Rossby waves.
Longwave Radiation   Another term for *infrared radiation*.
Low   Another term for *cyclone*.

M

Mammatus   A feature on parts of some cumulonimbus clouds marked by round, downward-extending protrusions.
Mature Stage   The stage in an air mass thunderstorm marked by heavy storm activity, with strong updrafts, lightning, and heavy precipitation.
Maunder Minimum   A period of Earth's history between about 1645 and 1715 characterized by minimal sunspot activity.

Mean Free Path   The average distance traveled by molecules before colliding with adjacent molecules; increases with altitude.
Mechanical Turbulence   (also called *forced convection*) Mixing of the air caused by horizontal movements (wind).
Medium-Range Forecast   Weather forecasts for predictions three to seven days in advance.
Mercury Barometer   The standard instrument for the measurement of atmospheric pressure.
Meridional Wind   Wind flowing north–south parallel to a line of longitude. Actual winds are seldom completely meridional, but usually have both a meridional and zonal component.
Mesocyclone   A rotating region within a cumulonimbus cloud where tornadoes often form.
Mesoscale   A scale of meteorological phenomena typically having horizontal extents of several tens of kilometers.
Mesoscale Convective Complex (MCC)   A type of mesoscale convective system having an oval or nearly circular shape.
Mesoscale Convective System (MCS)   A general clustering of thunderstorms.
Mesosphere   Region of the atmosphere from about 50 km to 80 km (30 to 50 mi), characterized by decreasing temperature with increasing altitude.
Meteorological Service of Canada   The official meteorological agency for Canada.
Meteorology   The science that studies the atmosphere.
Microburst   A small but severe downburst, whose wind shear is capable of causing air crashes.
Microscale   The smallest scale of meteorological phenomena, such as that which might surround a leaf.
Microwave Radiation   Electromagnetic radiation with wavelengths between about 0.1 and 300 μm. Weather radars use microwave radiation for imaging.
Midlatitude Cyclone   A low pressure system characterized by the presence of frontal boundaries.
Mie Scattering   Scattering of visible radiation caused by particulates.
Milankovitch Cycles   Variations in Earth's orbital characteristics having periodicities of tens of thousands of years.
Millibar   A unit of atmospheric pressure, abbreviated as mb. Sea level pressure is about 1013 mb.
Mirage   The apparent displacement of an object's true position due to refraction.
Mixed Layer   That part of the lower atmosphere in which vertical motion (convection) is strong enough for even dispersal of pollutants.
Mixing Ratio   A measure of atmospheric moisture: the mass of water vapor per unit mass of dry air, usually expressed in grams per kilogram (g/kg).
Moist Adiabatic Lapse Rate   Another term for the *Saturated Adiabatic Lapse Rate* (*SALR*).
Monsoon   A regional circulation pattern in which there is a seasonal reversal of wind and pressure, generally characterized by onshore flow during the summer and offshore flow during the winter.
Monsoon Depressions   Areas of low pressure superimposed in the southeasterly air flow out of the Bay of Bengal.
Mountain Breeze   A breeze that flows down a hill at night.

N

Nacreous Clouds   Multicolored, pearlescent clouds found in the stratosphere. Also called *mother-of-pearl clouds*, these consist of ice crystals or supercooled water.
National Weather Service   The official meteorological agency for the United States.
NCDC   Stands for National Climate Data Center.
NCEP   The initials of the National Centers for Environmental Prediction.
Nested Grid Model   A particular numerical weather prediction model.

Neutral Stability A condition in which a lifted parcel of air does not return to its original position nor continue to rise. Neutral stability occurs when the environmental lapse rate is equal to the appropriate adiabatic rate, so that temperatures inside the parcel match those of the surroundings.

Newton's Second Law An expression of the conservation of momentum that is stated as: net force equals mass times acceleration ($F = ma$).

NEXRAD Stands for Next Generation Weather Radar, a network of Doppler radar units established by the U.S. National Weather Service.

NHC Stands for National Hurricane Center.

Nimbostratus A low, layered cloud that yields light precipitation.

NOAA Stands for National Oceanic and Atmospheric Administration.

Noctilucent Cloud A type of cloud that exists in the mesosphere, visible just after sunset (or before sunrise), when the surface and the lower atmosphere are in Earth's shadow.

Nonselective Scattering Scattering of radiation in which all wavelengths are scattered about equally. This type of scattering causes clouds to appear white.

North Atlantic Oscillation *See* Arctic Oscillation.

Northeaster or Nor'easter A winter weather condition of the Atlantic Coast of the United States and Canada associated with the passage of midlatitude cyclones. The strong northeasterly winds are usually coupled with blizzard conditions.

NSSFC Stands for National Severe Storms Forecast Center.

Nuclear Fusion The thermonuclear process in which extreme heat and pressure cause atoms to combine, forming a different (heavier) element. A small part of the original mass is converted to tremendous quantities of energy and released to the environment.

Numerical Weather Prediction Weather prediction based on equations representing physical processes (as opposed to statistical relations).

NWS The initials of the National Weather Service.

O

Obliquity The degree of tilt of Earth's axis relative to the ecliptic plane, currently about 23.5°.

Occluded Front A front found in the late stages of a midlatitude cyclone.

Ocean Current The horizontal movement of surface waters caused by prevailing winds.

Omega High A pressure pattern depicted on upper-level weather maps by a pattern resembling the Greek letter Ω.

Orographic Lifting Rising motions caused by airflow over a mountain range or other topographic barrier.

Outgassing The emission of gases that accompanies volcanic eruptions.

Overrunning Warm air sliding over a dense cold air mass; the characteristic flow associated with a warm front.

Oxides of Carbon A general class of air pollutants consisting of oxygen and carbon.

Ozone Molecules consisting of three oxygen atoms, most abundant in the middle and upper stratosphere.

Ozone Hole Ozone depletions found at high latitudes (especially over Antarctica) in the spring of each year.

Ozone Layer The portion of the stratosphere where ozone is relatively abundant, reaching a few parts per million.

P

Pacific Decadal Oscillation An alternating pattern of sea surface temperature in the Pacific that reverses itself over periods of several decades.

Particulates *See* aerosols.

Pascal The standard unit of pressure in most scientific applications, equal to $1 N/m^2$.

Perihelion Earth's closest approach to the Sun (January 4).

Permanent Gases Those gases whose relative abundance is constant within the homosphere.

Persistence Forecast A weather forecast made by assuming some existing trend continues into the future.

Photochemical Smog Secondary air pollutants formed by chemical reactions in the presence of sunlight.

Photodissociation Splitting of molecules into atoms or submolecules by radiation. For example, in the thermosphere, ultraviolet radiation dissociates molecular oxygen (O_2) into atomic oxygen (O).

Photosphere That part of the Sun that emits most of the energy reaching Earth. It is the "visible" part of the Sun, a layer representing about 0.05 percent of the solar radius.

Photosynthesis The growth process of green plants, whereby water and carbon dioxide are converted to carbohydrate, releasing oxygen.

PM$_{2.5}$ Designation given to particulates smaller than 2.5 μm in diameter. Major attention has recently been given to this class of particulates as possibly the most damaging to human health.

PM$_{10}$ Designation given to particulates with diameters smaller than 10 μm, which are believed to have major health consequences for humans.

Polar Easterlies Low-level winds originating in the polar highs, a feature of the general circulation of the atmosphere, often very weak or absent.

Polar Front Transition zone between cold polar air and warmer air of the midlatitudes.

Polar Front Theory The theory postulated in the early part of the twentieth century describing the formation, development, and dissipation of midlatitude cyclones. Many of the features of the theory are still considered valid.

Polar Highs Low-level anticyclones of the Arctic and Antarctic. A feature of the general circulation of the atmosphere, often absent or weakly developed.

Polar Jet Stream A jet stream found in the upper troposphere above the polar front, a result of the strong temperature contrast across the front.

Polaris The North Star.

Potential Energy Energy possessed by virtue of an object's position above some reference level. Potential energy is available for conversion to kinetic energy.

Potential Instability The condition in which a layer of air can become statically unstable if lifted sufficiently.

Power The rate at which work is done or energy expended. The standard unit is the watt, equal to 1 J/sec.

Precession The wobble of Earth's axis that has a periodicity of about 27,000 years.

Precipitable Water Vapor A measure of the total water vapor content of the atmosphere. The depth of water that would result if all the water in the column were to condense. Global average precipitable water vapor is about 2.5 cm (1 in).

Precipitation Liquid water or ice that falls to Earth's surface. Rain is considered precipitation, but dew is not.

Precipitation Fog A type of fog that develops when falling raindrops evaporate enough water vapor into the air to saturate it.

Pressure Force exerted per unit area. In most sciences the standard unit of measurement is the pascal (Pa), equal to $1 N/m^2$. In daily meteorological applications, however, the millibar (mb) is frequently used in the United States and the kilopascal in Canada.

Pressure Gradient Force A force that arises from spatial variation in pressure. Acting alone, the pressure gradient force would cause air to blow from an area of high pressure toward an area of low pressure. The vertical pressure gradient force is always present but is nearly balanced by gravity most of the time. Much weaker horizontal pressure gradients are the ultimate cause of wind.

Primary Pollutants Substances that pollute the atmosphere upon release. *See* secondary pollutants.

Psychrometer An instrument for measuring atmospheric moisture.

Pyranometer An instrument for measuring solar radiation.

R

Radar   A device that uses microwave radiation for imaging the atmosphere.

Radiation   Another term for *electromagnetic radiation*.

Radiation Fog   A low-level cloud formed diabatically when the atmosphere loses heat by radiation upward.

Radiosonde   An instrument package carried by balloon, used to measure vertical profiles of temperature, moisture, and pressure. Measurements are radioed to the ground from the instrument cluster.

Rain   Precipitation arriving at the surface in the form of liquid drops, usually between 0.5 and 5 mm (0.02 and 0.2 in.). Outside of the tropics, rain usually begins in the ice stage and melts before reaching the surface. Rain that freezes on contact with the surface, forming a layer of ice, is called *freezing rain*.

Rain Shadow   An area on the lee (downwind) side of a mountain barrier having relatively low precipitation.

Rainbow   A wide, sweeping band of light caused by refraction of sunlight by rain drops.

Rawinsonde   A radiosonde tracked by radar to provide wind information.

Rayleigh Scattering   The scattering of radiation by agents substantially smaller than the radiation's wavelength. In the case of the atmosphere, this applies to the scattering of visible radiation by air molecules.

Reflection   A process in which radiation arriving at a surface bounces back, without being absorbed or transmitted. Reflection does not heat the reflector, because there is no net energy transfer to the surface.

Refraction   The bending of light within a medium or as it passes from one medium to another. Refraction results from density differences within/between the transfer media.

Relative Humidity   The measure of the amount of water vapor in the air as a fraction of saturation, often expressed as a percentage. Because the saturation point is temperature-dependent, relative humidity depends on both the moisture content and the temperature of the air.

Return Stroke   Synonymous with *lightning stroke*.

Ridge   An elongated axis of high pressure.

Riming   The growth of a falling ice particle as it collides with nearby water droplets that freeze onto the particle.

Roll Cloud   A rotating cloud often found within the leading edge of a gust front.

Rossby Waves   Also known as *long waves*. Waves in the midlatitude westerlies having wavelengths on the order of thousands of kilometers. Often a series of Rossby waves circle the planet, forming a pattern of ridges and troughs.

S

Saffir-Simpson Scale   A scheme for classifying the intensity of hurricanes.

Santa Ana Wind   A local name for a foehn wind in California.

Saturated Adiabatic Lapse Rate (SALR)   Rate of temperature change for a rising saturated parcel of air. The value ranges from about 4 to 10 °C/km, depending mainly on temperature.

Saturation   The maximum amount of water that can exist in the atmosphere as a vapor. More precisely, saturation occurs when a flat surface of pure water is in equilibrium with the overlying atmosphere. The evaporation rate equals the condensation rate, so the vapor content of the air is unchanging. The saturation point increases with increasing temperature.

Saturation Mixing Ratio   The mixing ratio of the atmosphere when it is saturated.

Saturation Specific Humidity   The specific humidity of the atmosphere when it is saturated.

Saturation Vapor Pressure   The vapor pressure of the atmosphere when it is saturated.

Scalar   A quantity or property that possesses magnitude but has no direction. Examples are temperature, pressure, and density.

Scattering   The dispersion or redirection of radiation by gases, dust, water drops, ice, and other particulates. Scattering does not heat the atmosphere, because there is no energy transfer to the scattering agent.

Sea Breeze   A flow of air from the water toward land along a coastal region.

Sea Level Pressure   The pressure that would presumably exist at a point if it were at sea level. This involves a conversion of observed surface air pressure.

Secondary Pollutants   Pollutants that form in the atmosphere from reactions involving anthropogenic or natural substances.

Semi-Desert   A type of dry climate that annually receives enough precipitation to distinguish it from a true desert.

Semipermanent Cell   Large area of high or low pressure present throughout the year, usually with size and location changing seasonally.

Sensible Heat   Heat transfer produced by the movement of warm air.

Severe Thunderstorm   A thunderstorm that produces either very strong winds, large hail, or tornadoes.

Severe Thunderstorm Warning   An advisory issued by a local office of the National Weather Service indicating that severe thunderstorms are occurring or imminent.

Severe Weather Warning   Advisory issued when severe weather is observed. In the case of a hurricane warning, the potential for landfall exists within a 24-hour period.

Severe Weather Watch   Advisory issued when atmospheric conditions are favorable for severe weather. In the case of a hurricane watch, the potential for landfall exists outside a 24-hour period.

Shelf Cloud   A portion of a severe thunderstorm cloud that protrudes ahead of the main portion of the cloud and above a gust front.

Short Wave   A small wave in the midlatitude westerlies. Often superimposed on Rossby waves, they move more quickly, and thus travel through the large-scale pattern.

Shortwave Radiation   Electromagnetic energy having wavelengths shorter than about 4 μm.

Siberian High   A semipermanent cell found in North Asia during winter.

Sleet   Precipitation in the form of ice pellets, resulting when raindrops freeze before reaching the surface.

Snow   Frozen, crystalline precipitation that forms and remains in the ice stage throughout its descent.

Solar Altitude   The angle between the horizon and the Sun. When the Sun is overhead, the solar altitude is 90°. *See* zenith angle.

Solar Constant   The amount of radiation reaching the top of the atmosphere when Earth is at its average distance from the Sun. This is not a pure constant, but rises and falls with changes in solar emission. Its value is about 1376 W/m^2.

Solar Declination   The latitude of overhead Sun; the place where one would go to find the Sun directly overhead at noon.

Solar Wind   A continuous stream of particles (mostly protons and electrons) emitted by the Sun, traveling about one-third to one-half the speed of light.

Solstices   The two times each year that mark the northern and southern limits of the latitude of overhead Sun. On the June Solstice (approximately June 21) Northern Hemisphere latitudes have their longest day of the year. On the December Solstice (approximately December 22), Southern Hemisphere latitudes have their longest day of the year.

Source Region   A large area of land or ocean of more-or-less uniform characteristics, above which an air mass can form.

Southern Oscillation   The reversal of surface pressure patterns over the tropical Pacific associated with El Niño events.

Specific Heat   The amount of energy required to raise the temperature of a given mass of a substance by a given amount.

Specific Humidity   A measure of atmospheric moisture. The mass of water vapor per unit mass of air, usually expressed in grams per kilogram (g/kg).

Speed   A scalar property representing the rate of motion.

Speed Convergence The compaction of air due to decreasing wind speed in the downwind direction.

Speed Divergence The spreading of air due to increasing wind speed in the downwind direction.

Squall Line A linear band of thunderstorms, often found several hundred kilometers ahead of a cold front.

Stable Air Air that, when displaced vertically, returns to its initial position. Stable air resists uplift.

Standard Atmosphere The mean structure of the atmosphere with regard to temperature and pressure.

Static Stability The condition of the atmosphere that inhibits or favors vertical displacement of air parcels.

Station Model A plotting on weather maps for individual locations depicting current temperature, dew point, pressure, and other meteorological information.

Stationary Front A transition zone between dissimilar air masses (a front) showing little or no tendency to move.

Steam Fog Fog that forms when cold air moves over a warmer water surface.

Stefan-Boltzmann Law A law for blackbody emission that states that the total energy emitted over all wavelengths is proportional to the fourth power of absolute temperature.

Steppe Another term for *semi-desert*.

Stepped Leader A narrow zone of ionized air that serves as a conduit for an initial lightning stroke.

Storm Surge A potentially damaging influx of coastal waters brought about by high winds and low pressures associated with hurricanes.

Stratocumulus A low, layered cloud having superimposed rows or cells of vertical development.

Stratopause Upper limit of the stratosphere; the transition between the stratosphere and mesosphere.

Stratosphere A layer of the atmosphere between about 16 and 50 km (10 and 30 mi), characterized by generally increasing temperature with increasing altitude.

Stratus A cloud with a layered structure.

Streamlines Lines that depict the path of wind. Air parcels are envisioned to flow along streamlines. *See* confluence *and* diffluence.

Structure The layering of the atmosphere combined with the reduction in density with altitude.

Stuve Diagram A particular type of thermodynamic diagram used for plotting temperature and moisture profiles.

Sublimation Change from a solid into a vapor without passing through the liquid phase. Also used to describe the reverse (vapor to solid).

Subpolar Low A belt of low pressure in the three-cell model, between the polar easterlies and midlatitude westerlies.

Subtropical High A semipermanent cell that occupies large areas of the midlatitude oceans, especially in the warm season.

Subtropical Jet A jet stream common in the upper troposphere on the poleward side of the Hadley cells, produced by the conservation of angular momentum.

Suction Vortex A zone of intense rotation within a large tornado that often causes the most devastation.

Sulfur Oxides A general class of pollutants consisting of sulfur and oxygen.

Sun Pillars Bands of light stretching vertically from the Sun caused by reflection off almost horizontally aligned ice crystals.

Sundogs Paired bright spots found 22° to the right or left of the Sun caused by ice crystal refraction.

Sunspots Magnetic storms of the Sun, appearing as dark (Earth-sized) spots on the photosphere.

Supercell Thunderstorm A very large thunderstorm formed from an extremely powerful updraft.

Supercooled Water Water existing in the liquid phase with a temperature less than 0 °C.

Supersaturation A relative humidity greater than 100%, when the atmosphere is more than saturated with water vapor. Requires a very clean atmosphere, where condensation nuclei are lacking.

Synoptic Scale The scale of meteorological phenomena having areas on the order of hundreds or thousands of square kilometers.

T

Teleconnection Relationship between weather or climate patterns at two widely separated locations.

Temperature An index of the average kinetic energy of the molecules comprising a substance.

Temperature Gradient Temperature change per unit distance. A strong temperature gradient implies that temperature changes rapidly over a short distance.

Temperature Inversion Condition in which temperature increases with increasing altitude.

Terminal Velocity The final speed obtained by an object falling through the atmosphere, when friction with the surrounding air balances the force of gravity.

Thermal Low Low-pressure cell produced by heating of the surface.

Thermistor An object whose electrical resistance changes with temperature, thus allowing temperature to be determined by measuring changes in electrical current.

Thermodynamic Diagram A diagram showing the relationship between pressure, temperature, density, and water vapor content, such that characteristics of air parcels can be determined as they *ascend* and *descend*.

Thermoelectric Effect A theory of lightning formation in which separation of charge is produced by positive ions migrating from warmer particles to colder ice crystals.

Thermohaline Circulation A movement of surface waters in the oceans due to variations in temperature and salt content.

Thermometer Instrument used to measure temperature.

Thermosphere Outermost reaches of the atmosphere, beginning at about 80 km (50 mi), characterized by increasing temperature with increasing altitude and by extremely low density.

Thornthwaite's Classification System The most widely used system for classifying general climate zones.

Threat Score A measure of precipitation forecast skill that considers the area correctly forecast relative to that under threat of precipitation.

Three-Cell Model A generalized description of global-scale circulation that calls for three large cells in each hemisphere. The cells rotate on a vertical plane with axes parallel to latitude lines, thereby moving heat and moisture in a north–south direction.

Thunder Sound produced when lightning discharges heat the surrounding air, causing pressure waves to emanate outward.

Tibetan Low A semipermanent cell found in southern Asia in summer.

Tipping-Bucket Gage A type of automated rain gage.

Tornado A rotating column of air with extreme horizontal winds.

Tornado Outbreak A severe storm that has spawned at least six tornadoes.

Tornado Warning An advisory issued by a local office of the National Weather Service indicating that a tornado is occurring or imminent.

Trade Wind Inversion A temperature inversion (layer of air having increasing temperature with altitude) commonly found at subtropical and tropical latitudes.

Trade Winds Prevailing lower troposphere winds of the tropics, associated with Hadley circulation. Strongest in the respective winter season, the trades blow from the northeast in the Northern Hemisphere and from the southeast in the Southern Hemisphere.

Transpiration Transfer of water to the atmosphere by vegetation, mostly by water evaporating and escaping plant tissues through leaf pores.

Tropic of Cancer A line of latitude at 23.5° N, the northern limit of solar declination.

Tropic of Capricorn   A line of latitude at 23.5° S, the southern limit of solar declination.

Tropical Depression   A closed zone of low pressure with wind speeds less than 60 km/hr (35 mph).

Tropical Disturbance   A disorganized group of thunderstorms with weak pressure gradients and little or no rotation.

Tropical Storm   A storm that originates in tropical regions and has wind speeds between 60 and 120 km/hr (35 and 70 mph)

Tropopause   Upper limit of the tropopause; the transition between the troposphere and stratosphere.

Troposphere   The lowest temperature layer of the atmosphere, from the surface to about 16 km, characterized by generally decreasing temperatures with increasing altitude.

Trough   An elongated axis of low pressure.

Turbidity   Loosely speaking, the "dustiness" of the atmosphere, including the effect of all particulates that reduce visibility.

U

Ultraviolet Radiation   Electromagnetic radiation at wavelengths too short to be visible, from about 0.001 to 0.4 μm.

Unstable Air   Air that experiences a buoyant force following a vertical displacement, causing it to rise. Uplift is promoted by instability.

Upwelling   Movement of ocean or lake water from lower levels toward the surface.

Urban Heat Island   Increased local temperatures that result from urbanization.

V

Valley Breeze   A low-level movement of air in an upslope direction, developing during the daylight hours as result of solar heating.

Vapor Pressure   A measure of atmospheric moisture, the partial pressure exerted by water vapor.

Variable Gases   Gases present in amounts that vary greatly in abundance, either vertically, horizontally, or seasonally. Water vapor is the most important variable gas.

Vault   An apparently empty area on a radar display, where moist air enters a supercell thunderstorm. Water droplets are abundant but are too small to provide a strong radar echo.

Vector   A quantity or property possessing both magnitude and direction. Examples are wind velocity, the Coriolis force, and the pressure gradient force.

Veering Wind   Wind that changes direction in a clockwise sense.

Velocity   A vector property that includes speed and motion of an object or substance.

Vertical Pressure Gradient Force   The upward-directed force that arises because pressure always decreases with increasing altitude.

Visible Radiation   Electromagnetic radiation between about 0.4 and 0.7 μm, which is detectable by the human eye.

Volatile Organic Compounds   Carbon-hydrogen molecules, both anthropogenic and naturally produced, which can be gaseous or particulate. A precursor to photochemical smog.

Vorticity   The turning of an object (such as an air parcel), usually with respect to the vertical direction. This is important to meteorology because of its association with areas of divergence and convergence.

W

Walker Circulation   An east–west circulation pattern of the tropics, characterized by several cells of rising and sinking air connected by horizontal motions along more-or-less parallel lines of latitude.

Wall Cloud   Thick cloud beneath a rotating thunderstorm, a place where severe weather often develops.

Warm Advection   Heat carried by airflow across isotherms from warm to cold.

Warm Core High   High-pressure cells with higher temperature than surrounding air. Also called *warm core anticyclones*, they originate from atmospheric motion, not from differences in heating.

Warm Core Low   Low-pressure cells that are warmer than surrounding air, produced by heating. Also called *warm core cyclones*.

Warm Front   Transition zone between two air masses of different temperature, produced when warm air advances on and overruns cold air.

Water Vapor   Water in its gaseous phase, not to be mistaken for small water droplets. Colorless and odorless, it seldom amounts to more than a few percent of the total atmospheric mass.

Waterspout   A rather weak whirlwind (narrow rotating column of air), forming over a water surface. Rising and condensing air makes the waterspout visible.

Watt   The SI unit of power, abbreviated as W, with dimensions of energy per unit time; 1 W = 1 J/sec.

Wave Cyclone   Low-pressure storm common in midlatitudes, typically formed in a favored position along a Rossby wave.

Wavelength   Distance between successive peaks of a wave, or successive troughs, or between any two corresponding points along a wavetrain.

Weather   Day-to-day conditions of the atmosphere.

Weather Forecast Office   A United States Weather Service facility that issues local and area forecasts.

Weighing-Bucket Gage   A type of automated gage for the measurement of precipitation.

West Greenland Drift   A branch of a current flowing southward in the North Atlantic.

Westerlies   Winds belts found in the middle latitudes of both hemispheres that have a strong west-to-east component.

Wet Bulb Depression   The difference between air temperature and wet bulb temperature. Large values (large differences between dry and wet bulb temperatures) are indicative of low humidity.

Wet Bulb Temperature   The minimum temperature achieved as water evaporates from a wick surrounding a thermometer's bulb.

Wien's Law   Law for blackbody emission that states that the wavelength of maximum emission is inversely proportional to absolute temperature.

Wind   The horizontal movement of air.

Wind Chill Temperature Index   An index of apparent temperature used for cold conditions that incorporates air temperature and wind speed.

Wind Profiler   A type of Doppler radar unit that provides information about vertical changes in wind speed and direction.

Wind Shear   Large change in wind direction or speed over a short distance.

Wind Vane   An instrument for measuring or indicating wind direction, often a pivoting arrow that continually points into the wind.

WMO   Stands for the World Meteorological Organization, the agency of the United Nations charged with collecting and disseminating meteorological data.

X

X-Ray Radiation   Electromagnetic radiation at wavelengths between about 0.000001 μm and about 0.001 μm.

Z

Zenith Angle   The angle between the Sun and the vertical direction. When the Sun is overhead, the zenith angle is zero. *See* solar altitude.

Zonal Wind (Flow)   Wind flowing east–west, parallel to a line of latitude. Actual winds are seldom completely zonal but usually have both a meridional and zonal component.

Zone Forecast   A weather forecast issued at regular intervals for a particular region.

Credits

Chapter 1

Photographs
Part 1 Opener, Brian Stablyk/Getty Images Inc. - Stone Allstock
Chapter 1 Opener, Edward Aguado
1-1, NASA/Johnson Space Center
1-3a,b, Space Science and Engineering/University of Wisconsin-Madison
1-5, George Holton/Photo Researchers, Inc.
1-7, Photo by Tom L. McKnight
1-10, Joe Towers/Corbis/Stock Market
1-12, Pekka Parviainen/Science Photo Library/Photo Researchers, Inc.
1-B1-1, NASA Goddard Laboratory for Atmospheres
1-B4-2, AP Wide World Photos
1-B4-3, Reuters/Dave Kaup/Landov LLC
1-B4-4, NASA/Johnson Space Center

Illustrations
1–4, 1–6, From National Oceanographic and Atmospheric Agency

Chapter 2

Photographs
Chapter 2 Opener, AP Wide World Photos
2-4a, Julian Baum & Nigel Henbest/Science Photo Library/Photo Researchers, Inc.
2-4b, National Optical Astronomy Observatories
2-8, NASA/Weatherstock
2-B3-2a, National Optical Astronomy Observatories

Illustrations
2–1, Modified from Nebel and Wright, *Environmental Science*, Sixth Edition, ©1998, Prentice Hall, Inc., Upper Saddle River, NJ
2–6, Modified from Chaisson and McMillan, *Astronomy Today*, Second Edition, ©1997, Prentice Hall, Inc., Upper Saddle River, NJ
2–B1–1; 2–B2–1 & 2b, From Chaisson and McMillan, *Astronomy Today*, Second Edition, © 1997, Prentice Hall, Inc., Upper Saddle River, NJ
2–B2–3, Data from Solar-Terrestrial Physics Division, National Geophysical Data Center, Boulder, Colorado, and John A. Eddy, "The Maunder Minimum," *Science*, Vol. 192, pp. 1189–1202.

Chapter 3

Photographs
Chapter 3 Opener, Mike R. Tuten
3-1, NASA/Goddard Space Flight Center/Laboratory for Atmospheres
3-4, NASA/Johnson Space Center
3-6, Larry Cameron/Photo Researchers, Inc.
3-11b, Robert J. Erwin/Photo Researchers, Inc.
3-20, Edward Aguado
3-23, NOVALYNX Corporation, Grass Valley, CA
3-24, Qualimetrics, Inc.
3-25, Edward Aguado

Illustrations
3–9b, J. N. Howard, *Proc. I.R.E.*, v. 47, 1959, and R. M. Goody and G. D. Robinson, *Quart. J. Roy. Meteorol. Soc.*, v. 77, 1951, as modified by R. G. Fleagle and J. A. Businger, *An Introduction to Atmospheric Physics*, Second Edition, © 1980, Academic Press, New York, NY
3–16, 3–18, Modified from Christopherson, *Geosystems, An Introduction to Physical Geography*, Third Edition, ©1997, Prentice Hall, Inc., Upper Saddle River, NJ
3–27b, From *The Atlas of Canada*, Natural Resources Canada.

Chapter 4

Photographs
Chapter 4 Opener, Jon Burbank/The Image Works
4-4, Charles D. Winters/Photo Researchers, Inc.
4-5a,c, Edward Aguado
4-22, Edward Aguado

Illustrations
4–5b, Modified from Moran and Morgan, *Meteorology, The Atmosphere and the Science of Weather*, Fifth Edition, © 1997, Prentice Hall, Upper Saddle River, NJ

Chapter 5

Photographs
Part 2 Opener, Charles A Mauzy/Getty Images Inc. - Stone Allstock
Chapter 5 Opener, © John Biever/NFL Photos
5-10, Michael Brisson
5-13a,b, NOVALYNX Corporation, Grass Valley, CA
5-17a, Rod Planck/Photo Researchers, Inc.
5-17b, F. Stuart Westmorland/Photo Researchers, Inc.
5-18, © Arjen & Jerrine Verkaik/SKYART
5-19, National Environmental Satellite, Data, and Information Center/NOAA
5-20, Getty Images, Inc.
5-B3-1, National Oceanic and Atmospheric Administration NOAA

Illustrations
5-8, National Climatic Data Center

Chapter 6

Photographs
Chapter 6 Opener, Howard Bluestein/Photo Researchers, Inc.
6-1, Copyright Howard B. Bluestein
6-3a-c Edward Aguado
6-4a, David Weintraub/Photo Researchers, Inc.
6-4b, George Ranalli/Photo Researchers, Inc.
6-16, Joyce Photographics/Photo Researchers, Inc.
6-17a, Edward Aguado
6-18, © Maurice Nimmo; Frank Lane Picture Agency/CORBIS
6-19, John Foster/Science Source/Photo Researchers, Inc.
6-20, Arjen & Jerrine Verkaik/SKYART
6-21, Arjen & Jerrine Verkaik/SKYART
6-22, Pekka Parviainen/Science Photo Library/Photo Researchers, Inc.
6-23, © Claudia Parks/CORBIS
6-24, ©Warren Faidley/Weatherstock
6-25, John Sohlden/Visuals Unlimited
6-26, David R. Frazier/Photo Researchers, Inc.
6-28, Day Williams/Photo Researchers, Inc.
6-29, Howard B. Bluestein/Photo Researchers, Inc.
6-30, Kent & Donna Dannen/Photo Researchers, Inc.
6-31, Richard J. Green/Photo Researchers, Inc.
6-32, ©Warren Faidley/Weatherstock
6-32b, Gene Rhoden/Visuals Unlimited
6-33, Polar Image
6-34, Pekka Parviainen/Science Photo Library/Photo Researchers, Inc.
6-35a-c, National Oceanic and Atmospheric Administration NOAA
6-B2-1a,b, Chuck O'Rear/Woodfin Camp & Associates

Illustrations
6-15, From Christopherson, *Geosystems*, 4th edition, ©2000, Prentice Hall, Upper Saddle River, NJ
6–17b, Adapted from Gedezelman, *The Science and Wonders of the Atmosphere*, © 1980, Wiley, Inc., New York, NY

Chapter 7

Photographs
Chapter 7 Opener, Howard B. Bluestein/Photo Researchers, Inc.
7-1, Courtesy of Spaceimaging.com
7-6, Doug Millar/Science Source/Photo Researchers, Inc.
7-9a, © Clyde H. Smith/Peter Arnold Inc.
7-9b, Scott Camazine/Photo Researchers, Inc.
7-9c, Richard C. Walters/Visuals Unlimited
7-11, Joe Traver/Getty Images, Inc - Liaison
7-13, Howard Bluestein/Photo Researchers, Inc.
7-17, Sycracuse Newspapers/The Image Works
7-18a, Edward Aguado
7-18b, Qualimetrics, Inc.
7-20, National Oceanic and Atmospheric Administration

Illustrations
7–2, Adapted from McDonald, *The Physics of Cloud Formation, Advances in Geophysics*, v. 5, Academic Press, New York, NY, as modified by Rogers, *A Short Course in Cloud Physics*, © 1979, Pergamon Press, New York, NY
7–14, Adapted from Browning and Foote, *Q.J.R. Meteorol. Soc.*, London, as modified by Eagleman, *Meteorology: The Atmosphere in Action*, Second Edition, © 1985, Wadworth, Inc., Belmont, CA
7–15, Adapted from NOAA
7–19, Courtesy of David Legates

Chapter 8

Photographs
Part 3 Opener, AP Wide World Photos
Chapter 8 Opener, C. S. Gray/The Image Works
8-1, © Dick Keen/Visuals Unlimited
8-3, Image produced by M. Jentoft-Nilsen, F. Hasler, D. Chesters (NASA/Goddard) and T. Nielsen (Univ. of Hawaii)./NASA Headquarters
8-9, NCDC at NOAA
8-16, O. Brown, R. Evans, and M. Carle, University of Miami Rosenstiel School of Marine and Atmospheric Science, Miami, Florida
8-18a,b, Tokai University Research and Information Center
8-23, NASA Earth Observing System
8-25, NASA/Johnson Space Center
8-28, NOAA/Science Photo Library/Photo Researchers, Inc.

Illustrations
8-4, Modified from Christopherson, *Geosystems, An Introduction to Physical Geography*, Third Edition, © 1997, Prentice Hall, Inc., Upper Saddle River, NJ
8-6, Adapted from Miller et al., *Elements of Meteorology*, © 1983, Merrill Publishing Co., Columbus, OH
8–14, Modified from Trujillo and Thurman, *Essentials of Oceanography*, Sixth Edition, © 1999, Prentice Hall, Inc., Upper Saddle River, NJ
8–27, Modified from Peixoto and Oort, *Physics of Climate*, New York: American Physics Inst., © 1992, p. 419
8–29, 8–30, 8–31, 8–34, 8–36, Climate Diagnostics Center/NOAA
8–32, Climate and Global Dynamics Division/NCAR
8–33, Climate Prediction Center/NOAA
8–35, Courtesy of NASA/JPL/California Institute of Technology

Chapter 9

Photographs
Chapter 9 Opener, Photo by Tom L. McKnight
9-7a, NOAA Central Library Photo Collection
9-7b, Weather Services International, a Landmark Company
9-B1-1b,d,f, National Oceanic and Atmospheric Administration/Seattle
9-B1-2b, National Oceanic and Atmospheric Administration/Seattle
9-B1-3b,d,f, National Oceanic and Atmospheric Administration/Seattle

Illustrations
9–1, Modified from Moran and Morgan, *Meteorology, The Atmosphere and the Science of Weather*, Fifth Edition, © 1997, Prentice Hall, Inc., Upper Saddle River, NJ
9–8, Modified from Christopherson, *Geosystems, An Introduction to Physical Geography*, Third Edition, © 1997, Prentice Hall, Inc., Upper Saddle River, NJ

Chapter 10

Photographs
Part 4 Opener, Ira Block/National Geographic Image Collection
Chapter 10 Opener, Airphoto - Jim Wark
10-11c, 10-12c, 10-13c, 10-14c, 10-18b, National Oceanic and Atmospheric Administration/Seattle

Illustrations
10–7, Adapted from Lutgens and Tarbuck, *The Atmosphere, An Introduction to Meteorology,* Ninth Edition, 2004, Prentice Hall, Upper Saddle River, New Jersey
10–8, Adapted from Petterson, *Weather Analysis and Forecasting,* Vol. 1, Second Edition, © 1956, McGraw-Hill, New York, NY

Chapter 11

Photographs
Chapter 11 Opener, Kennan Ward © 2003
11-1, NOAA Central Library Photo Collection
11-3, Ed Degginger/Color-Pic, Inc.
11-5, Daniel L. Osborne, University of Alaska/Detlev Van Ravenswaay/Photo Researchers, Inc.
11-6, Figure reprinted by permission from Nature: Figure 2c, vol. 416, pp. 152-154, 2002; V.P. Pasko, M.A. Stanley, J.D. Mathews, U.S. Inan and T.G. Wood. Nature Publishing Group/McMillian Publishers Ltd.
11-8, Larry Miller/Photo Researchers, Inc.
11-9, National Oceanic and Atmospheric Administration/Seattle
11-12, Weather Services International, a Landmark Company
11-13, NOAA
11-15, 11–17, Howard B. Bluestein, Professor of Meteorology
11-19b, National Oceanic and Atmospheric Administration/Seattle
11-24a, Weatherstock/Peter Arnold, Inc.
11-24b, Arjen & Jerrine Verkaik/SKYART
11-26, Arjen & Jerrine Verkaik/SKYART
11-36, Howard B. Bluestein, Professor of Meteorology
11-37, Courtesy of the Environmental Remote Sensing Center, University of Wisconsin-Madison, http://www.ersc.wisc.edu
11-40, Photo by Jeff Mitchell REUTERS/Landov LLC
11-41, National Oceanic and Atmospheric Administration
11-B5-1a, Herb Stein/Center for Severe Weather
11-B5-1b, National Center for Atmospheric Research/University Corporation for Atmospheric Research/National Science Foundation.

Illustrations
11–2, 11–14, 11–19a, Adapted from Wallace and Hobbs, *Atmospheric Science, An Introductory Survey,* © 1977, Academic Press, New York, NY
11–4, Adapted from Zajac and Weaver, Cooperative Institute for Research in the Atmosphere, Colorado State University.
11–7, Adapted from Gedzelman, *The Science and Wonders of the Atmosphere,* © 1980, Wiley, Inc., New York, NY
11–10, Image from Roger Edwards/NOAA Storm Prediction Center
11–13, 11–B4–1, National Weather Service/NOAA
11–18, Adapted from Djuric, *Weather Analysis,* © 1994, Prentice Hall, Inc., Upper Saddle River, NJ
11–22, 11–23, Orville, R., 2002, "The North American Lightning Detection Network (NALDN) - First Results: 1998-2000," *Monthly Weather Review,* v. 130, No. 8, pp. 2098-2109.
11–32, 11–33, Source: NOAA
11–27, From Bluestein, *Tornado Alley,* 1999, Oxford University Press
11–28, 11–29, Adapted from Brandes, in Church et al., eds., *The Tornado: Its Structure, Dynamics, Prediction, and Hazards,* © 1993, American Geophysical Union, Washington, DC
11–30, Modified from McKnight, *Physical Geography, A Landscape Appreciation,* © 1996, Prentice Hall, Inc., Upper Saddle River, NJ
11–31, Modified from Christopherson, *Geosystems: An Introduction to Physical Geography,* Sixth Edition, 2006, Prentice Hall, Upper Saddle River, NJ
11–34, National Severe Storms Laboratory/NOAA
11–35, Adapted from Fujita, *Journal of Atmospheric Sciences,* © 1981, American Meteorological Society, Boston, MA, as modified by Battan, *Fundamentals of Meteorology,* © 1984, Prentice Hall, Inc., Upper Saddle River, NJ
11–38, Storm Prediction Center/NOAA
11–39, National Weather Service

Chapter 12

Photographs
Chapter 12 Opener, NASA. Image and Animation produced by the Scientific Visualization Studio at NASA GFC.
12-1, 12–7, NASA Headquarters
12-8, Tropical Rainfall Measuring Mission/NASA/Goddard Space Flight Center
12-15, Marvin Nauman/FEMA
12-16a,b, National Geographic Image Collection
12-B3-2 and 3, NOAA
12-B3-4, Rick Wilking/Reuters/Corbis/Bettmann
12-B5-1, National Oceanic and Atmospheric Administration/Seattle

Illustrations
12–2, Modified from McKnight, *Physical Geography, A Landscape Appreciation,* Fifth Edition, © 1996, Prentice Hall, Inc., Upper Saddle River, NJ
12–3, 12–10, Modified from Moran and Morgan, *Meteorology: The Atmosphere and the Science of Weather,* Fifth Edition, © 1997, Prentice Hall, Inc., Upper Saddle River, NJ
12–4, Rauber et al., *Severe and Hazardous Weather,* 2002, Kendall/Hunt Publ., Dubuque, IA
12–5, Adapted from Miller, *Science,* v. 157, 1967, American Association for the Advancement of Science, as modified by Eagleman, *Meteorology: The Atmosphere in Action,* Second Edition, © 1985, Wadsworth, Inc., Belmont, CA
12–6, Anthes et al., "Three-Dimensional Particle Trajectories in a Model Hurricane," *Weatherwise,* v. 24, 1971
12–7, NASA Earth Observatory
12–8, NASA Tropical Rainfall Measuring Mission (TRMM)
12–11, 12–12, 12–17, 12–B2–1, 12–B2–2, 12–B3–1, National Hurricane Center/NOAA
12–14, Adapted from Novlan and Gray, *Monthly Weather Review,* © 1974, American Meteorological Society, Boston, MA, as modified by unpublished National Weather Service Training Center pamphlet
12–B3–2, 12–B3–3, National Weather Service/NOAA

Chapter 13

Photographs
Part 5 Opener, James Nubile/The Image Works
Chapter 13 Opener, AP Wide World Photos
13-1a-c, 13-4, Edward Aguado
13-20a-c, National Oceanic and Atmospheric Administration/Seattle
13-21, Weather Services International, a Landmark Company
13-B2-1, Christina Russo

Illustrations
13–3, Adapted from DeFelice, *Introduction to Meteorological Instrumentation and Measurement,* ©1998, Prentice Hall, Inc., Upper Saddle River, NJ
13–5, Adapted from Olson et al., *Weather and Forecasting,* v. 10, 1995
13–6, 13–7, Adapted from Mesinger, *Bulletin of the American Meteorological Society,* v. 77, 1996
13–9, 13–10, Source: NOAA
13–21, Weather Services International (Billerica, MA)
Appx 1, 3, Adapted from Olson et al., *Weather and Forecasting,* v. 10, 1995

Chapter 14

Photographs
Chapter 14 Opener, Paula Bronstein/Getty Images, Inc - Liaison
14-2, Adam Hart-Davis/Science Photo Library/Photo Researchers, Inc.
14-4, Martin Bond/Science Photo Library/Photo Researchers, Inc.
14-5, Tony Craddock/Science Photo Library/Photo Researchers, Inc.
14-7, copyright Blue Fier 2005
14-9b, Edward Aguado
14-B1-1a, Carnegie Library of Pittsburgh
14-B1-1b, Photo by Arthur G. Smith
14-B3-1a, The MapFactory/Photo Researchers, Inc.

Illustrations
14–1, From Nebel and Wright, *Environmental Science,* Sixth Edition, © 1998, Prentice Hall, Inc., Upper Saddle River, NJ

14–3, Modified from Moran and Morgan, *Meteorology: The Atmosphere and the Science of Weather,* Fifth Edition, © 1997, Prentice Hall, Inc., Upper Saddle River, NJ
14–8, 14–9a, Adapted from Oke, *Boundary Layer Climates,* © 1978, Methuen & Co., Ltd., New York, NY
14–10, Adapted from Hidore and Oliver, *Climatology, An Atmospheric Science,* © 1993, Macmillan Publishing Co., New York
14–B2–1, From Nebel and Wright, *Environmental Science,* Fifth Edition, © 1993, Prentice Hall, Inc., Upper Saddle River, NJ
14–B2–2, From Environmental Protection Agency.
14–B3–2, Adapted from Blumenthal et al., *Atmospheric Environment,* vol. 12, © 1978.

Chapter 15

Photographs
Part 6 Opener, Robin Scagell/Photo Researchers, Inc.
Chapter 15 Opener, Karl Weidmann/Photo Researchers, Inc.

Illustrations
15–1, 15–2, 15–3, 15–4, 15–5, 15–7, 15–8, 15–9, 15–10, 15–11, 15–12, 15–13, 15–14, 15–15, 15–16, Modified from McKnight, *Physical Geography, A Landscape Appreciation,* Fifth Edition, © 1996, Prentice Hall, Inc., Upper Saddle River, NJ
15–6, Source: NOAA
15–B1–1, Adapted from Critchfield, *General Climatology,* Third Edition, © 1974, Prentice Hall, Inc., Upper Saddle River, NJ

Chapter 16

Photographs
Chapter 16 Opener, Dennis Flaherty/Photo Researchers, Inc.
16-9, Phil Durkee/The Naval Postgraduate School
16-11, Simon Fraser/Science Photo Library/Photo Researchers, Inc.
16-12, NASA/Johnson Space Center
16-16, Stephen J. Krasemann/Photo Researchers, Inc.
16-17, James C. Knox

Illustrations
16–2, Modified from Hamblin, *Earth's Dynamic Systems, Eighth Edition,* © 1998, Prentice Hall, Inc., Upper Saddle River, NJ
16–3, P. J. Bartlein, 1997. "Past Environmental Changes: Characteristic Features of Quaternary Climate Variations." In B. Huntley, W. Cramer, A. V. Morgan, H. C. Prentice, and J. R. M. Allen (eds.), Past and Future Rapid Environmental Changes: The Spatial and Evolutionary Responses of Terrestrial Biota. *NATO ASI Series,* v. 147, pp. 11–29. Berlin: Springer-Verlag
16–4, Modified from McKnight, *Physical Geology, A Landscape Appreciation,* Fifth Edition, © 1996, Prentice Hall, Inc., Upper Saddle River, NJ
16–5, D. W. Stahle, M. K. Cleaveland, D. B. Blanton, M. D. Therrell, and D. A. Gay, 1998. "The Lost Colony and Jamestown Droughts." *Science,* v. 280, 24 April 1998, 564–567
16–6, Reprinted with permission from Richard Kerr, *Science* 307: 828-829 (11 February 2005). Copyright 2005 AAAS.
16–7, Modified from R. B. Alley, *Nature,* 392 (1998), 335–337
16–8, Modified from Chaisson and McMillan, *Astronomy Today,* Second Edition, © 1997, Prentice Hall, Inc., Upper Saddle River, NJ
16–10, NASA
16–13 & 14, Adapted from *Climate Change, The IPCC Scientific Assessment,* Houghton, Jenkins, and Ephraums, eds., © 1990, Cambridge University Press, Cambridge, England
16–B1–1, Source: NOAA
16–B2–1, National Climate Data Center/NOAA

Chapter 17

Photographs
Part 7 Opener, Galen Rowell/Mountain Light/The Stock Connection
Chapter 17 Opener, Kent Wood/Photo Researchers, Inc.
17-3, Polar Image
17-6, Pekka Parviainen/Science Photo Library/Photo Researchers, Inc.
17-7, Jim Steinberg/Photo Researchers, Inc.
17-11b, Arjen & Jerrine Verkaik/SKYART
17-12, 17-14, Warren Faidley/Weatherstock
17-13, Pekka Parviainen/Science Photo Library/Photo Researchers, Inc.

Index